STATE-OF-THE-ART TECHNOLOGIES IN FOOD SCIENCE

Human Health, Emerging Issues, and Specialty Topics

STATE-OF-THE-ART TECHNOLOGIES IN FOOD SCIENCE

Human Health, Emerging Issues, and Specialty Topics

Edited by

Murlidhar Meghwal
Megh R. Goyal

APPLE ACADEMIC PRESS

Apple Academic Press Inc.
3333 Mistwell Crescent
Oakville, ON L6L 0A2
Canada

Apple Academic Press Inc.
9 Spinnaker Way
Waretown, NJ 08758
USA

© 2019 by Apple Academic Press, Inc.

First issued in paperback 2021

Exclusive worldwide distribution by CRC Press, a member of Taylor & Francis Group
No claim to original U.S. Government works

ISBN 13: 978-1-77-463052-5 (pbk)
ISBN 13: 978-1-77-188616-1 (hbk)

Library and Archives Canada Cataloguing in Publication

State-of-the-art technologies in food science : human health, emerging issues, and specialty topics / edited by Murlidhar Meghwal, Megh R. Goyal.

(Innovations in agricultural and biological engineering)
Includes bibliographical references and index.
Issued in print and electronic formats.
ISBN 978-1-77188-616-1 (hardcover).--ISBN 978-1-315-16527-1 (PDF)

1. Functional foods. 2. Nutrition. 3. Health promotion. 4. Foodborne diseases--Prevention. I. Meghwal, Murlidhar, editor II. Goyal, Megh Raj, editor III. Series: Innovations in agricultural and biological engineering

| QP144.F85S73 2018 | 613.2 | C2018-900910-1 | C2018-900911-X |

CIP data on file with US Library of Congress

Apple Academic Press also publishes its books in a variety of electronic formats. Some content that appears in print may not be available in electronic format. For information about Apple Academic Press products, visit our website at **www.appleacademicpress.com** and the CRC Press website at **www.crcpress.com**

CONTENTS

LIST OF CONTRIBUTORS

Note: [*] after the author on first page of each chapter implies the ***corresponding author.***

R. Anjana
Biotechnologist, Pineapple Research Station, Kerala Agricultural University, P.O. Vazhakulam, Ernakulam 686670, Kerala, India, Tel.: +91-9946619746, E-mail: anjanasumathi@gmail.com

K. A. Athmaselvi*
Assistant Professor, Department of Food Process Engineering, School of Bioengineering, SRM University, Kattankulathur, Chennai 603203, Tamil Nadu, India, Tel.: +91-9600007823, E-mail: athmaphd@gmail.com

Divya Bansal
Associate Professor, Shri Ram Institute of Technology-Pharmacy, Jabalpur, Madhya Pradesh, India, Tel.: +91-9302958233, E-mail: drdivyabansal1126@gmail.com

Navneet Singh Deora*
Associate Specialist, Nestle R&D Centre India Pvt. Ltd. Applied Science Department CP-12A, Sector 8, IMT Manesar, Gurgaon 122050, India, Tel.: +91-7042307007
E-mail: navneet.deora@rd.nestle.com

K. Divyashree
Researcher, Food Technology, Centre for Emerging Technologies, Jain University, Jakkasandra 562112, Ramanagara, Karnataka, India, Tel.: +91-9164286583, E-mail: divyashreefst@gmail.com

Nazneen Dubey
Associate Professor, Shri Ram Institute of Technology—Pharmacy, Jabalpur, Madhya Pradesh, India, Tel.: +91-9407113473, E-mail: nazneenshabbir@gmail.com

Manju Gaare*
Assistant Professor, Department of Dairy and Food Microbiology, GN Patel College of Dairy and Food Technology, Sardarkrushi Nagar Dantiwada Agricultural University, Sardarkrushi Nagar, Dantiwada 385506, Banaskantha District, Gujarat, India, Tel.: +91-8971769001, E-mail: manjugdsc@gmail.com

Aditya Ganeshpurkar*
Assistant Professor, Shri Ram Institute of Technology-Pharmacy, Jabalpur, Madhya Pradesh, India, Tel.: +91-9993821431, E-mail: adityaganeshpurkar@gmail.com

Deepika Goswami
Scientist, ICAR—Central Institute of Post-Harvest Engineering and Technology (ICAR-CIPHET), Ludhiana 141004, Punjab, Inavndia, E-mail: deepikagoswami@rediffmail.com

Shaik Abdul Hussain
Scientist, Dairy Technology Division, ICAR—National Dairy Research Institute, Karnal 132001, Haryana, India, Tel.: +91-9896668983, E-mail: abdulndri@gmail.com

P. P. Joy*
Professor and Head, Pineapple Research Station, Kerala Agricultural University,
PO Vazhakulam, Ernakulam 686670, Kerala, India, Tel.: +91-9446010905,
E-mail: joy.pp@kau.in, joyppkau@gmail.com

Ravi Kumar Kadeppagari
Professor, Centre for Incubation, Innovation, Research and Consultancy, Thathaguni, Bangalore
560082, India, http://ciirc.jyothyit.ac.in/, Tel.: +91-9739204027, E-mail: ravikadeps@gmail.com

C. Kumar
Assistant Professor, Department of Physics, Government Arts and Science College for Men,
Nandanam, Chennai, 600035, Tamil Nadu, India, Tel.: +91-9600039748,
E-mail: vckumar60@yahoo.co.in

Deepak Kumar
Postdoctoral Research Associate, Agricultural and Biological Engineering, University of Illinois,
Urbana - Champaign, IL, 61801, USA, Tel.: +001-2173001929, E-mail: kumard@illinois.edu

Nitin Kumar
Assistant Professor, Department of Food Engineering National Institute of Food Technology
Entrepreneurship & Management, Kundli - 131028, Sonepat, Haryana, India
Email: nitinkumar.iit@gmail.com

S. P. Jeevan Kumar*
Scientist, ICAR—Indian Institute of Seed Science, Maunath Bhanjan, Uttar Pradesh 275103, India,
Tel.: +91-547-2530326, E-mail: jeevaniitkgp@gmail.com

Lakshita Maherda
MBBS Student, Sardar Patel Medical College, Bikaner, Mobile: +91-9413107109,
E-mail: maherdalakshita@gmail.com

Harshad M. Mandge
Assistant Professor, College of Horticulture, Banda University of Agriculture and Technology,
Banda-210001, Uttar Pradesh, India, E-mail: mandgeharshad@gmail.com

Murlidhar Meghwal*
Assistant Professor, Department of Food Science and Technology, National Institute of Food
Technology Entrepreneurship & Management, Kundli - 131028, Sonepat, Haryana, India;
Mobile: +91-9739204027; Email: murli.murthi@gmail.com

Irena Mladenova
Associate Professor and Head, Department of Hygiene, Epidemiology and Infectious Diseases,
Trakia University, Medical Faculty, Stara Zagora, 11 Armeiska Str., Bulgaria,
Mobile: +35-9897324472, E-mail: imladenova@yahoo.com

Amit Kumar Mukherjee*
Assistant Professor, Department of Food Technology, Haldia Institute of Technology,
P. O. HIT (Hatiberia), Haldia 721657, West Bengal, India. Mobile: +91-9477290235,
E-mail: mukherjee2001@gmail.com

P. Poojitha
M. Tech. Student, Department of Food Process Engineering, School of Bioengineering, SRM
University, Kattankulathur, Chennai 603203, Tamil Nadu, India, Tel.: +91-7418956148,
E-mail: poojithapushparaj@gmail.com

S. Rajendra Prasad
Director, ICAR—Indian Institute of Seed Science, Maunath Bhanjan 275103, Uttar Pradesh, India,
Tel.: +91-9415108377, E-mail: srprasad1989@yahoo.co.in.

Hradesh Rajput*
PhD Student, Sam Higginbottom Institute of Agriculture, Technology & Sciences (Formerly
Allahabad Agricultural Institute) Deemed-To-Be University, Allahabad 211007, Uttar Pradesh, India,
Tel.: +91-9454183802; E-mail: hrdesh802@gmail.com

T. A. Rashida-Rajuva
Food Technologist, Pineapple Research Station, Kerala Agricultural University,
PO Vazhakulam, Ernakulam 686670, Kerala, India, Tel.: +91-9349903270,
E-mail: rashh.ta@gmail.com

Anjana Ratheesh
Biotechnologist, Pineapple Research Station, Kerala Agricultural University,
PO Vazhakulam, Ernakulam 686670, Kerala, India, Tel.: +91-9946619746,
E-mail: anjanasumathi@gmail.com

Jagbir Rehal
Assistant Fruit & Vegetable Technologist, Food Science and Technology Division, Punjab
Agricultural University, Ludhiana 141004, Punjab, India, Tel.: +91-9417751567,
E-mail: jagbir@pau.edu.in

Anshu Singh
PhD Research Scholar, Indian Institute of Technology, Kharagpur, West Bengal 721302, India,
E-mail: anshusingh.biotech@gmail.com

Ashish Kumar Singh
Principal Scientist, Dairy Technology Division, ICAR—National Dairy Research Institute (NDRI),
Karnal 132001, Haryana, India, Tel.: +91-184-2259291, E-mail: akndri@gmail.com

H. T. Sowmya
Researcher, Food Technology, Center for Emerging Technologies, Jain Global Campus,
Jain University, Jain Global Campus, Jakkasandra, Kanakapura Main Road, Ramanagara District
562112, Karnataka, India, Tel.: +91-9739204027, E-mail: sowmyaht300@gmail.com

LIST OF ABBREVIATIONS

ACE	angiotensin-I-converting enzyme
AMD	age-related macular degeneration
AMS	automatic milking systems
AODF	antioxidant dietary fiber
ATCC	American Type Culture Collection
BPH	benign prostatic hyperplasia
CAD	coronary artery disease
CDC	Centers for Disease Control and Prevention
CFU	colony forming unit
CHD	coronary heart disease
CNS	central nervous system
CPE	*Clostridium perfringens* enterotoxin
CVD	cardiovascular disease
CYP	cytochrome P 450
DAG	diacylglycerol
DMS	differential mobility spectrometry
DNA	deoxyribo nucleic acid
DPPH	2,2-diphenyl-1-picrylhydrazyl
DRI	dietary reference intake
DV	daily value
EDTA	ethylene diamine tetraacetic acid
EHEC	entero hemorrhagic *E. coli*
EHO	environment health officers
EIA	enzyme immuno assay
EIEC	entero-invasive *E. coli*
EO	essential oil
EPA	Environmental Protection Agency
EPEC	entero-pathogenic *E. coli*
ETEC	entero-toxigenic *E. coli*
FBD	foodborne disease
FBI	foodborne illness
FBP	foodborne pathogens

FCM	flow cytometry
FDA	Food and Drug Administration
FFA	free fatty acids
FID	flame ionization detector
FSA	Food Standards Agency
FSIS	Food Safety and Inspection Service
FSMO	Food Safety Modernization Act
GABA	γ-amino butyric acid
GAG	glycosaminoglycans
GBF	germinated barley foodstuff
GBS	Guillain–Barré syndrome
GC-O	gas chromatography–olfactometry
GHP	good hygienic practices
GIT	gastrointestinal tract
GM	genetically modified
GM1	monosialotetrahexosylganglioside
GMP	good manufacturing practice
HACCP	hazard analysis critical control point
HDL-C	high-density lipoprotein cholesterol
HMG-CoA	3-hydroxy-3-methyl-glutaryl-coenzyme A
HPP	high-pressure processing
IMS	ion mobility spectrometry
ISO	International Organization for Standardization
LAB	lactic acid bacteria
LDL	low-density lipoprotein
LP	low pressure
MAG	monoacylglycerols
MCC	multicapillary columns
MS	mass spectrometer
NCDC	National Collection of Dairy Cultures
NDF	neutral detergent fibers
NDM	nonfat dry mil
NMFS	National Marine Fisheries Service
NSP	nonsoluble polysaccharide
PC	phosphatidylcholine
PCR	polymerase chain reaction
PE	phosphatidylethanolamine
PFGE	pulsed field gel electrophoresis

PL	photoluminescence
PME	pectin methyl esterase
PP	potato peels
PSV	protein storage vacuoles
PUFA	polyunsaturated fatty acid
RA	rheumatoid arthritis
RDA	recommended dietary allowance
SAH	S-adenosylhomocysteine
SAM	S-adenosylmethionine
SEA	Staphylococcal enterotoxin A
SEB	Staphylococcal enterotoxin B
SEC	Staphylococcal enterotoxin C
SED	Staphylococcal enterotoxin D
SEE	Staphylococcal enterotoxin E
SFI	solid fat index
SOD	superoxide dismutase
SSP	seed storage protein
TAG	triacylglycerols
TBARS	thiobarbituric acid-reactive substances
TCP	T-cell lymphoma
TG	triglycerides
TOF	time-of-flight
TPC	thyroid papillary carcinoma
VLDL	very low-density lipoprotein
WWF	whole wheat flour

LIST OF SYMBOLS

α (alpha)	first lowercase letter of the Greek alphabet
β (beta)	second lowercase letter of the Greek alphabet
Δ (delta)	fourth lowercase letter of the Greek alphabet
γ (gamma)	third lowercase letter of the Greek alphabet
ω (omega)	twenty-fourth (the last) lowercase letter of the Greek alphabet
μm	1 micrometer = 0.000001 m.
E	electrode
H_2O_2	hydrogen peroxide
I	current (A)
λ	lambda
L	distance between electrodes (m)
R	resistance (Ω)
σ	electrical conductivity (S·m^{-1})
s	second
∇V	voltage gradient (V·cm^{-1})
V	voltage (V)

PREFACE 1
BY MURLIDHAR MEGHWAL

Food and health are a most interrelated and very important part of day-to-day human life because everybody has to eat to survive and to maintain good health. In recent times there is an increasing interest in health benefits that are especially drawn from fruits and vegetables and food products based on them. There are various phytochemical, flavonoids, fibers, macronutrients, and micronutrients, minerals, etc., that are very good for health and essential to keep good health.

This book, *State-of-the-Art Technologies in Food Science: Human Health, Emerging Issues, and Specialty Topics,* provides a global perspective of present-age frontiers in food and health research, innovation, and emerging trends. It includes selected recent emerging trends and issues of food related to health management and health-related issues. The book volume explores topics of food for better health, functional foods, nutraceutical foods, and food science, which can help to provide solutions for the different issues, problems, and complexities related to the food crisis worldwide with the help of limited resources and technologies.

This book volume is divided into three major parts that include the following topics:

Part I – Foods for Human Health Promotion and Prevention of Diseases discusses fruits, vegetables, and grains: their peels and fiber for better human health, health prospects of bioactive peptides derived from seed storage proteins, mushrooms as novel source of antihyperlipidemic agents, emerging foodborne illnesses and their prevention.

Part II – Specific Fruits, Spices, and Dairy-Based Functional Foods for Human Health looks at the functional medicinal values of fenugreek, fruits as a functional food, and functional fermented dairy products.

Part III – Issues, Challenges, and Specialty Topics in Food Science focuses mainly on whole wheat flour: stability issues and challenges, physicochemical properties and quality of food lipids, methods for food analysis and quality control, and interventions of ohmic heating technology in foods.

The targeted audience for this book includes health practitioners, food specialists and nutrition producers and suppliers, yoga practitioners, practicing food process engineers, body builders, food technologists, researchers, lecturers, teachers, professors, food professionals, students of these fields, and all those who have inclination for food-processing sector. The book not only covers the practical aspect but also has a lot of basic information and also it is instructive. Therefore, students in undergraduate, graduate courses, postgraduate and post-doctoral researchers will be benefitted by this book. In order for the book to be useful to engineers, coverage of each topic is comprehensive enough to serve as an overview of the most recent and relevant research and technology. Numerous references are included at the end of each chapter supporting our knowledge.

The editors wish to acknowledge all the individuals who have contributed to this book.

Murlidhar Meghwal, PhD
Editor

PREFACE 2
BY MEGH R. GOYAL

25 grams of soy protein a day,
As part of a diet low in saturated fat and cholesterol,
May reduce the risk of heart disease, [USFDA 21CFR101.82].
— Ramabhau Patil, PhD

Let food be thy medicine and medicine be thy food—
Who said anything about medicine? Let's eat"
— attributed to Hippocrates
[https://en.wikipedia.org/wiki/Hippocrates]

We all know food is essential for our survival. The increasing world population and the continuous climate change result in the reduction of agricultural land for food production. Subsequently this urges modern food science and technology to develop sustainable food production systems and improve nutritional value of food products, while keeping the cost as low as possible. Quality and nutritional value of foods are highly dependent on the environment, agricultural practices, production conditions, and consumer preferences, which all may provide different effects for human health. One of the main challenges of the food science and technology is to optimize food production to have minimum environmental footprint, lower production costs, and improving quality and nutritional value.

Therefore, we introduce this book volume *State-of-the-Art Technologies in Food Science: Human Health, Emerging Issues, and Specialty Topics,* under the book series Innovations in Agricultural and Biological Engineering. This book covers mainly the current scenario of the research and case studies on: (1) **Foods for human health promotion and prevention of diseases** with emphasis on—Peels and fiber of fruits, vegetables, and grains, health prospects of bioactive peptides derived from seed storage proteins, mushrooms as novel source of antihyperlipidemic agents, emerging foodborne illnesses and their prevention; (2) **Functional foods such as fenugreek**, fruits, dairy products; (3) **Specialty topics in food science** that includes stability issues and challenges of wheat flour, physicochemical properties and quality of food lipids, methods for food analysis and quality control, Interventions of ohmic heating technology in foods.

This book volume sheds light on the potential of foods for human health for different technological aspects, and it contributes to the ocean of knowledge on food science and technology. We hope that this compendium will be useful for the students and researchers as well as the persons working with the food, nutraceuticals, and herbal industries.

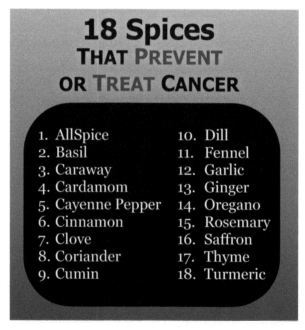

Source: [http://preventdisease.com/news/13/062713_18-Spices-Scientifically-Proven-To-Prevent-and-Treat-Cancer.shtml]

The contribution by all cooperating authors to this book volume has been most valuable in the compilation. Their names are mentioned in each chapter and in the list of contributors. We appreciate you all for having patience with our editorial skills. This book would not have been written without the valuable cooperation of these investigators, many of whom are renowned scientists who have worked in the field of food engineering and food science throughout their professional careers. I am glad to introduce Dr. Murlidhar Meghwal. With several awards and recognitions including from the President of India, Dr. Meghwal brings his expertise and innovative ideas in this book. Without his support, leadership qualities, and extraordinary work on food technology applications, readers will not have this quality book.

We would also like to thank the editorial staff, Sandy Jones Sickels, vice president, and Ashish Kumar, publisher and president at Apple Academic Press, Inc., for making their valiant effort to publish the book when the diminishing water and food resources are a major issue worldwide. Special thanks are due to the AAP Production Staff for typesetting the entire manuscript and for the quality production of this book.

We request the reader to offer constructive suggestions that may help to improve the next edition.

We express our admiration and gratitude to our beloved families and colleagues for understanding and collaboration during the preparation of this book volume. As an educator, there is a piece of advice to one and all in the world: *Permit that our almighty God, our Creator, provider of all and excellent Teacher, feed our life with Healthy Food Products and His Grace—; and Get married to your profession—"*

Megh R. Goyal, PhD, PE
Senior Editor-in-Chief

ABOUT THE LEAD EDITOR

Murlidhar Meghwal, PhD, is a distinguished researcher, engineer, teacher, and Assistant Professor in the Department of Food Science and Technology at the National Institute of Food Technology Entrepreneurship and Management, Kundli, Sonepat, Haryana, India. He is the lead editor for this book volume.

He received his B Tech degree (Agricultural Engineering) in 2008 from the College of Agricultural Engineering Bapatla, Acharya N. G. Ranga Agricultural University, Hyderabad, India; his M Tech degree is in the field of Dairy and Food Engineering in 2010; and he has also received his Ph D degree in Food Process Engineering in 2014 from the Indian Institute of Technology Kharagpur, West Bengal, India.

He worked for one year as a research associate at INDUS Kolkata for the development of a quicker and industrial-level parboiling system for paddy and rice milling. In his Ph D research, he worked on ambient and cryogenic grinding of fenugreek and black pepper by using different grinders to select a suitable grinder.

Currently, Dr. Meghwal is working on developing inexpensive, disposable, and biodegradable food containers using agricultural wastes; quality improvement, quality attribute optimization, and storage study of kokum (*Garcinia indica Choisy*); and freeze drying of milk. He has written six books and many research publications in food process engineering. He has attended many national and international seminars and conferences. He is reviewer and member of editorial boards of reputed journals.

He is a recipient of the Bharat Scout Award from the President of India as well as the Bharat Scouts Award from the Governor. He received a meritorious Foundation for Academic Excellence and Access (FAEA—New Delhi) Scholarship for his full undergraduate studies from 2004 to 2008.

He also received a Senior Research Fellowship awarded by the Ministry of Human Resources Development (MHRD), Government of India, during 2011–2014; and Scholarship of the Ministry of Human Resources Development (MHRD), Government of India, research during 2008–2010.

He is a good sportsperson, mentor, social activist, critical reviewer, thinker, fluent writer and well-wishing friend to all.

Readers may contact him at: murli.murthi@gmail.com

ABOUT THE SENIOR-EDITOR-IN-CHIEF

Megh R Goyal, PhD, PE
*Retired Professor in Agricultural and Biomedical
Engineering, University of Puerto Rico,
Mayaguez Campus Senior Acquisitions Editor,
Biomedical Engineering and Agricultural Science,
Apple Academic Press, Inc.*

Megh R. Goyal, PhD, PE, is, currently a retired
professor of agricultural and biomedical engineering from the General
Engineering Department at the College of Engineering at the University
of Puerto Rico–Mayaguez Campus; and Senior Acquisitions Editor and
Senior Technical Editor-in-Chief for Agricultural and Biomedical Engineering for Apple Academic Press Inc.

Dr. Goyal received his BSc degree in Engineering in 1971 from Punjab
Agricultural University, Ludhiana, India; his MSc degree in 1977, and
PhD degree in 1979 from the Ohio State University, Columbus, Ohio,
USA. He also earned a Master of Divinity degree in 2001 from the Puerto
Rico Evangelical Seminary, Hato Rey, Puerto Rico, USA.

Since 1971, he has worked as a Soil Conservation Inspector (1971);
Research Assistant at Haryana Agricultural University (1972–1975) and
Ohio State University (1975–1979); Research Agricultural Engineer/
Professor at the Department of Agricultural Engineering of UPRM (1979–
1997); and Professor of Agricultural and Biomedical Engineering in the
General Engineering Department of UPRM (1997–2012). He spent a one-
year sabbatical leave in 2002–2003 at the Biomedical Engineering Depart-
ment of Florida International University, Miami, USA.

Dr. Goyal was the first agricultural engineer to receive the professional
license in agricultural engineering in 1986 from the College of Engineers and
Surveyors of Puerto Rico. In 2005, he was proclaimed the "Father of Irrigation Engineering in Puerto Rico for the Twentieth Century" by the American Society of Agricultural and Biological Engineers, Puerto Rico Section,

for his pioneering work on micro irrigation, evapotranspiration, agroclimatology, and soil and water engineering. During his professional career of 45 years, he has received many awards, including Scientist of the Year, Blue Ribbon Extension Award, Research Paper Award, Nolan Mitchell Young Extension Worker Award, Agricultural Engineer of the Year, Citations by Mayors of Juana Diaz and Ponce, Membership Grand Prize for the American Society of Agricultural Engineers Campaign, Felix Castro Rodriguez Academic Excellence Award, Rashtrya Ratan Award and Bharat Excellence Award and Gold Medal, Domingo Marrero Navarro Prize, Adopted Son of Moca, Irrigation Protagonist of the University of Puerto Rico–Mayaguez, and Man of Drip Irrigation by the Mayor of Municipalities of Mayaguez/Caguas/Ponce and Senate/Secretary of Agriculture of ELA, Puerto Rico.

Dr. Goyal has authored more than 200 journal articles and edited more than 52 books. Apple Academic Press Inc. has published many books edited by him, many under his several books series, Innovations and Challenges in Micro Irrigation, Innovations in Agricultural & Biological Engineering, and Research Advances in Sustainable Micro Irrigation.

Readers may contact him at goyalmegh@gmail.com.

OTHER BOOKS ON AGRICULTURAL AND BIOLOGICAL ENGINEERING BY APPLE ACADEMIC PRESS, INC.

Management of Drip/Trickle or Micro Irrigation
Megh R. Goyal, PhD, PE, Senior Editor-in-Chief

Evapotranspiration: Principles and Applications for Water Management
Megh R. Goyal, PhD, PE, and Eric W. Harmsen, Editors

Book Series: Research Advances in Sustainable Micro Irrigation
Senior Editor-in-Chief: Megh R. Goyal, PhD, PE
Volume 1: Sustainable Micro Irrigation: Principles and Practices
Volume 2: Sustainable Practices in Surface and Subsurface Micro Irrigation
Volume 3: Sustainable Micro Irrigation Management for Trees and Vines
Volume 4: Management, Performance, and Applications of Micro Irrigation Systems
Volume 5: Applications of Furrow and Micro Irrigation in Arid and Semi-Arid Regions
Volume 6: Best Management Practices for Drip Irrigated Crops
Volume 7: Closed Circuit Micro Irrigation Design: Theory and Applications
Volume 8: Wastewater Management for Irrigation: Principles and Practices
Volume 9: Water and Fertigation Management in Micro Irrigation
Volume 10: Innovation in Micro Irrigation Technology

Book Series: Innovations and Challenges in Micro Irrigation
Senior Editor-in-Chief: Megh R. Goyal, PhD, PE

- Micro Irrigation Engineering for Horticultural Crops: Policy Options, Scheduling and Design
- Micro Irrigation Management Technological Advances and Their Applications
- Micro Irrigation Scheduling and Practices
- Performance Evaluation of Micro Irrigation Management: Principles and Practices
- *Potential of Solar Energy and Emerging Technologies in Sustainable Micro Irrigation*
- Principles and Management of Clogging in Micro Irrigation
- Sustainable Micro Irrigation Design Systems for Agricultural Crops: Methods and Practices
- Engineering Interventions in Sustainable Trickle Irrigation: Water Requirements, Uniformity, Fertigation, and Crop Performance

Book Series: Innovations in Agricultural & Biological Engineering
Senior Editor-in-Chief: Megh R. Goyal, PhD, P.E.

- Dairy Engineering: Advanced Technologies and Their Applications
- Developing Technologies in Food Science: Status, Applications, and Challenges
- Emerging Technologies in Agricultural Engineering
- Engineering Practices for Agricultural Production and Water Conservation: An Inter-disciplinary Approach
- Flood Assessment: Modeling and Parameterization
- Food Engineering: Emerging Issues, Modeling, and Applications
- Food Process Engineering: Emerging Trends in Research and Their Applications
- Food Technology: Applied Research and Production Techniques
- Engineering Interventions in Agricultural Processing
- Modeling Methods and Practices in Soil and Water Engineering

- Processing Technologies for Milk and Dairy Products: Methods Application and Energy Usage
- Soil and Water Engineering: Principles and Applications of Modeling
- Soil Salinity Management in Agriculture: Technological Advances and Applications
- Technological Interventions in the Processing of Fruits and Vegetables
- Technological Interventions in Management of Irrigated Agriculture
- Technological Interventions in Dairy Science: Innovative Approaches in Processing, Preservation, and Analysis of Milk Products
- Novel Dairy Processing Technologies: Techniques, Management, and Energy Conservation
- Sustainable Biological Systems for Agriculture: Emerging Issues in Nanotechnology, Biofertilizers, Wastewater, and Farm Machines
- State-of-the-Art Technologies in Food Science: Human Health, Emerging Issues and Specialty Topics
- Scientific and Technical Terms in Bioengineering and Biological Engineering
- Engineering Practices for Management of Soil Salinity: Agricultural, Physiological, and Adaptive Approaches
- Processing of Fruits and Vegetables: From Farm to Fork
- Engineering Interventions in Foods and Plants
- Technological Processes for Marine Foods, From Water to Fork: Bioactive Compounds, Industrial Applications and Genomics

EDITORIAL

Under book series titled *Innovations in Agricultural and Biological Engineering*, Apple Academic Press Inc., (AAP) will publish subsequent volumes in the specialty areas defined by *American Society of Agricultural and Biological Engineers* (<asabe.org>), over a span of 8 to 10 years. AAP wants to be principal source of books in Agricultural & Biological Engineering. **We need book proposals from the readers in area of their expertise.**

The mission of this series is to provide knowledge and techniques for agricultural and biological engineers (ABEs). The series aims to offer high-quality reference and academic content in **Agricultural and Biological Engineering** (ABE) that is accessible to academicians, researchers, scientists, university faculty, and university-level students and professionals around the world.

Agricultural and Biological Engineers (ABEs) ensure that the world has the necessities of life including safe and plentiful food, clean air and water, renewable fuel and energy, safe working conditions, and a healthy environment by employing knowledge and expertise of sciences, both pure and applied, and engineering principles. Biological engineering applies engineering practices to problems and opportunities presented by living things and the natural environment in agriculture. *"ABE embraces a variety of following specialty areas* (asabe.org)", such as: Aquacultural Engineering, Biological Engineering, Energy, Farm Machinery and Power Engineering, Food and Process Engineering, Forest Engineering, Information & Electrical Technologies engineering, Natural Resources, Nursery and Greenhouse Engineering, Safety and Health, and Structures and Environment.

For this book series, we welcome chapters on the following specialty areas (but not limited to):

1. Academia to industry to end-user loop in agricultural engineering
2. Agricultural mechanization
3. Aquaculture engineering
4. Biological engineering in agriculture

5. Biotechnology applications in agricultural engineering
6. Energy source engineering
7. Food and bioprocess engineering
8. Forest engineering
9. Hill land agriculture
10. Human factors in engineering
11. Information and electrical technologies
12. Irrigation and drainage engineering
13. Nanotechnology applications in agricultural engineering
14. Natural resources engineering
15. Nursery and greenhouse engineering
16. Potential of phytochemicals from agricultural and wild plants for human health
17. Power systems and machinery design
18. GPS and remote sensing potential in agricultural engineering
19. Robot engineering in agriculture
20. Simulation and computer modeling
21. Smart engineering applications in agriculture
22. Soil and water engineering
23. Structures and environment engineering
24. Waste management and recycling
25. Any other focus area

For more information on this series, readers may contact:

Ashish Kumar, Publisher and President
Sandy Sickels, Vice President
Apple Academic Press, Inc.
Fax: 866–222–9549
E-mail: ashish@appleacademicpress.com
http://www.appleacademicpress.com/
publishwithus.php

Megh R Goyal, PhD PE
Book Series
Senior Editor-in-Chief,
Innovations in Agricultural and Biological Engineering
E-mail: goyalmegh@gmail.com

PART I

Foods for Human Health Promotion and Prevention of Diseases

CHAPTER 1

FRUITS, VEGETABLES, AND GRAINS: THEIR PEELS AND FIBER FOR BETTER HUMAN HEALTH

MURLIDHAR MEGHWAL[1*], K. DIVYASHREE[2], and NITIN KUMAR[1]

[1]*Assistant Professor, Department of Food Science and Technology, National Institute of Food Technology Entrepreneurship & Management, Kundli - 131028, Sonepat, Haryana, India Mobile: +91 9739204027*

[2]*Centre for Emerging Technologies, Jain University, Jakkasandra 562112, Ramanagara, Karnataka, India Mobile: +91-9164286583, E-mail: divyashreefst@gmail.com*

*Corresponding author. E-mail: murli.murthi@gmail.com

CONTENTS

1.1   INTRODUCTION

Fruits and vegetables (FAV) are important parts of our healthy diet. In addition to delicious taste and flavor, they can reduce risk of several chronic diseases.[29] Fruits and vegetables contain a wide range of compounds including antioxidant,[32] vitamins, minerals, phenolics, thiols, carotenoids, tocopherols, and glucosinolates. They also exert chemo-protective effects through a variety of mechanisms.[23] Pretreatments used in the processing of FAV, such as peeling, have great importance in terms of nutritional and economic aspects. Peeling removes the peel when it is not edible or even when it is edible as the presence of peel is undesirable in the final product.[25]

Peel is also known as *rind* or *skin* and it is the outer protective layer of a fruit or vegetable, which can be peeled off. The rind is usually the botanical exocarp, although the term "exocarp" also includes the hard cases of nuts which are not called peels since they cannot be peeled off by hand or peeler, but rather are called shells because of their hardness. A fruit with a thick peel, such as citrus fruit, is called a *hesperidium*. In *hesperidiums*, the inner layer (also called *albedo* or *pith*) is peeled off together with the outer layer (called flavedo) and altogether they are called the peel. The flavedo and albedo are the exocarp and the mesocarp, respectively. The juicy layer inside the peel (containing the seeds) is the endocarp.

The aim of peeling is not only removing the peel but also minimizing the losses related to the tissues and nutritional value.[25] In some raw fruits, the peel has neutral flavor, such as in grapes and apples, and they are very rich in essential oils which give distinctive aroma to the fruit. In some fruits such as guava, the peel is firmly cohesive to its pulp and, indeed, in some fruits it turns tastier than the flesh as the fruit ripens.[3] The peel is a rich source of rough dietary fibers,[4,5] also known as nonsoluble polysaccharide, such as hemicellulose, pectin, tannins, gum, and so forth that add bulkiness to the food and help in treating constipation by reducing gastrointestinal transit time. Peel is low in calories, sugar, and fats, and is free from cholesterol.[18,19]

The main objective of this study is to identify the peels, skin, and outer coverings of FAVs and fibers of some grains; and to study in detail about their health benefits to humans. Authors also studied alternate ways to reduce the risk of diseases by utilizing the peels, which are considered as waste by the consumers, thus, creating awareness about the benefits of the fruits and vegetable fibers, skins, peels as well as fibers from grains.[18,19]

1.2 PEELING AND PEELING EQUIPMENT (PEELERS)

Decreasing losses and increasing the processing efficiency of fruits and vegetables is a matter of interest for the managers of food industries. Peeling as the preliminary and main stage of postharvest processing is currently conducted by mechanical, chemical, and thermal methods. Peeling is removal of the peel when it is not edible, or even when it is edible but its presence is undesirable in the final product.[2,10]

1.2.1 PEELING METHODS OF FRUITS AND VEGETABLES

The peeling methods are generally classified into mechanical, abrasion, chemical, thermal, and enzymatic peeling, as shown in Table 1.1 having the information from several articles.[2,3,10,11,15,32] Manual peeling is commonly used for some kinds of fruits and vegetables such as mango.[7,22] The requirement to develop new methods and tools for peeling that can be mechanized or automated has led to the versatile current peeling methods, machinery, and equipment.[2,10,11] There are several types of peelers that are available in the market. Such peelers are developed based on the geometry of the fruits.

1.2.1.1 MECHANICAL PEELING

Mechanical peelers are classified on the basis of the type of mechanism that is incorporated into the peeling system. Commercial mechanical peelers include abrasive devices, devices with drums, rollers, knifes or blades, and milling cutters. Generally, the quantity of losses in this kind of peeling is high, but the quality of the final peeled vegetables in terms of freshness is good.[10,11]

1.2.1.2 THERMAL PEELING

Thermal peeling is used for thick-skinned vegetables. This method can be performed by wet heat (steam peeling), dry heat (flame, infrared, hot gases peeling), thermal blast peeling, freeze-thaw, vapor explosion (vacuum peeling). It is preferred due to its high level of automation, precise control of time, temperature and pressure by electronic devices to minimize the peeling losses, and due to reduced environmental pollution as compared to chemical peeling. However, especially dry heat causes a cauterizing of the surface, wound areas, and small pieces of charred skin, which if not removed, gives a poor appearance to vegetables, especially canned ones.[10,11]

1.2.1.3   CHEMICAL PEELING

In the chemical peeling method, skin can be softened from the underlying tissues by submerging vegetables in hot alkali solution. The quantity of solution and the period of time are different for different kinds and varieties of vegetables. This method includes caustic (lye) and enzymatic peeling. This method reduces the losses but it causes harmful effects on the flesh of vegetables and also is not environmental-friendly.[10]

TABLE 1.1   Peeling Methods Based on Product and Intended Use. (*Source:* Adopted by collecting information from several articles.)[2,3,10,11,15,32]

Peeling methods	Fruits/Vegetables
1. Mechanical methods	
a) Knife	Apple
b) Abrasion	Potatoes
c) Combination of abrasion and lye	Potatoes
2. Thermal peeling	
a) Steam or hot water	Tomatoes, peach, carrot
b) Flame (dry heat)	Pepper
c) Thermal blast peeling	Pumpkin
d) Freeze-thaw	Tomatoes, peach
e) Vapor explosion (vacuum peeling)	Tomatoes
3. Chemical peeling	
a) Caustic (lye)	Carrot, pimiento peppers, tomatoes
4. Infrared irradiation peeling	Tomatoes
5. Enzymatic peeling	Apple

1.3   PEELS OF EDIBLE FRUITS

1.3.1   APPLE PEEL (MALUS DOMESTICA)

Apple is the fourth most important fruit crop worldwide, after citrus, vitis, and banana. The high content of phenolic compounds and antioxidant activity of apple peels [3,6] indicate that they are the valuable source of antioxidants and can impart health benefits when consumed.[29] Some studies show that antioxidant and antiproliferative activities of unpeeled apples were greater than those of peeled apples.[3,32] It was also known that the concentration of total phenolic compounds was much higher in the peel of apples than in the flesh. Both these facts suggest that apple peels may possess more bioactivity than the flesh.[23] The flavonoids such as quercitin, proanthocyanidin present in the peel of an apple along with pectin can help in reducing the blood pressure. Compounds, such as anthocyanins and quercetin, which are found in apple peel, have antihypertensive effects and are beneficial for decreasing the blood pressure level.[3,6,7]

1.3.2   ORANGE PEEL (CITRUS SINENSIS)

Citrus fruits have peculiar fragrance partly due to flavonoids and limonoids present in the peel and these fruits are good sources of vitamin C and flavonoids. The orange peels are rich in nutrients that can be used as drugs or as food supplements too.[29] Peel of citrus fruit has numerous glands that contain oil that is typically recovered as major a by-product. Each citrus fruit has its own typical set of compounds that comprise the oil that are responsible for its flavor and aroma in the products such as carbonated drinks, ice creams, cakes, air-fresheners, and perfumes.[28] The antioxidant/radical scavenging capacity and reducing power ability of different extracts of orange peel showed the highest yield of total phenolic content, total flavonoid content, chelating, and antioxidant activities (2,2-diphenyl-1-picrylhydrazyl (DPPH) scavenging activity).[14] Health benefits of orange peels are as follows:

- Dried orange peels can be used as homemade bath oils.
- Orange peels are rich in vitamins A and C; both are natural antioxidants that boost the overall health of the immune system and help fight against infection, colds, and flu.

• Orange peels have more phytonutrients and flavonoids than the inner pulp, endowing it with anti-inflammatory properties that can aid in digestion and relieve gastrointestinal problems such as acidity, heartburn, and the digestion of fatty foods.[6]

1.3.3 MANGO PEEL (MANGIFERA INDICA)

Mango belongs to the genus *Angifera*, consisting of numerous species of tropical fruiting trees in the flowering plant family *Anacardiaceae*. Mango fruit contains essential vitamins and dietary minerals. The antioxidant vitamins A, C, and E comprise 25, 76, and 9% of the dietary reference intake (DRI), respectively, in a 165-g serving. Mango peel and pulp contain other phytonutrients such as the pigment antioxidants, carotenoids, polyphenols, and omega-3 and omega-6 polyunsaturated fatty acids.[7] The edible mango peel has considerable value as a source of dietary fiber and antioxidant pigments.[4,5] Antioxidants of the peel and pulp include carotenoids, such as the pro-vitamin A compound, beta-carotene, lutein, and alpha-carotene.[13,22]

Mango peels can be fed fresh, dried, or ensiled to ruminants.[7] Due to the high sugar content, they are palatable and considered as an energy feed, but the high moisture and acidity of fresh peels may limit their use in ruminants. Because of their low protein content, addition of a source of nitrogen or protein is necessary to allow efficient utilization of the energy in the diet. In order to produce good silage, mango peels were mixed with rice straw and legume to facilitate fermentation. Dried mango peels up to 10% in the diet of finishing pigs had no deleterious effect on feed conversion ratio or performance and economized feeding cost.[22,34]

1.3.4 POMEGRANATE PEEL (PUNICA GRANATUM)

Pomegranate is a fruit native to the Middle East, belonging to the *Punicaceae* family. Pomegranate peels are considered inedible parts or by-product obtained through juice processing and it is characterized by the significant presence of ellagitannins and polyphenols, gallic acid and ellagic acid as well as flavonoids—associated with biological properties such as antioxidant and antimicrobial agents.[24] Pomegranate peel is rich in antioxidant, antiviral, anticancer, and antitumor properties, and these antioxidants are highly capable of protecting low-density lipoprotein (LDL) cholesterol against oxidation, lower risk of cancer and heart disease. It attracts attention due to its apparent wound healing properties and immune-modulating activity. Pomegranate skin powder can be used in a variety of recipes or blended into favorite smoothie for flavor and antioxidants. It is incorporated in *idli* and other ready-to-cook products to provide more nutritious meal.[12]

Phenolic compounds such as flavonoids and tannins present in pomegranate are responsible for its exceptional healing qualities.[29] Traditionally, aqueous pomegranate peels extract is obtained by boiling the fruit for 10–40 min. The extract has been used to treat diarrhea, dysentery, and dental plaque, in addition to being used as a douche and enema agent. Similarly, diarrhea, intestinal worms, bleeding noses, and ulcers have been treated in the Indian Ayurvedic system.[16]

1.3.5   BANANA (MUSA PARADISIACA)

Banana fruit has nutritional properties and also acclaimed therapeutic uses. Banana peel is known by its local and traditional use to promote wound healing, mainly from burns, and helps to overcome or prevent a substantial number of illnesses, such as depression. In traditional medicine, banana peel has a history of utility to promote wound healing mainly by burns when used topically. Peels of ripe bananas can be used to make a poultice for wounds, which is wrapped around an injury to reduce pain or swelling. As the inside of the peel has antiseptic properties, it can be wrapped directly around wounds or cuts in an emergency. Indeed, among the numerous sources of bioactive compounds, banana peel could be considered one of the complex plant matrices rich in high-value compounds. It is rich in phytochemical compounds, mainly antioxidants while ripe banana peel contains the anthocyanins, delphinidin, cyaniding, and catecholamines.[30,32]

Banana peel constitutes about 30% of fresh banana by weight, which is rich in dietary fiber,[4,5] proteins, essential amino acids, polyunsaturated fatty acids, and potassium.[29] Banana skins can also be utilized for the extraction of banana oil that can be used for food flavoring. They are also good source of lignin, pectin, cellulose, hemicellulose, and galacturoninc acid. Banana peel can also be used in wine, ethanol production, and as a base material for pectin extraction. Peel ash can be used as fertilizer for banana plants and as a source of alkali for soap production. Micronutrients (Fe and Zn) were found in higher concentration in peels compared to pulps, so that they could be a good feed material for cattle and poultry.[26]

1.3.6   PINEAPPLE (ANANAS COMOSUS)

Pineapple peel is rich in cellulose, hemicelluloses, and other carbohydrates. [3,6] Ensilaging of pineapple peels produces methane as a biogas. Anaerobic digestion occurs and the digested slurry is used as animal, poultry, and fish feeds or organic fertilizers.[33] Dietary fiber powder prepared from pineapple shell has 70.6% total dietary fiber with better sensory properties than the commercial dietary fibers from apple and citrus fruits.[5,18,19] Pineapple peel and leaf showed a high antioxidant activity with high phenolic compounds whereas the peel has been used for the alkali extraction of ferulic acid.[3,31,32]

1.4   PEELS OF EDIBLE VEGETABLES

Some of the peels of edible vegetables are discussed in the following sections.

1.4.1   POTATO PEEL

The main results show that there is a huge potential for potato peel (PP) extract as an antioxidant in food systems due to its high phenol content. In addition, PP powder could serve as a partial flour replacement in dough up to 10 g/100 g of flour weight without causing significant changes in sensory properties. Potato peel waste can serve as a solid substrate for fermentation.[1,15]

Raw PPs have high moisture and carbohydrate contents, but the overall protein and lipid contents are generally low. High content of starch makes it a good basis for fermentation. In addition, PPs contain a variety of valuable compounds, including phenols, dietary fibers, unsaturated fatty acids, and amides.[5] Potato dietary fiber is able to bind bile acids in vitro and can be part of the mechanism that lowers plasma cholesterol.[4,15]

1.4.2   TOMATO PEEL

Tomato skin and seeds, which are discarded, are rich in total phenolics, total flavonoids, lycopene, ascorbic acid, and antioxidant activity compared to their pulp. Therefore, it is important to consume tomatoes along with their skin and seeds, in order to attain maximum health benefits. The peel fraction of tomato waste contains lycopene up to five times more than the pulp. The protective effects of lycopene have been shown on oxidative stress, cardiovascular disease, hypertension, atherosclerosis, cancers, diabetes, and others. The intended use of lycopene extract from tomato is as a food color in dairy products, nonalcoholic-flavored drinks, cereals and cereal products, bread and baked goods and spreads, to provide color shades from yellow to red. Lycopene extract from tomato may also be used in food supplements.

1.4.3   BEETROOT

Beetroots contain a group of highly bioactive pigments known as betalains. It is a rich source of phytochemical compounds that include ascorbic acid, carotenoids, phenolic acids, and flavonoids. It has been primarily driven by the discovery that sources of dietary nitrate may have important implications for managing cardiovascular health. Betalains and beetroot extracts have emerged as potent anti-inflammatory agents.[17] Beetroot supplementation might serve as a useful strategy to strengthen endogenous antioxidant defense, helping to protect cellular components from oxidative damage, several studies have now established that beetroot supplementation is an effective means of enhancing athletic performance.[9,17]

1.4.4   CARROT PEEL

Carrots are excellent source of beta-carotene, vitamins A and K, and are also high in fiber and good source of potassium. The highest concentration of phytonutrients is found in skin or immediate underneath. The skin is important for vision, reproduction (sperm production), maintenance of epithelial integrity, growth and development.[2]

1.4.5   CUCUMBER PEEL

The cucumber (*Cucumis sativus*) is a widely cultivated plant in the gourd family *Cucurbitaceae*. Cucumber peels contain mainly silica, a chemical that helps to build collagen, which is vital for making skin last longer than ever. The dark green skin contains the majority of antioxidants, insoluble fiber, and potassium.[18,19] The cucumber peel also holds the most of its vitamin K. Cucumber peel extract has a bactericidal antimicrobial property and has also high content of silicon and sulfur that provide high nourishments and nutrients required for hair growth. The presence of minerals such as magnesium, potassium, silicon, and sulfur greatly helps in regulating blood pressure and maintaining body temperature.[21]

1.4.6   EGGPLANT PEEL

Eggplant is ranked as one of the top 10 vegetables in terms of the oxygen radical scavenging capacity due to its phenolic constituents. Anthocyanins, an important group of naturally occurring pigments of red and/ or purple colored fruits, are the main phenolic compounds in eggplant peel. An eggplant's purple hue comes from a powerful antioxidant called nasunin, which helps protect against cancerous development especially in the brain and other parts of the nervous system. Nasunin is also believed to have antiaging properties. Eggplant skin is also rich in chlorogenic acid, a phytochemical that boasts antioxidant and anti-inflammatory properties, and also promotes glucose tolerance. Although the eggplant interior contains chlorogenic acid, it is more prevalent in the skin.[20]

1.5   GRAIN FIBER

Fiber is indigestible carbohydrate found in plant foods, such as fruits, vegetables, and grain products. Dietary fiber is the edible part of plants, or analogous carbohydrates, that are resistant to digestion and absorption in the human small intestine with complete or partial fermentation in the large intestine.[5] Dietary fiber is an important part of a healthy diet. It helps to move food and waste efficiently through the digestive system.[5,18,19]

For instance, fiber prevents constipation, hemorrhoids, and diverticulosis. Fiber is also linked to prevent colon and breast cancer. In addition, fiber may help to lower the LDL cholesterol and the total cholesterol, thus, reducing the risk of heart disease. Furthermore, fiber can help to lower blood sugar and, therefore, can better help manage diabetes. Both soluble and insoluble fibers are undigested; they are therefore not absorbed into the bloodstream. Instead of being used for energy, fiber is excreted from our bodies. Soluble fiber forms a gel when mixed with liquid, while insoluble fiber does not. Insoluble fiber passes through our intestines largely intact.[18,19]

1.5.1   GRAINS: EDIBLE WITH HUSK AND FIBER

1.5.1.1   OAT FIBER FROM OAT HULL

Oat hulls comprise approximately 30% by weight of the grain. Traditionally, oat hulls were discarded during processing or used as animal feeds. Recently, it has become an important ingredient that can be incorporated in several food formulations due to its high fiber content; the content is higher than wheat or corn bran.[18,19]

Oat fibers have shown various health benefits. Numerous studies indicated that oat fiber was effective in controlling glycemic responses and improving intestinal regularity. Also studies showed that the addition of oat fiber to the diet induced a decrease in blood urea and renal nitrogen excretion relative to the control, indicating a potential for oat fiber diet therapy in chronic renal disease. Oat fiber is effective in improving intestinal regularity, indicating that it has a role in relieving constipation and/or diarrhea.[8] Oat fiber has desirable water holding capacity, thus mixes easily into dough, and does not affect color and flavor. A minimally extracted oat fiber with improved water binding capacity is desirable for manufacturing of crackers, cookies, and ice cream cones.[18,19]

1.5.1.2   BARLEY FIBER

Barley is a unique cereal grain as soluble and insoluble fibers are distributed throughout the mature seed. The fiber in barley is found in the hull, pericarp, and cell walls of the aleurone and starchy endosperm. The hull fiber has equal parts of cellulose and arabinoxylan, and about 20% lignin. Ingestion of barley grain or derived products is associated with increased fecal weight, accelerated transit time, increased cholesterol and fat excretion, and decrease in gallstones. Germinated barley foodstuff (GBF) derived from the aleurone and scutellum fractions of brewers' spent grains mainly consist of low-lignified hemicellulose and glutamine-rich protein. Barley-soluble fiber is also associated with increased fat and cholesterol excretion from the digestive system and prebiotic effects. The Food and Drug Administration (FDA) concluded that consuming whole grain barley and/ or dry milled barley products that provide at least 3 g of β-glucan-soluble fiber per day is effective in lowering blood total and LDL cholesterol.[8,18,19]

1.5.1.3   SUGAR BEET FIBER

Dietary fiber in sugar beet comes exclusively from its cell walls and is devoid of resistant starch or other reserve polysaccharides.[5] Sugar beet pulp has high dietary fiber content, typically > 75%, and is also known for its high soluble fiber content.[5] Sugar beet fiber is claimed to offer nutritional benefits to consumers as well as manufacturing and functional advantages to food processors. Moisture retention, good texture, and mouthfeel are the main technical properties of the beet fibers which are proposed with a variety of particle sizes for easy blending with other ingredients. The beet fiber also has the advantage of not containing phytic acid and gluten.[8,17]

1.5.1.4   OAT β-GLUCAN

The beneficial cholesterol and glucose effects of oat are attributed primarily to water-soluble fiber, called β-glucan. Oat is a good source of different dietary fibers such as β-glucan, arabinoxylans, and cellulose. β-glucan is located mainly in the endospermic cell walls and in the sub-aleurone layer of oats.[5,8]

1.5.1.5   PSYLLIUM

Psyllium is an excellent source of natural soluble fiber and contains, experimentally, eight times more soluble fiber than oat bran on a per weight basis.

It is also used in cosmetics and as an antitussive, anti-inflammatory, and an immune stimulant. It has also been used traditionally in food products such as in bread, honey, marmalade, soup, or mixed with wheat flour as a thickener in the making of chocolates and jellies. Recent uses of psyllium are in the production of ice cream as a thickener, sherbet, and yogurt. Psyllium has proved to be highly effective in the treatment of constipation and the maintenance of bowel regularity. Its stool-bulking activity principally results from the water holding property of the resident polysaccharide, but it has a range of properties such as high non-starch polysaccharide content, high viscosity on hydration, and, uniquely, the ability to retain some structure in the presence of significant microbial fermentation.[8]

1.6 USES OF PEEL AND FIBERS OF FRUITS AND VEGETABLES

- Depending on the thickness and taste, fruit peel is sometimes eaten as part of fruit, such as apples.
- In some cases, the peel is unpleasant or inedible, in which case it is removed and discarded, such as with bananas or grapefruits.
- The peel of some fruits—for example, pomegranates—is high in tannins and other polyphenols, and is employed in the production of dyes.
- The peel of citrus fruits is bitter and generally not eaten raw, but may be used in cooking.
- In gastronomy, the outermost, colored part of the peel is called the zest, which can be scraped off and used for its tangy flavor.
- The fleshy white part of the peel, bitter when raw in most species, is used as succade or is prepared with sugar to make marmalade or fruit soup.

1.7 USES OF GRAIN FIBER

- The introduction of fiber-rich foods, including whole grain breads and cereals as well as fruits and vegetables, early in a child's life, may promote acceptance and continued consumption of these foods later in life.

- Certain fibers such as inulin, polydextrose, and oligofructose can be used to enhance the inherent fiber content of certain foods or be added to foods that typically do not contain fiber in order to help consumers increase their fiber consumption.
- Fiber plays an important role in normal laxation, which is related primarily to fiber's effect on stool weight.
- Cereal fibers, such as bran, are most effective in increasing stool weight and decreasing transit time, since these fibers are partially fermented in the large intestine.
- A diet adequate in fiber is believed to reduce the risk of diverticular disease, which is prevalent in older adults.

1.8   PRECAUTIONS TO BE TAKEN WHILE USING SOME PEELS

Some of the precautions while using some peels are discussed in the following sections.

1.8.1   ORANGE PEEL

Orange peel contains synephrine, which produces stimulant effects similar to those of ephedra, which is associated with increased risks of stroke and heart attack. Some general stimulant effects experienced from synephrine are feeling nervous or restless as well as trouble in sleeping. It is also associated with cardiovascular system effects that include high blood pressure or hypertension, arrhythmias, that is, irregular heart rhythms, a fast heart rate or tachycardia, and fainting, heart palpitations as well as the chest pain. Another potential side effect of orange peel extract is the sensation of significant weakness or paralysis on one side of the body and vision problems and may produce migraine or cluster headaches.

1.8.2   BEETROOT PEEL

The consumption of beetroot provides many health benefits, but at the same time its consumption also brings many side effects. A certain percentage of the population suffers from red urine and coloration of blood, which take place after consuming beetroot. This condition of coloration of blood is called beeturia. A concentrated extract of betaine can also cause nausea, stomach upset, and diarrhea in some people and also increases the total cholesterol level in the body. The high oxalate content in beetroot increases the possibility of kidney stone formation in the body.[17] Drinking beetroot juice can cause a feeling of tightness in the throat and can even make speaking difficult. This happens due to the overconsumption of beetroot.[17]

1.8.3   CARROT PEEL

Carrot, when taken in excess, can cause certain side effects such as skin rashes, diarrhea, anaphylactic reactions, hives, and swelling; these allergies are caused due to the allergen present in carrot pollen. Carrots leave the skin abnormally yellow to orange in color, when taken in large amounts. This discoloration is caused by beta-carotene, a carotenoid content in carrots, and this discoloration can be seen on palms, hands, face, and soles of the feet.[2]

1.8.4   TOMATO PEEL

The major element found in fresh raw tomato is the carotenoid pigment "lycopene." It is a chemical compound that is supposed to keep cancer at bay. But excessive intake of this phytochemical can interfere with the regular activities of our immune system and slow it down. As a result, our body loses its ability to protect itself from several common microbial (bacterial, fungal, and viral) diseases. At the same time, it also becomes incapable of repairing the existing physical damages. Tomato seeds are rich in calcium and oxalate compounds may lead to kidney problems. Lycopene present in the seeds of the vegetable can cause abnormalities in the male prostate gland, affecting the reproductive system, these abnormality causes severe pain, erectile dysfunction, difficulty in urination, and so on and may sometimes it leads to prostate cancer. Excessive lycopene can give us allergic reactions. Some symptoms of lycopene allergy include itching, rashes, hives, chest constriction, swollen lips, burning sensation in eyes, and so on. Prolonged and consistent consumption of tomato can alter skin color also which might take on a slight orange tint.

1.9 SUMMARY

The waste materials such as peels, seeds and stones produced by processing of fruits and vegetables can be successfully used as a source of phytochemicals and antioxidants. The new aspects concerning the use of these wastes as by-products for further exploitation on the production of food additives or supplements with high nutritional value have gained increasing interest because these are high value products and their recovery may be economically attractive. The by-products represent important source of sugars, minerals, organic acids, dietary fiber and phenolics, which have a wide range of actions which include antitumoral, antiviral, antibacterial, cardioprotective and anti-mutagenic activities.

If effective utilization of food residues is to occur, food manufacturers should invest in specialized secondary industry to utilize these residues. Efforts are needed to develop new technologies and to institute suitable measures to promote waste reclamation. This can only be achieved if food residues are considered as complementary resources rather than as undesirable wastes. Many of the skins/seeds of fruits and vegetables are thrown in the garbage or fed to livestock. Fruits and vegetables wastes and by-products, which are formed in great amounts during industrial processing, represent a serious problem, as they exert an influence on the environment and need to be managed and/or utilized. On the other hand, they are very rich in bioactive components, which are considered to have a beneficial effect on health.

KEYWORDS

- allergy
- anti-inflammatory
- antioxidant
- antitumor
- apple
- banana
- barley
- beetroot
- carotenoids
- chemoprotective
- cholesterol
- chronic diseases
- citrus
- economic

- edible
- fiber
- flavor
- food industries
- fruits
- glucosinolates
- grains
- health
- healthy diet
- humans
- juice
- juicy layer
- mango
- minerals
- nutritional value
- oat
- peeling
- phenolics
- pomegranate
- potato
- processing
- psyllium
- pulp
- skin
- taste
- thermal
- thiols
- tocopherols
- tomato
- vegetables
- vitamins

REFERENCES

1. Al-Weshahy, A.; Rao, V. A. Eds. Potato Peel as a Source of Important Phytochemical Antioxidant Nutraceuticals and Their Role in Human Health—A Review. https://www.intechopen.com/; 2012; pp 101–205.
2. Aydın, O.; Bayındırl, L.; Artık, N. Effect of Peeling Methods on Quality of Carrots. *GIDA.* **2010,** *35*(3), 169–175.
3. Baluja, Z.; Kaur, S. Antihypertensive Properties of an Apple Peel—Can Apple a Day Keep a Doctor Away? *Bull. Pharm. Med. Sci.* **2014,** *1*, 19–91.
4. Berkeley. *Dietary Fiber*; University of California: Berkeley, 2012; pp 23–187.
5. Burkitt, D. P.; Walker, A. R.; Painter, N. S. Dietary Fiber and Disease. *J. Am. Med. Assoc.* **1974,** *229*, 1068–1074.
6. Chanda, S.; Baravalia Y.; Kaneria, M.; Rakholiya, K. Fruit and Vegetable Peels— Strong Natural Source of Antimicrobics. In *Current Research, Technology and Education Topics in Applied Microbiology and Microbial Biotechnology;* Mendez V., Ed.; FORMATEX Research Center: Badajoz, Spain, 2010; pp 209–380.
7. Chidan, K. C. S.; Mythily, R.; Chandraju, S. Utilization of Mango Peels (*Mangifera indica*) for the Extraction of Sugars. *Pharma Chem.* **2012,** *4*(6), 2422–2426.

8. Cho, S. S.; Samuel, P. *Fiber Ingredients Food Applications and Health Benefits;* CRC Press/Taylor & Francis Group: Boca Raton, FL, 2009; pp 249–393.

9. Clifford, T.; Howatson, G.; West, D. J.; Stevenson, E. J. The Potential Benefits of Red Beetroot Supplementation in Health and Disease. *Nutrients* **2015,** *7*, 2801–2822.

10. Emadi, B. *Experiment Studies and Modeling of Innovative Peeling Processes for Tough-Skinned Vegetables*; Queensland University of Technology: Brisbane, Queensland, 2005; pp 38–189.

11. Emadi, B.; Abbaspour-Fard, M. H.; Yarlagadda, P. K. D. V. *An Innovative Mechanical Peeling Method of Vegetables*; Queensland University of Technology: Brisbane, Queensland, 2007; pp 23–109.

12. Fathima, S.; Puraikalan, Y. D. Development of Food Products Using Pomegranate Skin. *Int. J. Sci. Res.* **2015,** *110*, 2319–2464.

13. Fowomola, M. A. Some Nutrients and Antinutrients Contents of Mango (*Magnifera indica*) Seed. *Afr. J. Food Sci.* **2012,** *4*(8), 472–476.

14. Hegazy A. E.; Ibrahium, M. I. Antioxidant Activities of Orange Peel Extracts. *World Appl. Sci. J.* **2012,** *18*(5), 684–688.

15. Igor, S.; Ruta, G. Industrial Potato Peel Waste Application: A Review. In: *Res. Rural Dev. Proceedings*; http://www2.llu.lv/; **2015,** *1*, 130–136.

16. Ismail, T.; Akhtar, P. S. S. Pomegranate Peel And Fruit Extracts: A Review of Potential Anti-Inflammatory and Anti-Infective Effects. *J. Ethnopharmacol.* **2012,** *143*, 397–405.

17. Jasna, M.; Sladjana, S.; Gordana, S.; Sonja, M.; Marko, V. Antioxidant and Antimicrobial Activities of Beet Root Pomace Extracts. *Czech. J. Food Sci.* **2011,** *29*(6), 575–585.

18. Joanne, S. Whole Grains and Human Health. *Nutr. Res. Rev.* **2004,** *17*, 1–12.

19. Joanne, S. Fiber and Prebiotics: Mechanisms and Health Benefits. *Nutrients* **2013,** *5*, 1417–1435.

20. Jung, E. B.; M. J.; Eun, K.; Jo, Y. H.; Lee, S. C. Antioxidant Activity of Different Parts of Eggplant. *J. Med. Plants Res.* **2011,** *5*(18), 4610–4615.

21. Jyoti, D. V.; Lakshmi, R.; Swetha, A. Biochemical, Antimicrobial and Organoleptic Studies of Cucumber (*Cucumis sativus*). *Int. J. Sci. Res.* **2014,** *3*, 662–664.

22. Kamila de, A. M.; Sergio M. C. M., R. U.; Giuseppina P. P. L. Sanitizer Effect in Mango Pulp and Peel Antioxidant Compounds. *Food Nutr. Sci.* **2014,** *5*, 929 –935.

23. Kellywolfe, X. W.; Rui, H. L. Antioxidant Activity of Apple Peels. *J. Agric. Food Chem.* **2003,** *51*, 609–614.

24. Livia, B. M.; Antonio R. G. M. Evaluation of Antioxidant and Antimicrobial Capacity of Pomegranate Peel Extract (*Punica granatum*) Under Different Drying Temperatures. *Chem. Eng. Trans.* **2015,** *44*, 2283–2416.

25. Luty, G.; Jerusa S.; Danilo, F. Effects of Peeling Methods on the Quality of Cubiu Fruits. *Cienc. Technol. Aliment.* (Campinas), **2012,** *32*(2), 255–260.

26. Mohapatra, D.; Mishra, S.; Sutar, N. Banana and its By-Product Utilization: An Overview. *J. Sci. Ind. Res.* **2010,** *69*, 323–329.

27. Nunes, X. P. *Phytochemicals as Nutraceuticals—Global Approaches to Their Role in Nutrition And Health*; InTech Publisher: Croatia, European Union, 2012; pp 12–48.

28. Ong, H. F. *Extraction of Essential Oil from Orange Peels*; University Malaysia: Pahang, 2012; pp 209–328.

29. Parashar, S.; Sharma, H.; Garg, M. Antimicrobial and Antioxidant Activities of Fruits and Vegetable Peels: A Review. *J. Pharmacogn. Phytochem.* **2014**, *3*(1), 160–164.

30. Pereiran, A. M. M. Banana (*Musa* spp) From Peel to Pulp: Ethnopharmacology, Source of Bioactive Compounds and its Relevance for Human Health. *J. Ethnopharmacol.* **2015**, *160*, 149–163.

31. Rudra, S. G.; Nishad, J.; Jakhar, N.; Kaur, C. Food Industry Waste: Mine of Nutraceuticals. *Int. J. Sci. Environ. Technol.* **2015**, *4*(1), 205–229.

32. Toor, R. K.; Savage, G. P. Antioxidant Activity in Different Fractions of Tomatoes. *Food Res. Int.* **2005**, *38*(5), 487–494.

33. Tortoe, C. J.; Paa-Nii, T. S.; Ted, M.; Matilde, T. T. Physicochemical, Proximate and Sensory Properties of Pineapple (*Ananas* sp.) Syrup Developed from ilts Organic Side-Stream. *Food Nutr. Sci.* **2013**, *4*, 163–168.

34. Wadhwa, M.; Bakshi, M. P. S. *Utilization of Fruit and Vegetable Wastes as Livestock Feed and as Substrates for Generation of Other Value-Added Products*; FAO-RAP Publication: Bangkok, Thailand, 2013; Vol. 4; pp 178–209.

CHAPTER 2

HEALTH PROSPECTS OF BIOACTIVE PEPTIDES DERIVED FROM SEED STORAGE PROTEINS

S. P. JEEVAN KUMAR[1,*], S. RAJENDRA PRASAD[2], AND ANSHU SINGH[3]

[1]*ICAR—Indian Institute of Seed Science, Maunath Bhanjan, Uttar Pradesh 275103, India, Tel.: +91-547-2530326*

[2]*ICAR—Indian Institute of Seed Science, Maunath Bhanjan, Uttar Pradesh 275103, India, Tel.: +91-9415108377, E-mail: srprasad1989@yahoo.co.in*

[3]*Indian Institute of Technology, Kharagpur, West Bengal 721302, India, E-mail: anshusingh.biotech@gmail.com*

Corresponding author. E-mail:jeevaniitkgp@gmail.com

CONTENTS

2.1   INTRODUCTION

In the course of plant evolution, seed is the most incredible event for its compliance toward the changing environment. Seed has a remarkable feature of storing products, which facilitate protection of tissues and nutritional supply for the developing embryo that enables the genotypes to adapt to the favorable conditions for germination. Seeds store a wide variety of products such as proteins, carbohydrates, lipids, antioxidants, alkaloids, antifungal terpene compounds, which serves not only the purpose of nourishment and protection to the seedling but also attracts other living organisms for seed dispersal.[17]

Seed proteins are mainly categorized into housekeeping proteins, which are responsible for normal plant cell metabolism and storage proteins. These abundant storage proteins play a key role in seed germination and growth. Proteins are classified into structural and biologically active proteins, which are present in minor proportion.[10] Structural proteins aid in building the architecture of cell (cell wall), whereas biological proteins confine for the protection of the cellular machinery such as enzymes, enzyme inhibitors and lectins, and so on. Apart from structural functions, seed storage proteins (SSPs) serve as a major source of dietary proteins for the humans. Their role is noteworthy as a prolific resource for essential amino acids. As human system is unable to synthesize essential amino acids which are indispensable for translation and maintenance of encoded proteins as well as other nitrogenous compounds, external source could satiate the requirement of essential amino acids. This external source is the diet that provides basic nutrition for the proper functioning of the body system.

The importance of SSPs as a source of essential amino acids is well-documented, but recently it has been observed that these dietary proteins possess other functionality in vivo through bioactive peptides. Several investigations on bioactive peptides of SSPs have associated it with positive health impact via regulating physiological functions (Fig. 2.1).[26]

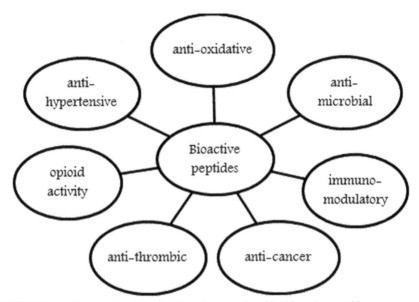

FIGURE 2.1   Various physiological functions regulated by bioactive peptides.

These unique properties of seed storability coupled with bioactive peptides have made seed a good host for the expression system of molecular farming objectives. In addition, due to the attributes such as antimicrobial and supplement of nitrogen in the form of amino acid, it can be incorporated in packaging to maintain food safety and quality.[9] However, despite possessing a huge potential in bioactive and industrial perspectives, the SSPs (plant source) have been undermined when compared with the animal proteins in terms of bioactive peptides.

In this chapter, scope of bioactive peptides and its potent applications for health as well as food industry have been elucidated which is paramount for further research.

2.2   SEED STORAGE PROTEINS

Seed storage proteins (SSPs) are important components in the human diet. It caters the protein requirement to more than half of the global population particularly for human and livestock.[25] It is economically viable and abundant in nature due to which these proteins have been explored over the animal source. Seed storage proteins are particularly important that determine not only the quality of protein but also protein content in the seed.[41] For instance, legumes contain meager amount of sulfur-containing amino acids (cysteine, methionine) whereas in cereals tryptophan, lysine, and threonine are low. These meager quantities of essential amino acids determine the quality of seed.[26] The seat of synthesis takes place either in cotyledon or in endosperm and stores the protein in the protein storage vacuoles (PSVs) called as protein bodies. In the mature seeds, the PSV was packed with the protein deposits which are dense in nature.[13] They have been classified into four types based on its solubility and extractability by Osborne:[31]

- Water soluble (albumins: 1.6S–2S)
- Dilute salt solutions (globulins: 7S–13S)
- Dilute acid or alkali (glutelins)
- Water/alcohol mixtures (prolamins)

Globulins occur as oligomers, prolamins form small aggregates and on the contrary glutelins form large disulfide-bonded aggregates, whose solubility is poor in water.[25] In dicots (e.g., legumes), albumins and globulins occupy a major portion whereas in monocots (e.g., cereals) glutelins and prolamines are predominant in nature.[7] These storage proteins, when undergo gastrointestinal (GI) digestion, release the bioactive peptides from the protein by the action of hydrolytic enzymes.

2.2.1 BIOACTIVE PEPTIDES

Bioactive peptides are specific fragments of protein, which influence the physiological functions with concomitant health improvement.[21] Bioactive peptides when released will perform various actions such as antimicrobial, anticancer, antithrombic, antihypertensive, antioxidant, and immunomodulatory.[26] Bioactive peptides and its assimilation depend on the length, size, and hydrophobicity of the peptide. For instance, mono and di peptides are actively transported via intestinal epithelium with the help of H^+ gradient, whereas oligopeptides would transport through intestinal epithelium either by para cellular channels or pinocytosis based on hydrophobicity and size of the peptide.[4] However, after absorption, the peptides might be degraded by the action of serum protease released from intestine. Hence, peptidase degradation-resistant mechanism would be inevitable to exert the physiological role on the body when administered through intravenous or oral mode.

2.2.1.1 PHYSIOLOGICAL EFFECTS OF BIOACTIVE PEPTIDES FROM SSPs

2.2.1.1.1 Immunomodulatory and Antitumor Activity of Bioactive Peptides

Immune system protects the body from foreign pathogens by executing the innate and acquired immunity processes. Bioactive peptides from plant source have demonstrated immune modulatory and also these work as an adjuvant to trigger the T-cell response for vaccine development.[26] In addition, some of the bioactive peptides work as an antitumor agent. Several peptides such as lunasin, soymetide-4, oryzatensin and soymetide-13 have been studied. Lunasin is a bioactive peptide derived from 2S albumin of soybean, barley, wheat, and amaranthus. It contains 43 amino acids with a cell adhesion motif of three peptides Arg-Gly-Asp followed by nine ASP units.[11]

Several studies substantiate that lunasin acts as antitumor agent, chemopreventive and adjuvant.[19] Lunasinis, a chromatin-binding peptide, interacts with nonacetylatedhistone H3 and H4, which inhibit acetylation of chromatin. As a result, it inhibits the carcinogen-induced fibroblast cells or colony suppression in RAS- transfected mouse.[37] Inhibition of acetylation of histones, a post-translational modification is the chief mechanism responsible for anticancer activity.[11] Recent investigations have revealed that the seed-based platforms are economically viable systems for vaccine production and molecular farming.[3] Tung et al.[47] have studied lunasin as adjuvant for the development of acquired immunity. Dendritic cells (DCs) were characterized after treatment with lunasin, which has not only resulted in increased levels of (CD86, CD40) but also cytokines (IL1B, IL6) and chemokines (CCL3, CCL4). Moreover, mice immunized with oval bumin and lunasin showed inhibition of ova expressed B- lymphomas. These results corroborate that the lunasin acts by the maturation of DC and augments the immune responses for antigens of vaccines.[47]

Soymetide-4 is a food-derived peptide that is agonist to N-formyl methionyl peptide (fMLP), which protects from the bacterial infections. Bacterial protein contains N-formyl methioninereside at its N-termini. On infection, it binds to the fMLP receptors located on the neutrophils and macrophages resulting into reactive oxygen species (ROS) generation and triggering of phagocytosis response from the host. In a similar fashion, soymetide-4 has low-affinity binding to the fMLP receptor but does not generate ROS by neutrophils. As it possesses lower affinity, short length and devoid of inflammation, it is considered to be safe as immune stimulating agent. In addition, Tsuruki et al.[46] have administered soymetide-4 orally, which showed antialopecia (prevention of hair loss) effect.

Bioactive peptides such as oryzatensin and wheat gluten lysate have showed immunomodulatory effect. Oryzatensin, derived from rice albumin tryptic digest, induced phagocytosis and ROS (superoxide anions) in human leukocytes in vitro.[21] Horuguchi et al. (2005) have observed the augmentation of natural killer (NK) cells when wheat hydrolysate was taken by the volunteers. NK cells are deprived in the elderly people and also are also important for autoimmune diseases, cancer, and viral infections. The effect of immune-enhancing activity is attributed due to immunomodulatory bioactive peptides in wheat gluten hydrolysate.[14] Therefore, from the illustration of above examples, it is well reasoned that bioactive peptides function as adjuvant for vaccine development and immunomodulatory effect.

2.2.1.1.2   Antihypertensive Activity

Hypertension (high blood pressure) implies elevation of blood pressure (BP) in the arteries and also a great risk to cause congestive heart failure, renal disease and stroke and coronary heart disease. The BP can be determined by systolic and diastolic blood pressure, which implies the contraction and relaxation phases of the cardiac cycle.[8] An individual is said to have hypertension, when systolic blood pressure is ≥ 140 mmHg or diastolic blood pressure is ≥ 90 mmHg.[48]

To understand the mechanism of antihypertensive activity of bioactive peptides, a brief illustration on the biochemistry is prerequisite. Blood pressure is elevated by the enzyme angiotensin-I-converting enzyme (ACE). It is aexo-carboxy peptidase and metallo-protease in nature. Two events are critical for the elevation of BP. Angiotensin-I (decapaptide) to angiotensin-II (octapeptide) and bradykinin (blood pressure-lowering peptide) act in a mutually exclusion fashion. Angiotensin-II formation, a strong vaso-constricting agent and also the breakdown of bradykinin (vasodilator), elevates the pressure in the arteries.[29] To combat the problem, the amino acids in the bioactive peptides particularly, aromatic amino acids C-termini bind to zinc of ACE which concomitantly inhibit the activity of enzyme. Study on rice, barley, and wheat have shown that they release angiotensin-converting enzyme (ACE)-inhibitory peptides.[5]

2.2.1.1.3 Antioxidant Activity of Bioactive Peptides

Free radicals are consistently produced in the human body due to incomplete reduction of oxygen. As a result, several ROS such as superoxide (O_2^-), hydroxyl (OH·), and hydrogen peroxide (H_2O_2) are generated and damages lipids, proteins, and DNA.[27] Although, antioxidants and enzyme mechanisms counterbalance the ROS species in the human body, still the cells have to be maintained by an external source of antioxidants. Bioactive peptides from enzymatic digest of soy β-conglycinin have demonstrated antioxidant activity against peroxidation of lipids. These peptides contain 5–16 amino acids with a hydrophobic moiety (Val, Leu) situated at N-terminal end whereas the linchpin amino acids (His, Tyr, and Pro) responsible for bioactivity are located in the sequences.[6] In another study, wheat germ protein hydrolysates showed antioxidant activity on par with α-tocopherol when digested by alcalase enzyme.[20] These studies corroborate the role of bioactive peptides in developing antioxidant activity. Table 2.1 depicts the various SSPs known for antioxidant peptide release.

TABLE 2.1 Antioxidant Peptides Obtained from Seed Proteins.

Antioxidant peptide source	References
Corn protein	[22]
Kidney bean protein	[45]
Pea protein	[35]
Soy protein	[33]
Sunflower protein	[36]

2.2.1.1.4   Bile Acid Sequestrant (Cholesterol Reducing Property)

Bile acids are derived from cholesterol (steroid), predominantly found in the juice of vertebrates and mammals. It emulsifies the lipids and helps in the digestion process. Bile acids are synthesized from cholesterol and conjugates with glycine or taurine in the liver to form bile salts. In the process of digestion, it is released into the duodenum, helps in digestion by forming emulsion of lipids and reabsorbs into the body through ileum. Dietary proteins possess the reducing ability of cholesterol. Bioactive peptides bind with the bile acid and hamper the reabsorption of bile acids (entero-hepatic circulation). As a result, the endogenous cholesterol (notably low-density lipids) is streamlined for the synthesis of bile acids that mitigates the serum cholesterol levels.

Dietary proteins have been explored for the property of cholesterol-reducing, particularly, soy-based peptides and isoflavones.[32] Lovati et al.[23] reported that a soy peptide (Leu-Pro-Tyr-Pro-Arg) showed a 25.4 and 30.6% reduction of total and LDL cholesterol levels in mice. Further research findings have demonstrated that the soy peptide derived from 11S globulin (pepsin hydrolysate) bind to cholic and deoxy cholic acids.[32] Nagaoka et al.[28] studied the mechanism of peptides for lowering the cholesterol. The soy protein hydrolysates with and devoid of phospholipids were administered to mice in vitro. From the results, it is hypothesized that the peptides exert its activity of lowering the serum cholesterol by binding bile acid (sequestrant) and hinders the reabsorption.[28]

2.3   PEPTIDES: BIOPROPERTIES AND ITS APPLICATIONS IN FOOD PRESERVATION

The increasing resistance to antibiotics by pathogens has incited to look for novel compounds. An attractive approach is to replace the antibiotics with natural origin molecules such as biopeptides.[2] These peptides have 1–50 amino acids with hydrophobic and cationic moieties that act as defense for the host organism against fungi, bacteria, parasites, and viruses.[12] Peptides with antimicrobial activity have potential applications in food-processing unit, food preservation, and biomedical appliances. Hence, the mode of action of peptides pertaining to antimicrobial activity has been discussed here for better understanding.

2.3.1 ANTIFUNGAL ACTIVITY

Molds are acquainted for resistance due to the consistent use of antifungal agents. Peptides offer as natural antifungal agent and have the potential to strike defense. Mechanism of antifungal activity of peptides can be accomplished by three modes such as peptides which act by cellular lysis, peptides that interact with cell internal targets by crossing fungal membrane (chitin, glucan), and the peptides which act by pore formation.[24,38]

2.3.1.1 CELLULAR LYTIC ACTION OF PEPTIDES

These peptides have a structural feature of amphipathic nature, which indicates that the cationic peptide consists of a hydrophilic region (positively charged head) that interacts with negatively charged moieties or water whereas the hydrophobic residue interacts with the lipids.[1] Amphipathic nature of peptides enables to interact with the amphipathic cell membranes, particularly bacteria but most of the animal cell membranes are devoid of net charge, as a result the cationic peptides remain unaffected.[42] Recent findings on Hispidalin, a cationic bioactive peptide derived from *Benincasa hispida* seeds, demonstrated potent antifungal activity that is comparable to commercial antifungal drugs.[34,40]

2.3.1.2 PEPTIDES ACTION VIA INTERCELLULAR TARGET

Strategy to strike defense by peptide against fungus is by interfering either on the synthesis of cell wall or its components such as chitin or glucan. Later, it may penetrate into the cell and interact with cytoplasmic membranes to inhibit the macromolecule synthesis, ultimately leading to death.[38] Lunatusin, a trypsin-stable antimicrobial peptide from Lima beans (*Phaseolus lunatus* L.), is an example of this category.[48]

2.3.1.3   PEPTIDES THROUGH PORE FORMATION

In this scheme, the peptides are organized in a selective fashion to form pores of different sizes through which the passage of ions and solutes take place.[16] When the peptide forms a pore on the membrane and interacts with cell membrane component (β-1, 3-glucan), it results in destabilization, damage of membranes, and increased permeability, which eventually leads to the death of the fungus.[15]

2.3.2   ANTIBACTERIAL ACTIVITY

Peptides' antibacterial activity has been demonstrated and till date is in debate about its probable mechanism. Hypothesis deduced has been illustrated to understand the mode of action of peptides for the antibacterial activity. The initial step involves the interaction of target cell and peptides through binding of cationic peptide and an anionic cell surface of the bacteria. In Gram-negative bacteria, the net negative charge is due to lipopolysaccharides and in Gram-positive bacteria it is lipoteichoic acid.[18] This interaction destabilizes the membrane by the removal of divalent cations such as Ca^{2+} and Mg^{2+} on the cell surface, which enables the peptide to interact with cell membranes.[34] Thus, the facilitated peptide exerts its action either by interacting with the cytoplasmic membrane or by directly participating in the inhibition of protein, nucleic acid and cell wall syntheses and enzyme activity. Hispidalin, a cationic bioactive peptide, was purified from the *Benincasa hispida* seeds and it showed antibacterial activity.[40]

2.3.3   ANTIVIRAL ACTIVITY

The antiviral activity of peptides has been attributed either to the entry of the host cell (adsorption) or to its effect on the envelope of a virus. The strategy mostly employed for antiviral activity is due to the interaction of a peptide either with the host or virus receptors, which hamper the entry of the virus.[9] For instance, peptides interact directly with viral-specific receptors of the host that inhibits the binding of virus to the cell membrane or intracellular binding.[18] Another strategy involves electrostatic association of negatively charged cell surface compounds on mammalian surface such as glycosaminoglycans (GAG) with virus, proteins and enzymes, respectively.[44]

As the insects are known to loom for resistance to the antibiotics, the natural origin, and eco-friendly molecules are quintessential. These antimicrobial natures of bioactive peptides have been envisaged to replace the conventional antibiotics. Applications of bioactive peptides in food industry have gained importance to mitigate the microbial load before packing or enhancement of shelf-life of the product. Incorporation of antimicrobial peptides in food packaging is called as active packaging, which is envisaged to reduce the microbial contamination and also enhance product shelf-life.[43] Antimicrobial peptides can be packed by three ways: direct incorporation of peptide in the polymer, coating of peptides on the surface of polymer, and immobilization of peptide in the polymer. The incorporated antimicrobial peptides diffuse from the packaging material to food and exert its biological role.[30,39]

2.4   CONCLUSIONS

Seed habitat is a unique property in gymnosperms and angiosperms, which helps to survive in unfavorable conditions. One of the main attributes of seed is SSPs, whose basic function is to nurture the embryo. In addition, it serves as a good protein source, and recent investigations reveal that the SSPs are good source for bioactive peptides.

Bioactive peptides have effect not only on physiological actions but antimicrobial nature also helps to incorporate in active packaging. Bioactive peptides perform various physiological functions such as antitumor activity, blood pressure, and cholesterol-reducing nature, immunomodulatory response revealing its potential as a strong agent for physiological responses. Bioactive peptides have gained great consensus particularly in the realm of antibiotic resistant microorganisms. Bioactive peptides are natural antimicrobial compounds which showed potent action on the resistant microorganisms.

Incorporation of peptides in the polymer is known as active packaging that reduces either the load of microbes or shelf-life of a product. Hence, bioactive peptides have tremendous potential in food industry and needs further research to explore novel bioactive peptides for various functions of human physiology and also in food industry.

2.5   SUMMARY

Seed storage proteins serve as a major protein source and also for bioactive peptides. Bioactive peptides are fragments of a protein, which manifest beneficial implications on humans. It has gained paramount impetus, due to its manifold use for health benefits, antimicrobial, antiviral, and antifungal activities. In plants, the bioactive peptides are ubiquitous in distribution, which serve as a part of innate response upon elicitation. In addition, the bioactive peptides derived from plant source have structural similarities with the insect and animal sources. In most of the cases, the bio-properties mechanism of the bioactive peptides starts by binding the target membrane that results in permeabilization and rupture of the membrane.

Recent research findings of the antimicrobial nature of peptides have potent applications in the food industry. Increasing pathogens resistance due to the wide use of antibiotics has heralded to emphasize on novel compounds with the similar functions. An ingeniousness approach is to find natural; eco-friendly molecules with same functions are desirable. Bioactive peptides with antimicrobial properties have been a potent alternative for its attributes. In addition, it can be incorporated into the packaging material that prolongs the food shelf-life. It is an innovative technology for the preservation of foods aiming either to reduce the microbial load before packing or prolong the desired product. In this chapter, scope of bioactive peptides and its potent applications for health as well as food industry have been elucidated which is paramount for further research.

KEYWORDS

- albumins
- angiotensin-converting enzyme-II
- angiotensin-I
- angiotensin-II
- antifungal activity
- antihypertensive activity
- antimicrobial peptides

- antioxidants
- antitumor activity
- antiviral activity
- bile acid sequestrant
- bioactive peptides
- bradykinin
- cysteine
- globulins
- glutelins
- glycosaminoglycans
- hydrogen peroxide
- hydroxyl
- lipoteichoic acid
- lunatusin
- lunasin
- methionine
- oryzatensin
- prolamins
- protein storage vacuoles
- reactive oxygen species (ROS)
- seed storage proteins
- soymetide-4
- sulfur-containing amino acids
- superoxide

REFERENCES

1. Avitabile, C.; Capparelli, R.; Rigano, M. M. Antimicrobial Peptides from Plants: Stabilization of the γ-Core of a Tomato Defensing by Intra Molecular Disulfide Bond. *J. Pept. Sci.* **2013,** *13*(4), 240–245.
2. Bechinger, B.; Lohner, K. Detergent-Like Actions of Linear Amphipathic Cationic Antimicrobial Peptides. *BBA Biomembr.* **2006,** *1758*(9), 1529–1539.
3. Boothe, J.; Nykiforuk, C.; Shen, Y.; Zaplachinski, S.; Szarka, S.; Kuhlman, P.; Murray, E.; Morck, D.; Moloney, M. M. Seed-Based Expression Systems for Plant Molecular Farming. *Plant Biotechnol J.* **2010,** *8*(5), 588–606.
4. Burton, P. S.; Conradi, R. A.; Ho, N. F. H.; Hilgers, A. R.; Borchardt, R. T. How Structural Features Influence the Biomembrane Permeability of Peptides. *J. Pharm. Sci.* **1996,** *85*, 1336–1340.
5. Cavazos, A.; de Mejia, E. G,; Identification of Bioactive Peptides from Cereal Storage Proteins and Their Potential Role in Prevention of Chronic Diseases. *Compr. Rev. Food Sci. Food Safety* **2013,** *12*, 364–384.
6. Chen, H. M.; Muramoto, K.; Yamauchi, F. Structural Analysis of Antioxidative Peptides from Soybean Beta-Conglycinin. *J. Agric. Food Chem.* **1995,** *43*, 574–578.
7. Derbyshire, E.; Wright, D. J.; Boulter, D. Legumin and Vicilin, Storage Proteins of Legume Seeds. *Phytochemistry* **1976,** *15*, 3–24.
8. Dodek, P. M.; Sackett, D. L.; Schechter, M. T. Systolic and Diastolic Learning: An Analogy to the Cardiac Cycle. *Can. Med. Assoc. J.* **1999,** *160*, 1475–1477.

9. Espitia, P. J. P.; Soares, N. F. F.; Coimbra, J. S. R. C.; Andrade, N. J.; Cruz, R. S.; Medeiros, E. A. A. Bioactive Peptides: Synthesis, Properties, and Applications in the Packaging and Preservation of Food. *Compr. Rev. Food Sci. Food Saf.* **2011,** *11*,187–204.

10. Fukushima, D. Recent Progress of Soybean Protein Foods: Chemistry, Technology, and Nutrition. *Food Rev. Int.* **1991,** *7*, 323–351.

11. Galvez A. F.; Chen N.; Macasieb, J.; de Lumen B. O. Chemopreventive Property of a Soybean Peptide (Lunasin) that Binds to Deacetylated Histones and Inhibit Acetylation. *Cancer Res.* **2001,** *61*, 7473–7478.

12. Hancock, R. E. W.; Sahl, H. G. Antimicrobial and Host-Defense Peptides as New Anti-Infective Therapeutic Strategies. *Nat. Biotechnol.* **2006,** *24*(12), 1551–1557.

13. Herman, E. M.; Larkins, B. A. Protein Storage Bodies and Vacuoles. *Plant Cell* **1999,** *11*, 601–614.

14. Horiguchi, N.; Horiguchi, H.; Suzuki, Y. Effect of Wheat Gluten Hydrolysate on the Immunity System in Healthy Human Subjects. *Biosci. Biochem.* **2005,** *69*, 2445–2449.

15. Hwang, B.; Hwang, J. S.; Lee, J.; Lee, D. G. Antifungal Properties and Mode of Action of Psacotheasin, A Novel Knottin-Type Peptide Derived from *Psacotheahilaris. Biochem. Biophys. Res. Commun.* **2010,** *400*(3), 352–357.

16. Jang, W. S.; Kim, H. K.; Lee, K. Y.; Kim, S. A.; Han, Y. S.; Lee, I. H. Antifungal Activity of Synthetic Peptide Derived from Halocidin, Antimicrobial Peptide from the Tunicate, *Halocynthia aurantium. FEBS Lett.* **2006,** *580*(5), 1490–1496.

17. Jasicka-Misiak, Z.; Jacek, L.; Nowakowska, E. M.; Wieczorek, P. P.; Młynarz, P.; Kafarski, P. Antifungal Activity of the Carrot Seed Oil and its Major Sesquiterpene Compounds. *Z. Naturforsch* **2004,** *59*, 791–796.

18. Jenssen, H.; Hamill, P.; Hancock, R. E. W. Peptide Antimicrobial Agents. *Clin. Microbiol. Rev.* **2006,** *19*(3), 491–511.

19. Jeong, H. J.; Jeong, J. B.; Kim, D. S.; Park, J. H.; Lee, J. B.; Kweon, D. H.; Chung, G. Y.; Seo, E. W.; De Lumen, B. O. The Cancer Preventive Peptide Lunasin from Wheat Inhibits Core Histone Acetylation. *Cancer Lett.* **2007,** *255*, 42–48.

20. Kong, B.; Xiong, Y. L. Antioxidant Activity of Zein Hydrolysates in a Liposome System and the Possible Mode of Action. *J. Agric. Food Chem.* **2006,** *54*, 6059–6068.

21. Korhonen, H.; Pihlanto, A. Bioactive Peptides: Production and Functionality. *Int. Dairy J.* **2006,** *16*, 945–960.

22. Li, H. M.; Hu, X.; Guo, P.; Fu, P.; Xu, L.; Zhang, X. Z. Antioxidant Properties and Possible Mode of Action of Corn Protein Peptides and Zein Peptides. *J. Food Biochem.* **2010,** *34*, 44–60.

23. Lovati, M. R.; Manzoni, C.; Gianazza, E.; Arnoldi, A.; Kurowska, E.; Carroll, K. K.; Sirtori, C. R. Soy Protein Peptides Regulate Cholesterol Homeostasis in hep g2 Cells. *J. Nutr.* **2000,** *130*, 2543–2549.

24. Lucca, A. J.; Walsh, T. J. Antifungal Peptides: Novel Therapeutic Compounds against Emerging Pathogens. *Antimicrob. Agents Chemother.* **1999,** *43*(1), 1–11.

25. Mandal, S.; Mandal, R. K. Seed Storage Proteins and Approaches for Improvement of Their Nutritional Quality by Genetic Engineering. *Curr. Sci.* **2000,** *78*(1), 79–85.

26. Marambe, P. W. M. L. H. K.; Wanasundara, J. P. D. Seed Storage Proteins as Sources of Bioactive Peptides. In: *Bioactive Molecules in Plant Foods;* Florence, O. U, Ed.; Nova Science Publishers: UK, 2012; pp 50–80.

27. Marczak, E. D.; Ohinatam, K; Lipkowski, A.W.; Yoshikawa, M. Arg-Ile-Tyr (RIY) Derived from Rapeseed Protein Decreases Food Intake and Gastric Emptying after Oral Administration in Mice. *Peptides* **2006**, *27*, 2065–2068.

28. Nagaoka, S.; Miwa, K.; Eto, M.; Kuzuya, Y.; Hori, G.; Yamamoto, K. Soy Protein Peptic Hydrolysate with Bound Phospholipids Decreases Micellar Solubility and Cholesterol Absorption in Rats and Caco2 Cells. *J. Nutr.* **1999**, *129*, 1725–1730.

29. Natesh, R.; Schwager, S. L. U.; Sturrock, E. D.; Acharya, K. R. Structural Details on the Binding of Antihypertensive Drugs Captopril and Enalaprilat to Human Testicular Angiotensin I-Converting Enzyme. *Nature* **2003**, *421*, 551–554.

30. Onaizi, S. A.; Leong, S. S. J. Tethering Antimicrobial Peptides: Current Status and Potential Challenges. *Biotechnol. Adv.* **2011**, *29*(1), 67–74.

31. Osborne, T. B. *The Vegetable Proteins;* Longmans Green & Co.: London, 1924; pp 6–15.

32. Pak, V. V.; Koo; M. S.; Kasymova, T. D.; Kwon, D. Y. Isolation and Identification of Peptides from Soy 11S-globulin with Hypocholesterolemic Activity. *Chem. Nat. Compd.* **2005**, *41*, 710–714.

33. Park, S.Y.; Lee, J. S.; Baek, H. H.; Lee, H. G. Purification and Characterization of Antioxidant Peptides from Soy Protein Hydrolysate. *J. Food Biochem.* **2010**, *34*, 120–132.

34. Powers, J. P. S.; Hancock, R. E. W The Relationship between Peptide Structure and Antibacterial Activity. *Peptides* **2003**, *24*(11), 1681– 1691.

35. Pownall, T. L.; Udenigwe, C. C.; Aluko, R. E. Amino Acid Composition and Antioxidant Properties of Pea Seed (*Pisum sativum* L.) Enzymatic Protein Hydrolysate Fractions. *J. Agric. Food Chem.* **2010**, *58*, 4712–4718.

36. Ren, A.; Zheng, X. Q.; Liu, X. L.; Liu, H. A. Purification and Characterization of Antioxidant Peptide from Sunflower Protein Hydrolysate. *Food Technol. Biotechnol.* **2010**, *48*, 519–523.

37. Roberts, P. R.; Burney, J. D.; Black, K. W.; Zaloga, G. P. Effects of Chain Length on Adsorption of Biologically Active Peptides from the Gastrointestinal Tract. *Digestion* **1999**, *60*, 332–337.

38. Rubinchik, E.; Dugourd, D.; Algara, T.; Pasetka, C.; Friedland, H. D. Antimicrobial and Antifungal Activities of a Novel Cationic Antimicrobial Peptide, Omiganan, in Experimental Skin Colonisation Models. *Int. J. Antimicrob. Agents* **2009**, *34*(5), 457–461.

39. Sarmadi, B. H.; Ismail, A. Antioxidative Peptides from Food Proteins: A Review. *Peptides* **2010**, *31*(10), 1949–1956.

40. Sharma, S.; Verma, H. N.; Sharma, N. K. Cationic Bioactive Peptide from the Seeds of *Benincasa hispida. Int. J. Pept.* **2014,** Article ID 156060; 12 pages.

41. Shewry, P. R.; Napier, J. A.; Tatham, A. S. Seed Storage Proteins: Structures and Biosynthesis. *Plant cell* **1995**, *7*, 945–956.

42. Slavokhotova, A. A.; Rogozhin, E. A.; Musolyamov, A. K. Novel Antifungal α-Hairpinin Peptide from *Stellaria* Media Seeds: Structure, Biosynthesis, Gene Structure and Evolution. *Plant Mol. Biol.* **2014**, *84*(1), 189–202.

43. Soares, N. F. F.; Pires, A. C. S.; Camilloto, G. P.; Santiago-Silva, P.; Espitia, P. J. P.; Silva, W. A. Recent Patents on Active Packaging for Food Application. *Recent Pat. Food, Nutr. Agric.* **2009**, *1*(1), 171–178.

44. Spillmann, D. Heparansulfate: Anchor for Viral Intruders? *Biochemie* **2001,** *83*(8), 811–817.
45. Tang, L.; Sun, J.; Zhang, H. C.; Zhang, C. S.; Yu, L. N.; Bi, J.; Zhu, F.; Liu, S. F.; Yang, Q. L. Evaluation of Physicochemical and Antioxidant Properties of Peanut Protein Hydrolysate. *Plos One* **2012,** *7*, 1–7.
46. Tsuruki, T.; Yoshikawa, M. Anti-Alopecia Effect of Gly–Leu–Phe, an Immunostimulating Peptide Derived from—Lactalbumin. *Biosci. Biotechnol. Biochem.* **2005,** *69*, 1633–1635.
47. Tung; Chun-Yu; Sun, J.; Zhou, B.; Robertson, M. J.; Li, F.; Yao, S.; Jaja, M.; Ling H.; David, E. L.; Hua-Chen, C. Activation of Dendritic Cell Function by Soy Peptide Lunasin as a Novel Vaccine Adjuvant. *Vaccine* **2014,** *32*, 541–545.
48. Wong, J. H.; Ng, T. B. Lunatusin, A Trypsin-Stable Antimicrobial Peptide from Lima Beans (*Phaseolus lunatus* L.). *Peptides* **2005,** 26(11), 2086–2092.

CHAPTER 3

MUSHROOMS AS NOVEL SOURCE OF ANTIHYPERLIPIDEMIC AGENTS

ADITYA GANESHPURKAR[1,*], DIVYA BANSAL[2], AND NAZNEEN DUBEY[3]

[1]Shri Ram Institute of Technology—Pharmacy, Jabalpur, Madhya Pradesh, India, Tel.: +91-9993821431

[2]Shri Ram Institute of Technology—Pharmacy, Jabalpur, Madhya Pradesh, India, Tel.: +91-9302958233
E-mail: drdivyabansal1126@gmail.com

[3]Shri Ram Institute of Technology—Pharmacy, Jabalpur, Madhya Pradesh, India, Tel.: +91-9407113473
E-mail: nazneenshabbir@gmail.com

*Corresponding author. E-mail: adityaganeshpurkar@gmail.com

CONTENTS

3.1 INTRODUCTION

Hyperlipidemia can be regarded as one of the major "socioeconomic" issues among the worldwide population. This pathological condition corroborates a strong association with cardiovascular complications, due to the intake of diet rich in fats and reduced physical activity.[19] Later continuous accumulation of low-density lipoproteins (LDLs) cholesterol in "extracellular subendothelial space of arteries" ultimately leads to situations such as hypertension, atherosclerosis, obesity, hypertension, and diabetes along with decreased functionality of vital organs like heart, kidney, and liver.[37]

Apart from alterations in normal metabolic processes, hyperlipidemia is also observed due to several diseases (e.g., diabetes mellitus, hypothyroidism, chronic renal failure, obesity, and alcohol consumption) and the side effects of some medications. The use of antihypertensive drugs also raises the level of lipids in the blood stream. For instance, loop diuretics tend to increase very low-density lipoprotein (VLDL)-cholesterol and low-density lipoprotein (LDL)-cholesterol. β-blockers increase plasma triglycerides and decrease high-density lipoprotein (HDL)-cholesterol. Prednisolone (a corticosteroid) increases total cholesterol and triglycerides by elevating low-density lipoprotein (LDL)-cholesterol. Similar effects have been observed by cyclosporine.[72]

This chapter explores the potential of mushrooms as novel source of antihyperlipidemic agents to control hyperlipidemia.

3.2 MUSHROOMS: TIME-TESTED MEDICINE

Since time immortal, humankind has been using roots, leaves, and fruits from plants in order to cure various ailments. Mushrooms have been regarded as "nutritious" and "tasty" food. Romans considered mushrooms as "food of the gods," which produced as a result of "bolts of lightning". Egyptians deemed mushrooms as "gift from the god Osiris," whereas the Chinese considered them as "elixir of life".[32] The Greek physician, Dioscorides recognized that *Laricifomes* (*Fomitopsis*) *officinalis* (Vill.) Kotl. and Pouzar (Fomitopsidaceae) could be used for the treatment of tuberculosis. A 5300-year-old mummified Ice Man carried *Piptoporus betulinus* (Bull.) P. Karst. (Fomitopsidaceae) and *Fomes fomentarius* (L.) J.J. Kickx (Polyporaceae) with him in his medicinal kit. The use of fungi and macro-fungi by mankind in brewing, baking, and fermentation technology is an effective way of waste treatment and in turn, bioactive and therapeutic agents are obtained.[30]

3.3 MUSHROOMS: A NOVEL SOURCE FOR THE DISCOVERY OF THERAPEUTIC AGENTS

These days, herbal remedies are used by about 80% of the population of the world as primary health care. They are also popular due to their low cost, safety, effectiveness, wide acceptability, and fewer side effects. Herbal constituents are believed to be compatible with human biological system. Mushrooms, as a food and medicine, have been known to humankind since the prehistoric era. Mushrooms have drawn a great attention of scientists throughout the world due to diverse and potent therapeutic activities. Various mushrooms with diverse pharmacological and therapeutic activities are listed in Table 3.1.

TABLE 3.1 Mushrooms with Diverse Pharmacological and Therapeutic Activities.

Activity	Constituents	References
Antimicrobial agents	5,8-epidioxy-5α,8α-ergosta-6,22-dien-3β-ol	[63]
	Oxalic acid	[9]
	Epicorazins	[5]
	Epipolythiopiperazine-2,5-diones	[5]
Antiviral agents	Ganoderic acid B	[18]
	Ganoderiol F	[18]
	Ganodermadiol	[53]
Antidiabetic agents	Ganoderan A and B	[31]
	Coriolan	[35]
	Tremellastin	[74]
	Dehydrotrametenolic acid	[61]
Anti-inflammatory agents	Ganoderic acids A, B, G	[44]
	Ergosterol	[84]
	ergosta-4-6-8(14),22-tetraen-3-one	[84]
	1-oleoyl-2-linoleoyl-3-palmitoylglycerol	[84]

Based on various scientific evidence, mushrooms can be regarded as "functional foods".[10,33] They are rich source of macro- and micronutrients such as carbohydrates, proteins, niacin, biotin, riboflavin, and minerals.[13, 21]

For instance, two preparations of *Lentinus edodes*, namely mycelium extract and lentinan have immunomodulatory and anticancer activity, which is expressed as a result of differentiation, maturation, and cell proliferation that ultimately increases the resistance of host against various types of cancers.[65,48]

3.4   ANTIHYPERLIPIDEMIC AGENTS FROM MUSHROOMS

3.4.1   CHRYSIN

Chrysin (5,7-dihydroxyflavone, Fig. 3.1) is one of the important natural flavonoids which are found in many plants, propolis, and honey.[76,59,40] Under various experiments, chrysin has demonstrated to prevent neuronal death,[36] and to improve cognitive decline.[64] Antidepressant, antiamyloidogenic, antiepileptic, and antiatherogenic effects of chrysin have been extensively reviewed and well documented.[54] Chrysin intake is also associated with antihypercholesterolemic effect. Administration of chrysin causes partial decrease in non-HDL-cholesterol and triglycerides. Thiobarbituric acid relative substances (TBARS) levels and restoration of SOD levels have also been observed.[83] Restoration of lipid parameters and hepatic marker enzyme markers is also observed.[7]

3.4.2   ERITADENINE

Eritadenine (Fig. 3.1) is a hypolipidemic agent that was earlier isolated from shiitake mushroom (*Lentinus edodes*).[42] Eritadenine consumption decreases phosphatidylcholine (PC): phosphatidylethanolamine (PE) ratio in hepatic microsomes and the S-adenosylmethionine (SAM): S-adenosylhomocysteine (SAH) ratio in the liver, along with decrease in plasma cholesterol concentration. Hypercholestremic action of eritadenine is probably due to modification of hepatic phospholipid metabolism.[66]

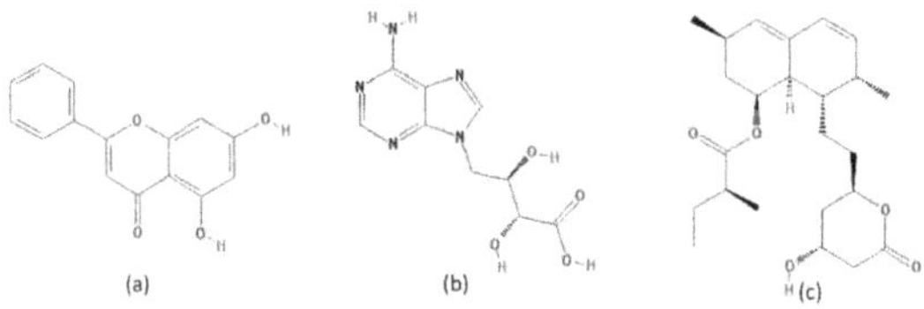

FIGURE 3.1 Antihyperlipidemic agents from mushrooms (a) chrysin; (b) eritadenine; (c) lovastatin.

3.4.3 LOVASTATIN

Lovastatin (Fig. 3.1) is a chemical inhibitor of 3-hydroxy-3-methylglutaryl-coenzyme. This enzyme plays an important role in conversion of HMG-CoA to mevalonate, that is an important substrate for biosynthesis of cholesterol. Lovasatain inhibits cholesterol biosynthesis by a competitively and reversibly blocking HMG-CoA reductase. Lovastatin itself is inactive. Its hydrolysis produces an active metabolite that blocks HMG-CoA reductase.[4,69]

3.5 ANTIHYPERLIPIDEMIC EFFECTS OF MUSHROOMS

3.5.1 PLEUROTUS OSTREATUS

This is one of the important forms of mushroom with medicinal and nutritional value. It is an integral part of many cuisines and is one of the largest edible medicinal mushrooms worldwide. Flavonoids, polyphenols and carotenoids are some of the bioactive constituents found in *P. ostreatus*.[38] Chrysin (5,7-dihydroxyflavone) is a bioactive flavonoid that is also found in it. Anandhi et al.[6] reported that antihyperlipidemic effect of *P. ostreatus* and its major constituent, chrysin, in Triton WR-1339 induced hypercholesterolemia in rats. It was observed that after oral administration of mushroom extract (500 mg/kg by wt.) and chrysin (200 mg/kg) there was a significant reduction in lipid profile, hepatic markers and blood glucose. Administration of chrysin demonstrated significant hypocholestremic effects. Lovastatin is a lipid lowering agent that is also found in *P. ostreatus*.[27] Similar protective effects of *P. ostreatus* have been observed in hereditary hypercholesterolemic rats. Consumption of mushroom reduced the increase in serum cholesterol at the end of 4th week of experiment and reduction in 40% cholesterol on 7th week[12] (Table 3.2).

TABLE 3.2   Antihyperlipidemic Agents from Mushrooms.

Source	Active constituent	Reference
Botryosphaeria rhodina	Botryosphaeran	[51]
Ganoderma lucidum	Ergosterol	[82]
Lentinus edodes	Eritadenine	[78]
Pleurotus citrinopileatus	Ergosterol, Nicotinic acid	[33]
Pleurotus ostreatus	Chrysin	[6]
Pleurotus ostreatus	Lovastatin	[28]
Pleurotus ostreatus	Lovastatin (Mevinolin)	[28]
Pleurotus saca		
Pleurotus sapidus		

3.5.2   PLEUROTUS FLORIDA

Another important culinary mushroom is *Pleurotus florida* that contains large number of myochemicals such as terpenoids, tannins, steroidal glycosides and carbohydrates.[23] *P. florida* has been extensively studied for antidiabetic,[24] analgesic,[26] antioxidant,[25] anticataract,[22] antihyper-cholesterolemic and anthelmintic[38] potential. Khan et al.[43] reported that administration of *P. florida* extract to Long-Evans rats caused reduction in thiobarbituric acid relative substances (TBARS) along with improvement in the levels of reduced glutathione levels. The hypercholesterolemic effect was also observed due to antioxidant constituents found in mushroom.

3.5.3   PLEUROTUS SAJOR-CAJU

Pleurotus sajor-caju can be regarded as one of the best source of high quality of protein.[56] Its administration increases blood hemoglobin levels.[56] *P. sajor-caju* has been investigated for antiviral[71] and anti-inflammatory[62] activities. Administration of *P. sajor-caju* extract caused restoration in the levels of TBARS. The effect predisposed due to antioxidative property of mushroom.[43]

3.5.4   PLEUROTUS FERULAE

It is an edible mushroom that is composed of highest amount of lovastatin in the fruiting bodies.[29] In a study, rats fed with 5% *P. ferulae* diet caused decrement in triglycerides, LD lipoproteins and plasma total cholesterol. Significant weight reduction was also observed in mushroom fed rats. Cholesterol excretion in feces was promoted. The protective effect of mushroom administration was also revealed by histopathological analysis.[3]

3.5.5   PLEUROTUS ERYNGII

Pleurotus eryngii (King oyster mushroom) is recognized for flavor. Alam et al.[2] reported that feeding 5% powder of the fruiting bodies of *P. eryngii* to hypercholesterolemic rats resulted in the decrement of triglyceride, total cholesterol, LDL, and total lipids. No adverse effect was observed on the vital functioning of body organs. Biochemical estimations revealed that administration of mushroom caused decrement of plasma β and pre-β-lipoprotein, whereas increase in α-lipoprotein. Such effects were also confirmed by histopathological analysis.

3.5.6   PLEUROTUS SALMONEOSTRAMINEUS

In a study on 6-week-old female, Sprague-Dawley albino rats, administration of 5% powder of *P. salmoneostramineus* in hypercholesterolemic rats resulted in a decrease in triglyceride, LDL, total lipids, total cholesterol, LDL/HDL ratio, and phospholipids. Excretion of cholesterol and lipids in feces was also observed. In lipoproteins, levels of α-lipoprotein were increased whereas β and pre-β-lipoprotein levels were decreased.[81]

3.5.7   PLEUROTUS NEBRODENSIS

Pleurotus nebrodensis (white sanctity mushroom) is one of the widely consumed mushrooms in China. It is a good source of dietary fibers, nutrients, and many other biologically active compounds. In female Sprague-Dawley rats, supplementation of *P. nebrodensis* led to restoration of lipid profile to normal. There was no alteration in the functioning of other vital organ systems.[1]

3.5.8 PLEUROTUS DJAMOR

Pleurotus djamor administration in experimental animals caused significant decrease in the levels of total cholesterol, LDL, VLDL, and free fatty acids.[58]

3.5.9 PLEUROTUS CITRINOPILEATUS

It is a popular edible mushroom and has been studied for antihypercholesterolemic effects in hamster rats. Administration of dry fruiting body and extracts in rats caused decrease in serum triglycerides and total cholesterol levels. Ergosterol and nicotinic acid were two identified constituents, which might be responsible for antihypercholesterolemic effect.[34]

3.5.10 AGARICUS BISPORUS

White button mushroom (*A. bisporus*) is one of the widely consumed mushrooms throughout the world. *A. bisporus* is a rich source of dietary fibers, polysaccharides, and antioxidants.[20,49,45,46] With respect to cholesterol-lowering effect on rats, serum cholesterol was lowered when supplemented with white button mushroom.[20]

In another independent study, streptozotocin-induced diabetic rats were administered with *A. bisporus* powder (200 mg/kg) for three weeks. There was reduction in blood glucose and triglycerides, after this treatment. Restoration in the activity of hepatic biomarker enzymes was observed. Decrease in total cholesterol and LDL was noted.[39] These effects might be due to the presence of lovastatin.[15]

3.5.11 AGARICUS BLAZEI MURRILL

Royal Sun Mushroom (*Agaricus blazei Murrill*) is rich source of proteins and carbohydrates. It is useful to fight against various diseases such as tumor, cancer, diabetes, and chronic hepatitis.[73] In a research study, administration of *A. blazei Murrill* to animals caused reduction in blood pressure as compared to control group. Hypotensive effect was observed due to the presence of gamma amniobutyric acid and angiotensin-converting enzyme inhibitory substance (unknown) in mushroom.[75] In a clinical study, administration of γ-aminobutyric acid-enriched *A. blazei* to mild hypertensive human subjects caused reduction in systolic and diastolic blood pressure.[70] Protein-bound polysaccharides from *A. blazei*, when administered to 90 female volunteers for 8 weeks, caused significant weight-reduction and hypocholesterolemic effect. Alteration in cholesterol absorption and metabolism were the post mechanisms of such effect.[47]

3.5.12   LENTINUS EDODES

Shiitake mushroom (*Lentinus edodes*) has multipharmacological effects and has been studied against cancer, fungal infection, bronchial inflammation, hypertension, hepatitis, and diabetes.[11] In an independent study, administration of *L. edodes* extract to mice on high-fat diet resulted in the reduction of total cholesterol and LDL in dose-dependent manner. The mRNA expression of cholesterol 7-α-hydroxylase 1 (CYP7A1) was decreased in mushroom treated group. Histopatcal analysis revealed suppression of atherosclerotic plaques due to high-fat diet. Eritadenine, n adenosine analog alkaloid in *L. edodes* is responsible for such effects.[78]

3.5.13   LENTINUS LEPIDEUS

It is one of the popular edible mushrooms in China and Japan. In a study performed by Yoon et al.,[80] administration of *L. lepideus* to albino rats caused reduction in total plasma cholesterol, lipids, LDL, and the ratio of LDL to HDL. However, there was no adverse effect on plasma biochemical and enzyme profiles. Such protective effects were also evident by histopathological studies.

3.5.14   GANODERMA LUCIDUM

Resihi mushroom (*Ganoderma lucidum*) is used as food in Asian subcontinent and is known to endorse health and longevity. It is regarded as "the mushroom of immortality." *G. lucidum* contains about more than 400 bio-active constituents. It has demonstrated various therapeutic effects (viz., anti-inflammatory, analgesic, antifibrotic, antidiabetic, chemopreventive, antitumor, antibacterial, antiviral, antiulcer and antiaging effects).[60] In an independent study performed by Kabir et al.,[40] the effect of administration of *G. lucidum* powder on cholesterol levels was observed. Total cholesterol levels in plasma as well as liver in *G. lucidum*-fed rats were significantly lower as compared to control group. The mechanism behind such effect was thought due to either inhibition in the synthesis of cholesterol or increased cholesterol metabolism.

3.5.15   CALVATIA CYATHIFORMIS

Purple-spored puffball mushroom (*Calvatia cyathiformis*) is a saprobic macrofungi. It is abundantly found in Australia and North America. In a study on male Wistar rats, administration of aqueous and methanol extract of *C. cyathiformis* to animals caused decrement of glycemic index and reduction in cholesterol.[67]

3.5.16   BOTRYOSPHAERIA RHODINA

Macrofungi *Botryosphaeria rhodina* is known to produce water-soluble exopolysaccharide confirmed as β-(1 $\rightarrow$ 3;1 $\rightarrow$ 6)-D-glucan, and termed "botryosphaeran"[8]. In a study, botryosphaeran has demonstrated anticlastogenic and anti-mutagenic activity[50]. In a study, administration of 12 mg botryosphaeran/kg body weight of the rats for 15 days decreased total plasma cholesterol by (18%) and LDL-cholesterol (by 27%).[50]

3.5.17   TERMITOMYCES MICROCARPUS

It is an edible and flavorful mushroom native to Asia and Africa and grows in clusters in the vicinity of bamboo stumps close to termite nests.[85] In a study by Nabubuya et al.,[55] diet containing 25, 45, and 60% air-dried mushroom flour mixed with the basal feed and 0.5% cholesterol were administered to animals. The addition of mushroom powder in diet caused a significant decrement in triglycerides, LDL cholesterol and total cholesterol. The mechanism behind lowering cholesterol level was possibly due to high quantities of fiber in the mushrooms.[55]

3.5.18   TREMELLA FUCIFORMIS

Tremella fuciformis is known as snow fungus or white jelly mushroom and is popular in China. Systematic studies have been performed to determine antihypercholesterolemic activity of *T. fuciformis*. Animals were supplemented with *T. fuciformis* dietary fiber (0, 5, and 10%) along with 1 g of cholesterol/100 g of diet with or without 0.7% Nebacitin for a period of 4 weeks. Administration of mushroom fibers caused decrease in total cholesterol, triglycerides, and serum LDL cholesterol. Excerption of lipids in feces was also promoted. The probable reason for hypocholesterolemic effect was increased excretion if fecal steroids and bile acids along with increased short chain fatty acid production.[16]

3.5.19   AURICULARIA AURICULAR

Jew's wood (*Auricularia auricular*) is an edible macro-fungus found worldwide. It has an important place in Chinese traditional medicine. So far, *A. auricular* has been studied for hypoglycemia,[68] anticoagulant[79] and antitumor activity.[52] In a study, cholesterol-lowering property of *A. auricular* was studied on four-week-old male ICR mice. Oral administration of ethanol extract of *A. auricular* (dose 150 mg/kg/daily) for 8 weeks resulted in improved antioxidant status, decrease in total cholesterol level and atherosclerosis index. There was a remarkable increase in levels of HDL-cholesterol and fecal excretion of bile acids.[14]

3.5.20   CORDYCEPS SINENSIS

Caterpillar fungus is an integral part of traditional Chinese and Tibetian medicine. Traditionally, this fungus is recommended for the treatment of "all illness" and as a "tonic".[14] In a study, water extract of *Cordyceps* fruiting bodies was studied for cholesterol-lowering effects in mice. Administration of water extract to mice caused decrement in increased serum lipid peroxide level. There was also suppression of increased aortic cholesteryl ester level. The protective effect was observed due to the prevention of deposition of cholesterol in aorta due to inhibition of free radical-mediated LDL oxidation.[77]

3.5.21   *VOLVARIELLA VOLVACEA*

Straw mushroom is an edible mushroom found abundantly in Asia. Antihypercholestremic effect of this mushroom was experimentally determined on male Golden Syrian hamsters. Administration of mushroom fruiting body (5%) and mushroom mycelium (5%) caused promotion in the excretion of fecal steroids and bile acids. Fruiting bodies demonstrated more significant cholesterol-lowering effect as compared to that of mycelium.[17]

3.6   SUMMARY

Mushrooms, by virtue of the presence of various constituents (e.g., chrysin, lovastatin, and eritadenine), can be regarded as best source of antihyperlipidemic agents. In addition to the aforementioned constituents, many mushroom polysaccharides have moderate to weak antihyperlipidemic effect when administered orally. Therefore, mushrooms can be regarded as "functional foods," intake of which is beneficial for humans. There are many other issues such as uncovering mechanism of action, safety, and toxicity of mushroom metabolites. Interdisciplinary studies from various science domains are necessary to deal with this issue.

KEYWORDS

- *Agaricus*
- *Auricularia auricular*
- *Botryosphaeria rhodina*
- *Calvatia cyathiformis*
- cholesterol
- chrysin
- eritadenine
- functional food
- *Ganoderma lucidum*
- hyperlipidemia

- *Lentinus edodes*
- *Lentinus lepideus*
- lovastatin
- metabolites
- mushrooms
- myochemicals
- nutraceuticals

- *Pleurotus citrinopileatus*
- *Pleurotus ostreatus*
- *Pleurotus saca*
- steroids
- *Termitomyces microcarpus*
- *Tremella fuciformis*
- triglycerides

REFERENCES

1. Alam, N.; Yoon, K. N.; Lee, J. S.; Cho, H. J.; Shim, M. J.; Lee T. S. Dietary Effect of *Pleurotus eryngii* on Biochemical Function and Histology in Hypercholesterolemic Rats. *Saudi J. Biol. Sci.* **2011,** *18*(4), 403–409.
2. Alam, N.; Yoon, K. N.; Lee, J. S; Lee, M. W.; Lee, T. S. *Pleurotus nebrodensis* Ameliorates Atherogenic Lipid and Histological Function in Hypercholesterolemic Rats. *Int. J. Pharmacol.* **2011,** 7, 455–462.
3. Alam, N.; Yoon, K. N.; Lee, T. S. Antihyperlipidemic Activities of *Pleurotus ferulae* on Biochemical and Histological Function in Hypercholesterolemic Rats. *J. Res. Med. Sci.* **2011,** *16*(6), 776–786.
4. Alberts, A. W. Discovery, Biochemistry and Biology of Lovastatin. *Am. J. Cardiol.* **1988,** *62*(15), 10J–15J.
5. Al-Fatimi, M. A. M. Isolation and Characterization of Antibiotic Compounds from *Ganoderma pfeifferi Bres.* and from *Podaxis pistillaris* (L.:Pers) Morse (*Isolierung und Charakterisierung antibiotisch wirksamer Verbindungen aus Ganoderma pfeifferi Bres. und aus Podaxis pistillaris (L.:Pers.) Morse*). Universität Greifswald, **2001,** *15*, 175. (in German).
6. Anandhi, R.; Annadurai, T.; Anitha, T. S.; Muralidharan, A. R.; Najmunnisha, K.; Nachiappan, V.; Thomas, P. A.; Geraldine, P. Antihypercholesterolemic and Antioxidative Effects of an Extract of the Oyster Mushroom, *Pleurotus ostreatus*, and its Major Constituent, Chrysin, in Triton WR-1339-induced Hypercholesterolemic Rats, Part I. *J. Physiol. Biochem.* **2013,** *69*(2), 313–315.
7. Anandhi, R.; Annadurai, T.; Anitha, T. S.; Muralidharan, A. R.; Najmunnisha, K.; Nachiappan V.; Thomas, P. A.; Geraldine, P. Antihypercholesterolemic and Antioxidative Effects of an Extract of the Oyster Mushroom, *Pleurotus ostreatus*, and its Major Constituent, Chrysin, in Triton WR-1339-induced Hypercholesterolemic Rats, Part II. *J. Physiol. Biochem.* **2013,** *69*(2), 316–323.
8. Barbosa, A. M.; Steluti, R. M.; Dekker, R. F.; Cardoso, M. S.; Corradi da Silva M. L. Structural Characterization of Botryosphaeran: A (1→3;1→6)-β-D-glucan Produced

by the Ascomyceteous Fungus, *Botryosphaeria* sp. *Carbohydr. Res.***2003**, *338*, 1691–1698.

9. Bender, S.; Dumitrache, C. N.; Backhaus, J.; Christie, G.; Cross, R. F.; Lonergan, G. T.; Baker, W. L. A Case for Caution in Assessing the Antibiotic Activity of Extracts of Culinary-medicinal Shiitake Mushroom [*Lentinus edodes* (Berk.) Singer] (Agaricomycetidae). *Int. J. Med. Mushrooms* **2003**, *5*, 31–35.

10. Bensky, D.; Gamble, A. *Chinese Materia Medica*, 2nd Ed.; Eastland Press: Seattle, 1993, pp 55–64.

11. Bisen, P. S.; Baghel, R. K.; Sanodiya, B. S.; Thakur, G. S.; Prasad, G. B. *Lentinus edodes*: A Macro Fungus with Pharmacological Activities. *Curr. Med. Chem.* **2010**, *17*(22), 2419–2430.

12. Bobek, P.; Ginter, E.; Jurcovicová, M.; Kuniak, L. Cholesterol-Lowering Effect of the Mushroom *Pleurotus ostreatus* in Hereditary Hypercholesterolemic Rats. *Ann. Nutr. Metab.* **1991**, *35*(4), 191–195.

13. Breene, W. Nutritional and Medicinal Value of Specialty Mushrooms. *J. Food Prod. Mark.* **1990**, *53*, 883–894.

14. Chen, G.; Luo, Y. C.; Ji, B. P., Li; B., Su; W., Xiao; Zhang, G.Z. Hypocholesterolemic Effects of *Auricularia auricula* Ethanol Extract in ICR Mice Fed a Cholesterol-enriched Diet. *J. Food Sci. Technol.* **2011**, *48*(6), 692–698.

15. Chen, S. Y.; Ho, K. J.; Hsieh, Y. J.; Wang, L. T.; Mau, J. L. Contents of Lovastatin, γ-Aminobutyric Acid and Ergothioneine in Mushroom Fruiting Bodies and Mycelia. *LWT—Food Sci. Technol.* **2012**, *47*, 274–278.

16. Cheng, H. H.; Hou, W. C.; Lu, M. L. Interactions of Lipid Metabolism and Intestinal Physiology with *Tremella fuciformis* Berk Edible Mushroom in Rats Fed a High-Cholesterol Diet with or without Nebacitin. *J. Agric. Food Chem.* **2002**, *50*(25), 7438–7443.

17. Cheung, P. C. Plasma and Hepatic Cholesterol Levels and Fecal Neutral Sterol Excretion are Altered in Hamsters Fed Straw Mushroom Diets. *J. Nutr.* **1998**, *128*(9), 1512–1516.

18. el-Mekkawy, S.; Meselhy; M. R., Nakamura; N., Tezuka, Y.; Hattori, M.; Kakiuchi, N.; Shimotohno, K.; Kawahata, T.; Otake, T. Anti-HIV-1 and Anti-HIV-1-Protease Substances from *Ganoderma lucidum*. *Phytochemistry* **1998**, *49*(6), 1651–1657.

19. Freedman, J. E. High-Fat Diets and Cardiovascular Disease: Are Nutritional Supplements Useful? *J. Am. Coll. Cardiol.* **2003**, *41*(10), 1750–1752.

20. Fukushima, M.; Nakano, M.; Morii, Y.; Ohashi, T.; Fujiwara, Y.; Sonoyama, K. Hepatic LDL Receptor mRNA in Rats is Increased by Dietary Mushroom (*Agaricus bisporus*) Fiber and Sugar Beet Fiber. *J. Nutr.* **2000**, *130*, 2151–2156.

21. Ganeshpurkar, A.; Rai, G.; Jain, A. P. Medicinal Mushrooms: Towards a New Horizon. *Pharmacogn. Rev.* **2010**, *4*(8), 127–135.

22. Ganeshpurkar, A.; Bhadoriya, S. S.; Pardhi, P.; Jain, A. P.; Rai, G. In Vitro Prevention of Cataract by Oyster Mushroom *Pleurotus florida* Extract on Isolated Goat Eye Lens. *Indian J. Pharmacol.* **2011**, *43*(6), 667–670.

23. Ganeshpurkar, A.; Bhadoriya, S. S.; Pardhi, P.; Rai, G. Investigation of Anthelmintic Potential of Oyster Mushroom *Pleurotus florida*. *Indian J. Pharmacol.* **2012**, *44*, 539–540.

24. Ganeshpurkar, A.; Kohli, S.; Rai, G. Antidiabetic Potential of Polysaccharides from The White Oyster Culinary-Medicinal Mushroom *Pleurotus florida* (higher Basidiomycetes). *Int. J. Med. Mushrooms* **2014**, *16*(3), 207–217.

25. Ganeshpurkar, A.; Pardhi, P.; Bhadoriya, S. S.; Jain, N.; Rai, G.; Jain, A. P. Antioxidant Potential of White Oyster Culinary – Medicinal Mushroom, *Pleurotus florida* (Higher Basidiomycetes). *Int. J. Med. Mushrooms.* **2015**, *17(5)*, 491–498.

26. Ganeshpurkar, A.; Rai, G. Experimental Evaluation of Analgesic and Anti-Inflammatory Potential of Oyster Mushroom *Pleurotus florida. Int. J. Med. Mushrooms.* **2013**, *45*(1), 66–70.

27. Gunde-Cimerman, N.; Cimerman, A. *Pleurotus* Fruiting Bodies Contain the Inhibitor of 3-hydroxy-3-methylglutaryl-coenzyme a Reductase-Lovastatin. *Exp. Mycol.* **1995**, *19*, 1–6.

28. Gunde-Cimerman, N.; Cimerman, A. *Pleurotus* Fruiting Bodies Contain the Inhibitor of 3-hydroxy-3-methylglutaryl-coenzyme a Reductase-Lovastatin. *Exp. Mycol.* **1995**, *19*(1), 1–6.

29. Gunde-Cimerman, N.; Plemenitas, A.; Cimerman, A. *Pleurotus* Fungi Produce Mevinolin, an Inhibitor of HMG CoA Reductase. *FEMS Microbiol. Lett.* **1993**, *113*(3), 333–337.

30. Hamlyn, P. F. Fungal Biotechnology. *Brit. Mycol. Soc. Newsl.* **1997**, *3*, 25–27.

31. Hikino, H.; Konno, C.; Mirin, Y.; Hayashi, T. Isolation and Hypoglycemic Activity of Ganoderans A and B, Glycans of *Ganoderma lucidum* Fruit Bodies. *Planta Med.* **1985**, *51*(4), 339–340.

32. Hobbs, C. *Medicinal Mushrooms: An Exploitation of Tradition, Healing and Culture.* Botanica Press: Santa Cruz, 1995, pp 51–55.

33. Hobbs, C. *Medicinal Mushrooms: An Exploitation of Tradition, Healing and Culture.* Botanica Press: Empire Grade, Santa Cruz, 1995, pp 57–62.

34. Hu, S. H.; Liang, Z. C., Chia; Y. C., Lien, J. L.; Chen, K. S.; Lee, M. Y.; Wang, J. C. Antihyperlipidemic and Antioxidant Effects of Extracts from *Pleurotus citrinopileatus. J. Agric. Food Chem.* **2006**, *54*(6), 2103–2110.

35. Ikuzawa, M.; Oguchi, Y.; Matsunaga, K.; Toyoda, N.; Furusho, T.; Fujii, T. Pharmaceutical Preparation Containing a Glycoprotein. *German Patent DE.* **1985**, *3*, 429–551.

36. Izuta, H.; Shimazawa, M.; Tazawa, S.; Araki, Y.; Mishima, S.; Hara, H. Protective Effects of Chinese Propolis and its Component, Chrysin, Against Neuronal Cell Death Via Inhibition of Mitochondrial Apoptosis Pathway in SH-SY5Y cells. *J. Agric. Food Chem.***2008**, *56*, 8944–8953.

37. Jain, K. S.; Kulkarni, R. R.; Jain, D. P. Current Drug Targets for Antihyperlipidemic Therapy. *Mini-Rev. Med. Chem.* **2010**, *10*(3), 232–262.

38. Jayakumar, T.; Thomas, P. A.; Geraldine, P. In-vitro Antioxidant Activities of an Ethanolic Extract of the Oyster Mushroom, *Pleurotus ostreatus. Innovative Food Sci. Emerging Technol.* **2009**, *10*, 228–234.

39. Jeong, S. C.; Jeong, Y. T.; Yang, B. K.; Islam, R.; Koyyalamudi, S. R.; Pang, G.; Cho, K. Y.; Song, C. H. White Button Mushroom (*Agaricus bisporus*) Lowers Blood Glucose and Cholesterol Levels in Diabetic And Hypercholesterolemic Rats. *Nutr. Res.* **2010**, *30*(1), 49–56.

40. Kabir ,Y.; Kimura, S.; Tamura T. Dietary effect of *Ganoderma lucidum* mushroom on Blood Pressure and Lipid Levels in Spontaneously Hypertensive Rats (SHR). *J. Nutr. Sci. Vitaminol.* **1988**, *34*(4), 433–438.

41. Kalogeropoulos, N.; Yanni, A. E.; Koutrotsios, G.; Aloupi, M. Bioactive Microconstituents and Antioxidant Properties of Wild Edible Mushrooms from the Island of Lesvos, Greece. *Food Chem. Toxicol.* **2013**, *55*, 378–385.

42. Kamiya, T.; Saito, Y.; Hashimoto, M.; Seki, H. Structure and Synthesis of Lentysine, A New Hypocholesterolemic Substance. *Tetrahedron Lett.* **1969**, *10*(53), 4729–4732.

43. Khan, M. A.; Rahman, M. M.; Tania, M.; Uddin, M. N.; Ahmed, S. *Pleurotus sajorcaju* and *Pleurotus florida* Mushrooms Improve Some Extent of the Antioxidant Systems in the Liver of Hypercholesterolemic Rats. *Open Nutraceutical J.* **2011**, *4*, 20–24.

44. Koyama, K.; Imaizumi, T.; Akiba, M.; Kinoshita, K.; Takahashi, K.; Suzuki, A.; Yano, S.; Horie, S.; Watanabe, K.; Naoi, Y. Antinociceptive Components of *Ganoderma lucidum*. *Planta Med.* **1997**, *63*(3), 224–227.

45. Koyyalamudi, S. R.; Jeong, S. C.; Cho, K. Y.; Pang, G. Vitamin B_{12} is the Active Corrinoid Produced in Cultivated White Button Mushrooms (*Agaricus bisporus*). *J. Agric. Food Chem.* **2009**, *57*, 6327–6333.

46. Koyyalamudi, S. R.; Jeong, S. C.; Song, C. H.; Cho, K. Y.; Pang, G. Vitamin D2 Formation and Bioavailability from *Agaricus bisporus* Button Mushrooms Treated with Ultraviolet Irradiation. *J. Agric. Food Chem.* **2009**, *57*, 3351–3355.

47. Kweon, M. H.; Kwon, S. T.; Kwon, S. H.; Ma, M. S.; Park, Y. I. Lowering Effects in Plasma Cholesterol and Body Weight by Mycelial Extracts of Two Mushrooms: *Agaricus blazei* and *Lentinus edodes*. *Korean. J. Microbiol. Biotechnol.* **2002**, *30*, 402–409.

48. Maeda, Y. Y.; Sakaizumi, M.; Moriwaki, K.; Yonekawa, H. Genetic Control of the Expression of Two Biological Activities of an Antitumor Polysaccharide, Lentinan. *Int. J. Immunopharmacol.* **1991**, *13*, 977.

49. Mattila, P.; Könkö, K.; Eurola, M.; Pihlava, J. M.; Astola, J.; Vahteristo, L.; Hietaniemi, V.; Kumpulainen, J.; Valtonen, M.; Piironen, V. Contents of Vitamins, Mineral Elements, and Some Phenolic Compounds in Cultivated Mushrooms. *J. Agric. Food Chem.* **2001**, *49*, 2343–2348.

50. Miranda, C. C.; Dekker, R. F.; Serpeloni, J. M.; Fonseca, E. A.; Cólus, I. M.; Barbosa, A. M. Anticlastogenic Activity Exhibited by Botryosphaeran, a new Exopolysaccharide Produced by *Botryosphaeria rhodina* MAMB-05. *Int. J. Biol. Macromol.* **2008**, *42*(2), 172–177.

51. Miranda-Nantes, C. C.; Fonseca, E. A.; Zaia, C. T.; Dekker, R. F.; Khaper, N.; Castro, I. A.; Barbosa, A. M. Hypoglycemic and Hypocholesterolemic Effects of Botryosphaeran from *Botryosphaeria rhodina* MAMB-05 in Diabetes-Induced and Hyperlipidemia Conditions in Rats. *Mycobiology*. **2011**, *39*(3), 187–193.

52. Mizuno, T.; Saito, H.; Nishitoba, T.; Kawagishi, H. Antitumor Active Substances from Mushrooms. *Food Rev. Int.* **1995**, *11*, 23–61.

53. Mothana, R. A.; Awadh Ali, N. A.; Jansen, R.; Wegner, U., Mentel, R., Lindequist, U. Antiviral Lanostanoid Triterpenes from the Fungus *Ganoderma pfeifferi*. *Fitoterapia* **2003**, *74*(1–2), 177–80.

54. Nabavi, S. F.; Braidy, N.; Habtemariam S.; Orhan, I. E.; Daglia, M.; Manayi, A.; Gortzi, O.; Nabavi, S. M. Neuroprotective Effects of Chrysin: From Chemistry to Medicine. *Neurochem. Int.* **2015**, *90*, 224–231.

55. Nabubuya, A.; Muyonga, J. H.; Kabasa, J. D. Nutritional and Hypocholesterolemic Properties of *Termitomyces microcarpus* Mushrooms. *Afr. J. Food, Agric., Nutr. Dev.* **2010**, *10*(3), 2235–2257.

56. Oyetayo, F. L.; Oyetayo, V. O. Assessment of Nutritional Quality of Wild and Cultivated *Pleurotus sajor-caju*. *J. Med. Food.* 12(5), 1149–1153.

57. Panda, A. K.; Swain, K. C. Traditional Uses and Medicinal Potential of *Cordyceps sinensis* of Sikkim. *J. Ayurveda. Integr. Med.* **2011**, *2*(1), 9–13.

58. Raman, J.; Nanjian, R.; Lakshmanan, H.; Ramesh, V.; Srikumar, R. Hypolipidemic Effect of *Pleurotus djamor* var. roseus in Experimentally Induced Hypercholesteromic Rats. *Res. J. Pharm., Biol. Chem. Sci.* **2014**, *5*, 581–88.

59. Rapta, P.; Misík, V.; Stasko, A.; Vrábel, I. Redox Intermediates of Flavonoids and Caffeic Acid Esters from Propolis: An EPR Spectroscopy and Cyclic Voltammetry Study. *Free Radical Bio. Med.* **1995**, *18*, 901–908.

60. Sanodiya, B. S.; Thakur, G. S.; Baghel, R. K.; Prasad, G. B.; Bisen, P. S. *Ganoderma lucidum*: A Potent Pharmacological Macrofungus. *Curr. Pharm. Biotechnol.* **2009**, *10*(8), 717–742.

61. Sato, M.; Tai, T.; Nunoura, Y.; Yajima, Y.; Kawashima, S.; Tanaka, K. Dehydrotrametenolic Acid Induces Preadipocyte Differentiation and Sensitizes Animal Models of Noninsulin-Dependent Diabetes Mellitus to Insulin. *Biol. Pharm. Bull.* **2002**, *25*(1), 81–86.

62. Silveira, M. L.; Smiderle, F. R.; Agostini, F.; Pereira, E. M.; Bonatti-Chaves, M.; Wisbeck, E.; Ruthes, A. C.; Sassaki, G. L.; Cipriani, T. R.; Furlan, S. A.; Iacomini, M. Exopolysaccharide Produced by *Pleurotus sajor-caju*: Its Chemical Structure and Anti-Inflammatory Activity. *Int. J. Biol. Macromol.* **2015**, *75*, 90–96.

63. Smania, A., Jr.; Delle Monache, F.; Smania, E. F. A.; Cuneo, R. S. Antibacterial Activity of Steroidal Compounds Isolated from *Ganoderma applanatum* (Pers.) Pat. (Aphyllophoromycetideae) Fruit Body. *Int. J. Med. Mushrooms.* **1999**, *1*, 325–330.

64. Souza, L. C.; Antunes, M. S.; Filho, C. B.; Del Fabbro, L.; de Gomes, M. G.; Goes, A. T.; Donato, F.; Prigol, M.; Boeira, S. P.; Jesse, C. R.; Flavonoid Chrysin Prevents Age-Related Cognitive Decline Via Attenuation of Oxidative Stress and Modulation of BDNF Levels in Aged Mouse Brain. *Pharmacol., Biochem. Behav.* **2015**, *134*, 22e30.

65. Suga, T.; Maeda, Y. Y.; Uchida, H.; Rokutanda, M.; Chihara, G. Macrophage-mediated Acute-Phase Transport Protein Production Induced by Lentinan. *Int. J. Immunopharmacol.* **1986**, *8*, 691.

66. Sugiyama, K.; Akachi, T.; Yamakawa, A. Hypocholesterolemic Action of Eritadenine is Mediated by a Modification of Hepatic Phospholipid Metabolism in Rats. *J. Nutr.* **1995**, *125*(8), 2134–2144.

67. Tamez de la O, E. J.; Garza-Ocañas, L.; Zanatta-Calderón, M. T.; Lujan-Rangél, R.; Garza-Ocañas, F.; Badillo-Castañeda, C. T.; Gómez, X. S. R. Hypoglycemic and Hypocholesterolemic Effects of Aqueous and Methanolic Extracts of *Lentinus lepideus*, *Calvatia cyathiformis* and *Ganoderma applanatum*, from Northeastern Mexico, in Wistar rats. *J. Med. Plants Res.* **2013**, *7*(11), 661–668.

68. Tanakeuchi, H.; He, P.; Mooi, L. Reductive Effect of Hot-Water Extracts from Woody Ear (*Auricularia auricula-judae Quel*) on Food Intake and Blood Glucose Concentration in Genetically Diabetic KK-AY mice. *J. Nutr. Sci. Vitaminol.* **2004**, *50*, 300–304.

69. Tobert, J. A. Lovastatin and Beyond: The History of the HMG-CoA Reductase Inhibitors. *Nat. Rev. Drug Discovery* **2003**, *2*(7), 517–26.

70. Toshiro, W.; Ayako, K.; Satoshi, I.; Kumar, M. T.; Shiro, N.; Keisuke, T. Antihypertensive Effect of Gamma-Aminobutyric Acid-Enriched *Agaricus blazei* on Mild Hypertensive Human Subjects. *J. Jpn. Soc. Food Sci. Technol. Nippon Shokuhin Kagaku Kogaku Kaishi* **2003**, *50*, 167–173.

71. Verma, S. M.; Prasad, R.; Kudada, N. Investigations of Anti-Viral Properties on Extract of *Pleurotus sajor caju*. *Ancient Sci. Life* **2001**, *21*(1), 34–37.

72. Walker, R.; Whittlesea, C. *Clinical Pharmacy and Therapeutics*; 5th Ed.; Churchill Livingstone: London, 2011; pp 32–35.

73. Wang, H.; Fu, Z.; Han, C. The Medicinal Values of Culinary-Medicinal Royal Sun Mushroom (*Agaricus blazei Murrill*). *J. Evidence-Based Complementary Altern. Med.* **2013**, 842619.

74. Wasser, S. P.; Tan, K. K.; Elisashvili, V. I. Hypoglycemic, Interferonogenous, and Immunomodulatory Activity of Tremellastin from the Submerged Culture of *Tremella mesenterica* Retz.:Fr. (Heterobasidiomycetes). *Int. J. Med. Mushrooms.*, **2002**, *4*, 215–227.

75. Watanabe, T. T.; Yamada, H.; Tamaka, S.; Jiang, T. K.; Mazumder Nagai, S.; Tsuji, K. Antihypertensive Effect of Gamma-Aminobutyric Acid-Enriched *Agaricus blazei* on Spontaneously Hypertensive Rats. *J. Jpn. Soc. Food Sci. Technol. Nippon Shokuhin Kagaku Kaishi* **2002**, *49*, 166–173.

76. Williams, C. A.; Harborne, J. B.; Newman, M.; Greenham, J.; Eagles, J. Chrysin and Other Leaf Exudate Flavonoids in the Genus *Pelargonium*. *Phytochemistry* **1997**, *46*, 1349–1353.

77. Yamaguchi, Y.; Kagota, S.; Nakamura, K.; Shinozuka, K.; Kunitomo, M. Inhibitory Effects of Water Extracts from Fruiting Bodies of Cultured *Cordyceps sinensis* on Raised Serum Lipid Peroxide Levels and Aortic Cholesterol Deposition in Atherosclerotic Mice. *Phytother. Res.* **2000**, *14*(8), 650–652.

78. Yang, H.; Hwang, I.; Kim, S.; Hong, E. J.; Jeung, E. B. *Lentinus edodes* Promotes Fat Removal in Hypercholesterolemic Mice. *Exp. Ther. Med.* **2013**, *6*(6), 1409–1413.

79. Yoon, S. J.; Yu, M. A.; Pyun, Y. R.; Hwang, J. K.; Chu, D. C. Nontoxic Mushroom *Auricularia auricula* Contains a Polysaccharide with Anticoagulant Activity Mediated by Antithrombin. *Thromb. Res.* **2003**, *112*,151–158.

80. Yoon, K. N.; Lee, J. S.; Kim, H. Y.; Lee, K. R.; Shin, P. G.; Cheong, J. C.; Yoo, Y. B.; Alam, N.; Ha, T. M.; Lee, T. S. Appraisal of Antihyperlipidemic Activities of *Lentinus lepideus* in Hypercholesterolemic Rats. *Mycobiology* **2011**, *39*(4), 283–289.

81. Yoon, K. N.; Alam, N.; Shim, M. J.; Lee, T. S. Hypolipidemic and Antiatherogenesis Effect of Culinary-Medicinal Pink Oyster Mushroom, *Pleurotus salmoneostramineus* L. Vass. (higher Basidiomycetes), in Hypercholesterolemic Rats. *Int. J. Med. Mushrooms* **2012**, *14*(1), 27–36.

82. Yuan, J. P.; Wang, J. H.; Liu, X.; Kuang, H. C.; Huang, X. N. Determination of Ergosterol in *Ganoderma* Spore Lipid from The Germinating Spores of *Ganoderma*

lucidum by High-Performance Liquid Chromatography. *J. Agric. Food Chem.* **2006,** *54*(17), 6172–6176.

83. Zarzecki, M. S.; Araujo, S. M.; Bortolotto, V. C.; de Paula, M. T.; Jesse, C. R.; Prigol, M. Hypolipidemic Action of Chrysin on Triton WR-1339-Induced Hyperlipidemia in Female C57BL/6 mice. *Toxicol. Rep.* **2014,** *1*, 200–208.

84. Zhang, Y.; Mills, G. L.; Nair, M. G. Cyclooxygenase Inhibitory and Antioxidant Compounds from the Mycelia of the Edible Mushroom *Grifola frondosa*. *J. Agric. Food Chem.* **2002,** *50*(26), 7581–5.

85. Zhishu, B.; Zheng, G.; Taihui, L. *The Macrofungus Flora of China's Guangdong Province;* (Chinese University Press), Columbia University Press: New York, 1993; pp 142.

CHAPTER 4

EMERGING FOODBORNE ILLNESSES AND THEIR PREVENTION

MURLIDHAR MEGHWAL[1,*], H. T. SOWMYA[2],
LAKSHITA MAHERDA[3], DEEPAK KUMAR[4],
RAVI KUMAR KADEPPAGARI[5]

[1]*Assistant Professor, Department of Food Science and Technology, National Institute of Food Technology Entrepreneurship & Management, Kundli - 131028, Sonepat, Haryana, India Email: murli.mdm@niftem.ac.in, murli.murthi@gmail.com Tel.: +91 9739204027*

[2]*Department of Food Technology, Center for Emerging Technologies, Jain Global Campus, Jain University, Jain Global Campus, Jakkasandra, Kanakapura Main Road, Ramanagara District, 562112, Karnataka, India, Tel.: +91-9739204027 E-mail: sowmyaht300@gmail.com*

[3]*Sardar Patel Medical College, Bikaner, India, Tel.: +919413107109, E-mail: maherdalakshita@gmail.com*

[4]*Agricultural and Biological Engineering, University of Illinois at Urbana, Champaign, IL 61801, USA, Tel.: +001-2173001929 E-mail: kumard@illinois.edu*

[5]*Centre for Incubation, Innovation, Research and Consultancy, Thathaguni, Bangalore 560082, India, http://ciirc.jyothyit.ac.in/, Tel.: +91-9739204027, E-mail: ravikadeps@gmail.com*

Corresponding author. E-mail: murli.murthi@gmail.com

CONTENTS

4.1   INTRODUCTION

Foodborne disease (FBD) (also referred as foodborne illness (FBI) or food-poisoning) is any illness that results from the consumption of food contaminated with pathogenic bacteria, viruses, or parasites. FBDs are major problem in the present fast-food scenario.[20] These problems are also caused by the food contaminated with harmful fungi or chemicals or toxins. Usually the harmful microbes do not make the contaminated food appear, smell, or taste different from the normal food when it was prepared fresh. Therefore, it is difficult to know whether the food is contaminated or not just by physical appearance when it was served fresh.[2,11,12] FBDs are acute illnesses that are associated with the consumption of toxic, polluted or spoiled food.[20] Such food contains enough pathogens or toxicants necessary to make a person sick.[2,5,11]

The WHO reports that approximately 30% of the population in industrialized countries suffers from FBDs each year, and more than one-third of the overall reported cases of *salmonella* occur in children under the age of 10 years.[9,11,12] According to WHO estimation, the worldwide food- and water-borne diarrheal diseases taken together kill about 2.2 million people annually.[2,12,15] FBI is a preventable public health challenge that causes an estimated 48 million illnesses each year in the United States [2,12,15,19] according to the Centers for Disease Control and Prevention (CDC). The CDC estimates that each year in the United States, 1 in 6 common Americans or 48 people get sick, 128,000 are hospitalized, and 3000 die of FBDs.[20]

The onset of symptoms may occur within minutes to weeks and often presents itself as flu-like symptoms such as nausea, vomiting, diarrhea, or fever. Since the symptoms are often flu-like, many people may not recognize that the illness is caused by harmful bacteria or other pathogens in food.[5,14] Everyone is at risk for getting an FBI. However, some people are at greater risk for experiencing a more serious illness or even death should they get a FBI.[12,19] Those at greater risk are infants, young children, pregnant women and their unborn babies, older adults, and people with weakened immune systems (such as those with HIV/AIDS, cancer, diabetes, kidney disease, and transplant patients). Some people may become ill after ingesting only a few harmful bacteria whereas others may remain symptom free after ingesting thousands of bacteria.[2,5,8,13]

FBDs are still not well controlled and outbreaks can cause health and economic losses. The main causes for FBI are unhygienic practices during food preparation, production handling, and crop harvesting. There are around 30 important foodborne pathogens that cause illness. The significant ones are *Salmonella* (nontyphoidal), *Campylobacter*, *Listeria* and *shiga*-toxin producing *Escherichia coli*.[20] These are monitored by national authorities, and outbreaks are assessed in depth to determine the trends of outbreaks and steps necessary to combat future outbreaks. FBD can be mild with recovery in days or severe resulting in hospitalization and death in certain cases.

Generally a healthy and normal person does not have much long-term effect on his/her health due to minor FBI, but even a minor FBI can be lethal for very young girl or boy, very old persons, pregnant women, and sick people. FBI reaction may occur in a few hours or up to several days after exposure. Infectious disease spreads through food or beverages are common.

There are more than 250 known FBIs and different diseases have different symptoms, so that there is no one "syndrome" that is FBI. However, when the microbe or toxin enters the body through the gastrointestinal (GI) tract, often causes the initial symptoms like nausea, vomiting, abdominal cramps, and diarrhea, which are common in many FBDs. Many microbes can spread in more than one way, so we cannot always know that a disease is a foodborne. The distinction matters, because public health authorities need to know how a particular disease is spreading to take the appropriate steps to stop it. For example, *E. coli* O157:H7 infections can spread through contaminated food or drinking water or swimming water or toddler to toddler at a day-care center.[12,17] Depending on the cause, the measures to stop future cases could range from removing contaminated food from stores, chlorinating a swimming pool, or closing a child day-care center. These are most prevalent diseases in poor and developing countries such as south East Asia and South Africa.[2,15]

The main objectives of this chapter is to study about FBI, to review the main causes of FBI, how the disease affects the health, what are the sources of contamination of food, and how to prevent FBIs.[12,15]

4.2   TYPES OF FBI AND CAUSATIVE AGENTS

4.2.1   TOXICATION OR POISONING

Causative pathogens that cause FBD are listed in Table 4.1.[4,5,10,12,14,17] Illness occurs as a consequence of ingestion of a preformed bacterial or mold toxin. Once the microorganisms have grown and produced toxin in the food, there is no need of viable cells during the consumption of food for illness to occur. *Staphylococcus* enterotoxin food poisoning is an example.[2,12,16]

4.2.2   INFECTION

Infection is caused due to the consumption of foods or liquids contaminated with bacteria, viruses, or parasites. These pathogens cause infection by multiplying in the lining of the intestine and/or other tissue invasion and releasing a toxin. It is necessary that pathogenic bacteria and viruses remain alive in the food or water during consumption. Viable cells, even if present in small numbers, have the potential to establish and multiply in the digestive tract to cause the illness. Salmonellosis and hepatitis-A are examples of such problems.[2,4,6]

4.2.3   INTOXICOINFECTION

It is caused by consuming foods or beverages already contaminated with a toxin. Sources of toxins are as follows:

- Certain bacteria (preformed toxins).
- Poisonous chemicals
- Natural toxins found in animals, plants, and fungi.

Generally, the bacterial cells either sporulate or die and release toxins to produce the symptoms. *Clostridium perfringens* gastroenteritis is an example.[2,4,6,13]

TABLE 4.1   Microbial FBI and Causative Pathogens. (*Source:* Adopted by Compiling Information from Several Articles.)[4,5,10,12,14,17]

Causative pathogens1	Incubation period	Symptoms	Foods	Source	References
Bacterial infection					
E. coli	3–4 days	Diarrhea, vomiting, mild fever	Undercooked ground beef, unpasteurized cider	Human and bovine intestinal tract	[2]
Campylobacter jejune	2–5 days	Diarrhea, vomiting, headache, fever, muscle pain	Poultry, dairy products, water	Intestinal tract of wild domestic animals	[12]
Clostridium perfringens	10–12 h	Abdominal pain, nausea, diarrhea, fever, headache, vomiting usually absent	Stews, gravies, beans	soil, animal, and human intestinal tracts	[2]

TABLE 4.1 *(Continued)*

Causative pathogens1	Incubation period	Symptoms	Foods	Source	References
Listeria monocytogenes	3–70 days	Flu-like, meningitis, encephalitis, spontaneous abortion	Unpasteurized milk, ice cream, ready-to-eat, lunchmeats	soil, water, damp environments, domestic/wild animals (esp. fowl)	[15]
Salmonella enteriditis	12–36 h	Abdominal cramps, headache, fever, nausea, diarrhea	Poultry, meat, eggs and egg products, sliced melons	Water, soil, insect, animals, and humans	[11]
Intoxication					
Clostridium botulinum	4 h–8 days	Vomiting, constipation, difficulty with vision, swallowing, speaking paralysis, death	baked potatoes, soups vide, garlic/oil mixtures, low-acid canned foods	present on almost all foods, soil, water	[13]
Bacillus cereus	30 min–6 h (Emetic) and 6–15 h (Diarrheal)	Nausea, vomiting, watery diarrhea	rice products, starchy foods, casseroles, puddings, soups	soil and dust, cereal crops	[12]
Staphylococcus aureus	1–7 h	Nausea, retching, abdominal cramps, diarrhea	ready-to-eat, reheated foods, dairy products	skin, hair, nose, throat, infected sores, animals	[2]

TABLE 4.1   *(Continued)*

Causative pathogens1	Incubation period	Symptoms	Foods	Source	References
Viruses (infection)					
Hepatitis-A	10–50 days	Sudden fever, vomiting, jaundice	water (ice), shellfish, ready-to-eat, fruit juices, vegetables	human intestinal/ urinary tracts	[15]
Norwalk virus	10–50 h	Nausea, diarrhea, headache, mild fever	water, shellfish, raw vegetables and fruits	human intestinal tract, water	[11]
Rotavirus	1–3 days	Vomiting, diarrhea, mild fever	ready-to-eat, water and ice	human intestinal tract, water	[2]
Parasites					
Giardia lamblia	3–25 days	Fatigue, nausea, gas, weight loss, abdominal cramps	water, ice, raw vegetables	beavers, bears, dogs, cats, humans	[12]
Anisakis simplex	hours–2 weeks	Tickle in throat, coughing up worms	undercooked, improp-erly frozen seafood	marine fish-bottom feeders	[11]
Cryptosporidium parvum	1–12 days	Severe diar-rhea, may have no symptoms	water, raw foods, unpas-teurized cider, ready-to-eat	humans, cattle, barn-wash	[15]
Cyclospora cayetanensis	days to weeks	Watery diarrhea, weight loss, bloating, cramps, vomiting, muscle aches	water, marine fish, raw milk, raw produce	humans, water	[15]

TABLE 4.1 *(Continued)*

Causative pathogens1	Incubation period	Symptoms	Foods	Source	References
Trichinella spiralis	2–28 days	Flu-like, swelling around eyes, extreme sweating, hemor-rhaging	undercooked pork, game	domestic pigs, bear, walrus	[2]

List of causative pathogens is given in a website of the Centers of Disease Control (the US Government): https://www.cdc.gov/foodborneburden/PDFs/11_228412_Pitts_factsheet_tables_remediated.pdf

4.2.4 BACTERIAL ILLNESS

Bacteria are tiny organisms that can cause infections of the GI tract. Not all bacteria are harmful to humans. Raw foods including meat, poultry, fish and shellfish, eggs, unpasteurized milk and dairy products, and fresh produce often contain bacteria that can cause FBIs.[3,4,12] Bacteria can contaminate food—making it harmful to eat—at any time during growth, harvesting or slaughter, processing, storage, and shipping. Bacteria multiply quickly when the temperature of food is around 37°C. Bacteria multiply more slowly when food is refrigerated, and freezing food can further slow or even stop.[2] The majority of FBIs are caused by harmful bacteria and viruses. Some parasites and chemicals also cause FBIs.[17,19] *Salmonella* causes 31% of food-related deaths, followed by *Listeria* (28%), *Campylobacter* (5%), and *E. coli* O157:H7 (3%), by multiplying rapidly on food and inside human body. CDC estimates around 6.5–33 million FBI cases each year and in which *E. coli* was found to be causing about 21,000 FBI cases each year, *Salmonella* causes FBI 2-4 million cases per year and campylobacter cause 1–6 million cases per year.[2,12,15]

Listeria monocytogens can grow well under refrigeration temperatures, although very slowly. This is the basis for seven-day date-marking requirements for refrigerated potentially hazardous foods. The illness due to this pathogen can be very severe for pregnant women. There were several outbreaks associated with fresh cheeses made from unpasteurized milk. The cheese (Queso fresco) is very popular in the Hispanic community. Outbreaks occur each year in the United States along the Mexican border. It is important for pregnant women to avoid these types of cheeses because listeriosis can lead to spontaneous abortions or fetal death even if the mother is asymptomatic. Death rate among new born children can be as high as 50% due to this illness.

The control measures include the prevention of cross-contamination and usage of pasteurized milk.[2,3] *Bacillus cereus* is a spore-forming bacterium that produces toxins that cause vomiting or diarrhea. Symptoms are generally mild and short-lived (up to 24 h). *B. cereus* is commonly found in the environment and in a variety of foods. Spores are able to survive harsh environments including normal cooking temperatures. It is a Gram-positive, motile (flagellated), spore-forming, and rod-shaped bacterium. It produces two types of toxins—emetic (vomiting) and diarrheal—causing two types of illness. The emetic toxin produced by the bacteria during the growth phase in the food leads to the emetic syndrome. The diarrheal syndrome is caused by diarrheal toxins produced during growth of the bacteria in the small intestine.

4.2.4.1   CLOSTRIDIUM BOTULINUM

Bacteria produce a toxin that causes botulinum disease that can cause paralysis, breathing failure and death. It is an indicator organism in canned foods, which have a low-acid and low-oxygen environment.[2] The foodborne intoxication can be caused by *C. perfringens* enterotoxin (CPE) produced in the GI tract by entero-toxigenic strains of *C. perfringens*. The organism is found in the soil, dust, water, sewage, marine sediments, decaying materials, intestinal tracts of humans and other animals. This organism is a spore-forming, anaerobic, Gram-positive bacillus. Food-poisoning strains have a variety of origins including human and animal feces, abattoirs, sewage, and flies. Spores produced by these organisms can resist boiling for 4 h or more. If the spores are present as contaminants on raw meat, they may resist boiling or steaming, and on slow cooling the spores will germinate and rapidly multiply.

Botulism is a type of foodborne intoxication caused due to the consumption of enterotoxins produced by the strains of *Clostridium botulinum*. It is an obligate, spore-forming anaerobe, and a Gram-positive bacillus. The strains are divided into proteolytic and nonproteolytic types depending on their ability to hydrolyze proteins. Botulinum can cause paralysis, breathing failure, and death. It is an indicator organism in canned foods which have a low-acid and low-oxygen environment. The intoxication is caused by botulinal toxins A, B, E, F and G, produced by C. botulinum type A, B, E, F and G, respectively, when the organism grows in the food. The C. botulinum types C and D produce toxins C and D that cause disease in animals. Type E strains are nonproteolytic whereas the rest are proteolytic. Spores of C. botulinum type A can survive temperatures of 120°C. [2,4,6,7]

4.2.4.2   SHIGELLA AND YERSINIA ENTEROCOLITICA

The Shigella bacterium will be present in the stools of people who are infected. If the infected people are not hygienic, they can contaminate the food materials. Water mixed with infected stools can also contaminate produce in the field. These bacterial infections are most common in children. Both Shigella and Yersinia can cause diarrheal and abdominal pain. Infection with Yersinia often occurs from raw or undercooked meat products.[1,2,3,5,7]

   Shigella bacterium can spread from person to person. These bacteria are present in the stools of people who are infected. If people who are infected and do not wash their hands thoroughly after using the bathroom, they can contaminate food that they handle or prepare. Water contaminated with infected stools can also contaminate produce in the field. These are bacterial infections most common in children. It can cause diarrheal and abdominal pain. Infection with Yesinia often occurs from raw or undercooked meat products.[2,4,5]

4.2.4.3   VIBRIO

It is a bacterium that can infect shellfish in coastal water. It is found in higher concentration in the summer and most outbreaks occur during summer from the consumption of raw oysters and clams.[2] The bacterium Campylobacter jejuni is commonly found in raw poultry meat and can be easily destroyed by cooking. It causes diarrhea, abdominal cramps, nausea, headache, and fevers usually lasting 2–7 days. The bacteria,

E. coli are potential food-poisoning pathogens which are widely distributed in low numbers in food environments.[2,3,7]

4.2.4.4 CAMPYLOBACTER, SALMONELLA, AND E. COLI

It is commonly found in raw poultry meat and can easily be destroyed by cooking. *Campylobacter* causes diarrhea, abdominal cramps, nausea, headache, and fevers usually lasting for 2–7 days. *E. coli* are potential food-poisoning pathogens which are widely distributed in low numbers in food environments.[2,3,5,16]

4.2.4.4.1 Salmonella

It is the second most common bacterial FBI after campylobacter. It is most common in children and CDC data show that more than one-third of *salmonella* cases occur in the children under age below 10. Symptoms usually include nausea, headache, fever, diarrheal, and abdominal cramps. This lasts for 2–7 days.[2,5]

4.2.4.4.2 Escherichia coli

It includes several different strains, only a few of which cause illness in humans. *E. coli O157:H7* is the strain that causes the most severe illness.[12] Common sources of *E. coli* include raw or undercooked hamburger, unpasteurized fruit juices and milk, and fresh produce. These are numerous strains of *E. coli* bacteria that can cause foodborne infection. But the most serious one is *E. coli* O157:47. It lives in the intestines of animals. Particularly cattle *E-coli* o157:H7 can cause severe diarrheal and abdominal cram.[2,4,5]

The infection of *Salmonella* is most common in the children and according to the CDC data, more than one-third of *Salmonella* cases occur in the children under 10 years. It is found in the intestines of birds, reptiles, and mammals. It can spread through the consumption of raw poultry, eggs, meat, and unwashed fruits. A person with this infection usually develops fever, diarrhea, and abdominal cramps which last for 2–7 days. Most people get recovered on their own and do not need medication. Some sick patients require antibiotics, intravenous fluids, and hospital admission. In case of people with weakened immune system, it can get into the bloodstream and cause severe illness and ultimately leading to death. Occasionally, people recovering from *Salmonella* infection can develop irritated eyes, painful joints, and pain during urination, a condition called Reiter's syndrome. Some infected people can have no symptoms at all, but become chronic carriers who can spread the disease to others. The strain, *E. coli O157:H7* causes the most severe illness. Common sources of *E. coli* include raw or undercooked hamburger, unpasteurized fruit juices and milk, and fresh produce. There are numerous strains of *E. coli* that cause FBI. Particularly cattle *E. coli* O157:H7 can cause severe diarrheal and abdominal cramps. The strains involved in the foodborne infection are grouped into the following: Entero pathogenic *E. coli* (EPEC), enterotoxigenic *E. coli* (ETEC), entero invasive *E. coli* (EIEC), and enterohemorrhagic *E. coli* (EHEC).[2]

4.2.4.5   STAPHYLOCOCCUS AUREUS

It is a type of foodborne intoxication that is caused by consumption of food contaminated with staphylococcal enterotoxins produced by certain strains of *Staphylococcus aureus* while growing in food. The bacterium, *S. aureus* produces a toxin that is heat stable, but not a fatal one. Around 50% of individuals are carriers of these bacteria in their mucous membranes or on skin or hair. This pathogen is associated with other illnesses of the skin that can be as mild as pimples and infected cuts to sepsis leading to death. Outbreaks have been associated with cold protein salads, meats, and cream pastries. The control measures to prevent *Staphylococcus* intoxication are proper time and temperature control and good personal hygiene. The organism produces the following five serologically different enterotoxins that are involved in foodborne intoxication: Staphylococcal enterotoxin-A (SEA), Staphylococcal enterotoxin-B (SEB), Staphylococcal enterotoxin-C (SEC), Staphylococcal enterotoxin-D (SED) and Staphylococcal enterotoxin-E (SEE).

4.2.5 VIRAL ILLNESS

Viruses are tiny capsules, much smaller than bacteria. Viruses cause infections that can lead to sickness. People can pass viruses to each other. Viruses are present in the stool or vomit of people who are infected. People who are infected with a virus may contaminate food and drinks, especially if they do not wash their hands thoroughly after using the bathroom.[4] Common sources of foodborne viruses include:[12,19]

- Food prepared by a person infected with a virus
- Shellfish from contaminated water
- Produce irrigated with contaminated water

The most common viruses are Noro viruses, hepatitis-A, hepatitis-E, and rotaviruses.[12]

Norovirus is also known as norwalk-like viruses or calico viruses are a group of viruses that cause inflammation of the stomach lining and intestines. They are most commonly spread through food handling.[12] CDC estimates that noroviruses are the most common known source of FBI, causing approximately 5.5 million cases each year.[12,13]

Hepatitis-A can cause diarrhea, nausea, fever, and inflammation of the liver. There is no effective treatment for hepatitis-A.[5,12]

4.2.6 WATER

Food contamination can occur at the source, such as seafood harvested from fecal contaminated waters, or by an infected food-handler through poor personal hygiene. On cruise vessels, the major cause of norovirus outbreaks has been attributed to person-to-person spread. These viruses cause inflammation of the stomach lining and intestines. They are most commonly spread through food handling. According to CDC estimation, noroviruses are most commonly known source of FBI, causing approximately 5.5 million cases each year.[12] Hepatitis-A virus also occurs when food or water are contaminated by feces. The contamination can occur at the source, for example, sewage contamination of shellfish beds. The reservoir for hepatitis-A is human. It can cause diarrhea, nausea, fever, and inflammation of the liver. There is no effective treatment for hepatitis-A.[12]

4.2.7 PARASITES

Parasites are the organisms that live inside another organism. Many FBIs in the world can also be caused by parasites. The *Cryptosporidium parvum* and *Giardia intestinalis* are parasites that are spread through water contaminated with the stools of people or animals that are infected. Another parasite *Trichinella spiralis* is a kind of roundworm. People may be infected with this parasite by consuming raw or undercooked pork. They are transmitted from the host to host through contaminated food or water or through oral contact with anything that has touched the feces of an infected animal or person.[5,12]

4.2.8 CHEMICALS

Harmful chemicals that cause severe illness may contaminate foods such as fish or shellfish. In addition, they may feed on algae that produce toxins, leading to high concentrations of toxins in their bodies. Some types of fish, including such as tuna, may be contaminated with bacteria that produce toxins if the fish are not properly refrigerated before they are cooked or served. Certain types of poisonous mushrooms, unwashed fruits, and vegetables contain high concentrations of pesticides.

4.3 CONTAMINATION OF FOODS

4.3.1 TYPES OF CONTAMINATION

4.3.1.1 PHYSICAL

Physical contaminants such as metal pieces, stones, wood, and glass pieces can contaminate the food. Commonly, contamination occurs during handling and transportation of food and in other ways it can occur through the machineries used for food manufacturing.[2]

4.3.1.2   CHEMICAL

Chemical contaminants in food may be naturally present or can get added during the processing of food. Examples of naturally occurring chemicals are mycotoxins (e.g., aflatoxin), scombrotoxin (histamine), ciguatoxin, mushroom toxins, and shellfish toxins. Cleansers, metal leaching (copper, lead, and cadmium), and pesticides can also contaminate the food.

4.3.1.3   BIOLOGICAL

These organisms are commonly associated with humans and with raw products they enter the food preparation site. Many of these microorganisms occur naturally in the environment where food is grown. Therefore, some contamination by these pathogens can be expected in any raw food. The level of contamination can be minimized by adequate control of handling and sound storage practices (hygiene, temperature, and time). Furthermore, most of the organisms of concern are killed or inactivated by proper cooking practices.[14] Bacteria and fungi present the greatest risk. First, both raw and cooked food can provide a fertile medium and support rapid growth of these organisms. Second, there are toxins of fungal and bacterial origin that are relatively heat stable and can remain at hazardous levels even after cooking. Therefore, the contamination levels in the raw food should be minimized even though bacteria, fungi, viruses, and parasites were killed during cooking.[2,8] Unlike bacteria and fungi, human-pathogenic viruses are unable to reproduce outside a living cell. In general, they cannot replicate in food, and can only be carried by it. Furthermore, most foodborne viruses affecting humans are limited to human hosts. This makes contamination by the unclean hands of infected food-handlers or contamination from human feces the prime risk factors.[8]

A range of helminthic and protozoan parasites can contaminate food. Many are zoonotic and can infect a range of animals leading to the direct contamination of the food. Some diseases are fecal origin whereas others are transmitted via consumption of contaminated flesh. Parasitic infections are commonly associated with undercooked meat products or contaminated ready to eat food. Some parasites in products that are intended to be eaten raw, marinated or partially cooked can be killed by suitable effective freezing techniques.[2,4,7,18]

4.4 PATHWAYS OF CONTAMINATION

4.4.1 FOOD-HANDLERS

A food-handler can be any person, who is involved in a food business. Food is the part of their course of work, or part of their duties. Prevention of infected food-handlers from the food business will help in the prevention of FBI. Screening of food-handlers routinely for sporadic illness and in an outbreak situation will also help in line with this. Treatment guidelines, if appropriate, for asymptomatic food-handlers, will eliminate the carriage.[9,14]

4.4.2 WATER

Drinking water comes from two main sources, that is, groundwater (wells and springs) and surface water (rivers, lakes, and reservoirs). Water is the important means of contamination and different methods were used for disinfecting water. Although smaller water utility companies use groundwater sources, most people in the United States live in big towns and cities, which get their water from surface water sources. Most suppliers' use a process that makes water passes through different steps, including flocculation, filtration, and disinfection. Flocculation is a process where chemicals added to the water make dirt and other contaminants clump together and settle at the bottom. The water is then passed through filters, which remove smaller particles, and then disinfected, most commonly with chlorine.[4]

4.4.3 PACKAGING

Packaging design and materials should provide adequate protection for products to minimize contamination, prevent damage, and accommodate proper labeling. Packaging materials or gases must be nontoxic and do not pose a threat to the safety and suitability of food under the specified conditions of storage and use. Whenever appropriate, reusable packaging should be suitably durable so that it is easy to clean and is easy to disinfect.

4.4.4   INSECTS AND RODENTS

Foods attract flies, cockroaches, and rodents to food-processing facilities. They transfer dirt and pathogens from contaminated areas to food through their body parts and excreta.

4.4.5   SEWAGE

Untreated sewage carries high microbial load and may contaminate water, food, or equipment through faulty plumbing. To prevent this kind of contamination, proper waste management is necessary. Rats and mice carry dirt and disease with their feet, fur and feces.

4.4.6   FOOD CONTACT SURFACES

Food contact surfaces are any type of surface that may come in direct contact with exposed food, for example, meat or poultry product come in contact with conveyor belt, table tops, saw blades, augers, and stuffers.[3]

4.4.7   AIR

Air is one of the best ways for the contamination of food. Air contains dirt dust and some of the microorganisms, which contaminate the food and spoil their quality and develop the disagreeable smell. Figure 4.1 shows the flow chart showing the overview of FBI.

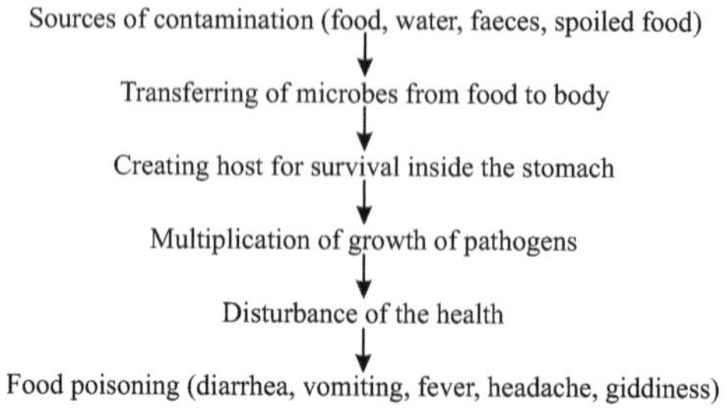

Sources of contamination (food, water, faeces, spoiled food)
↓
Transferring of microbes from food to body
↓
Creating host for survival inside the stomach
↓
Multiplication of growth of pathogens
↓
Disturbance of the health
↓
Food poisoning (diarrhea, vomiting, fever, headache, giddiness)

FIGURE 4.1   Overview of how food contamination occurs.

4.4.8 SEQUENCES FOR CONTAMINATION

Several events should happen in sequence. Understanding these sequences is helpful to investigate the cause (source and means of transmission) of a FBD.[17] It also helps recognize how the sequence can be broken in order to stop a FBD. Initially, there has to be a source of a pathogen. Next, the pathogen has to contaminate a food. Consumption of the food contaminated with a pathogenic virus can lead to viral infection. For bacterial pathogens (and toxicogenic molds), the contaminated food has to support growth and be exposed for a certain period of time at a suitable temperature to enable the pathogens to grow. However, for some potent pathogens (such as *E. coli* O157:H7), growth may not be necessary to cause a foodborne infection. For intoxication, growth should reach a sufficient level to produce enough toxins so that when the food is consumed, an individual develops the symptoms. For bacterial infection, viable cells of a pathogen need to be consumed in sufficient numbers, which vary greatly with pathogens, to survive stomach acidity, establish in the digestive tract, and cause illness. For toxic infection, viable cells should be consumed either in very high numbers (for those that cannot multiply in the digestive tract, such as *C. perfringens*) or in reasonable numbers (for those that can multiply in the digestive tract, such as *Vibrio cholera*) so that toxins released by them in the digestive tract can produce the symptoms.[4,14]

4.4.9 PREVENTING CONTAMINATION

Washing, rinsing and sanitizing cutting boards, knives, utensils, and countertops after contact with raw meat will prevent contamination. Raw meat should be stored below low temperature and away from all ready-to-eat foods. Wash, rinse, and sanitize food-contact equipment (slicers, knives, cutting boards) for at least every 4 h. Maintaining good hygienic practices by the people who handles and consume the food will also help to the greater extent.[3]

4.4.10 FOODS THOSE ARE ASSOCIATED WITH FBI

Raw foods of animal origin are the most likely to be contaminated, for example, raw meat and poultry, raw eggs, unpasteurized milk, and raw shellfish. Due to filter-feeding, shellfish strain microbes from the sea over many months and they are particularly likely to be contaminated if there are any pathogens in the seawater. Foods that mingle the products of many individual animals, such as bulk raw milk, pooled raw eggs, or ground beef, are particularly hazardous because a pathogen present in any one of the animals may contaminate the whole batch. A single hamburger may contain meat from hundreds of animals.[19] A single restaurant omelet may contain eggs from hundreds of chickens. A glass of raw milk may contain milk from hundreds of cows. A broiler chicken carcass can be exposed to the drippings and juices of many thousands of other birds that went through the same cold water tank after slaughter. Fruits and vegetables consumed raw are a particular concern. Washing can decrease but not eliminate contamination, so that the consumers can do little to protect themselves. Recently, a number of outbreaks have been traced to fresh fruits and vegetables that were processed under less than sanitary conditions. These outbreaks show that the quality of the water used for washing and chilling the produce after it is harvested is critical. Using water that is not clean can contaminate many boxes of produce. Fresh manure used to fertilize vegetables can also contaminate them. Alfalfa sprouts and other raw sprouts pose a particular challenge, as the conditions under which they are sprouted are ideal for growing microbes as well as sprouts, and because they are eaten without further cooking. That means that a few bacteria present on the seeds can grow to high number of pathogens on the sprouts. Unpasteurized fruit juice can also be contaminated if there are pathogens in or on the fruit that is used to make it.[7,14]

4.5   MOST VULNERABLE AGE GROUPS TO FBI

4.5.1   YOUNG CHILDREN (INFANTS AND SCHOOL-AGED)

Children face higher risks when exposed to foodborne pathogens because their less-developed immune systems have a limited ability to fight infections.[8,19] Also, their lower body weight reduces the amount of a pathogen needed to cause illness. Children under 5 experience higher rates of laboratory-confirmed infections from 8 of the 10 major foodborne pathogens, both bacteria and parasites, tracked by the public health system. Even though young children are more likely than the general population to get a diagnosis of FBI, the incidence of many pathogens remains higher for the young than for all other ages combined. Significant reductions in foodborne infections among young children are likely to be necessary to achieve the Healthy People 2020 objectives for several of its targeted pathogens.[4,9,13,19]

4.5.2   OLDER ADULTS (ELDERLY)

As people age, their immune system and other organs become sluggish in recognizing and ridding the body of harmful bacteria and other pathogens that cause infections,[4] such as FBI. Many older adults have been diagnosed with one or more chronic conditions, such as diabetes, arthritis, cancer, or cardiovascular disease, and are taking at process and or the side effects of some medications may also weaken the immune system. In addition, stomach acid decreases as people get older, and stomach acid plays an important role in reducing the number of bacteria in the intestinal tract and the risk of illness.[8,19]

4.5.3   INDIVIDUALS WITH COMPROMISED IMMUNE SYSTEMS (PREGNANT)

Changes during pregnancy alter the mother's immune system, making pregnant women more susceptible to FBI. Harmful bacteria can also cross the placenta and infect an unborn baby whose immune system is underdeveloped and not able to fight infection. FBI during pregnancy is serious and can lead to miscarriage, premature delivery, stillbirth, sickness, or the death of a newborn baby.[4] *Listeria monocytogenes* infections pose an increased risk for pregnant women and newborns. Pregnant women are about 10 times more likely than the general population to get a *Listeria* infection; about 1 in 7 cases occur during pregnancy. Illness can lead to miscarriage or stillbirth, premature delivery, or illness/death of the newborn. Many other severe complications and long-term effects can result.[1,4,8]

4.5.4 PEOPLE WITH IMMUNE SYSTEM WEAKENED BY DISEASE OR MEDICAL TREATMENT

The immune system is the body's natural reaction or response to "foreign invasion." In healthy people, a properly functioning immune system readily fights off harmful bacteria and other pathogens that cause infection. However, the immune system of transplant patients and people with certain illness (such as HIV/AIDS, cancer, and diabetes) are often weakened from the disease process and or the side effects of some treatments, making them susceptible to many types of infections like those that can be brought on by harmful bacteria that cause FBI. In addition, diabetics may lead to a slowing of the rate at which food passes through the stomach and intestines, allowing harmful foodborne pathogens an opportunity to multiply.[4,14,19]

4.6 MODES OF TRANSMISSION OF FBIs

4.6.1 POOR CONSUMER HANDLING OF FOOD

As a means to address domestic and catering food hygiene practices, the Food Hygiene Campaign will be a valuable adjunct to other pathogen-specific work within the FDS. Food preparation in the home is a vitally important yet highly variable step of the food chain, with different practices, knowledge, facilities, and habits in the United Kingdom's 26 million households. Since 2001, the Food Hygiene Campaign has promoted the simple 4Cs principles of good food hygiene (see the graphic below) that, if adhered to, should prevent the majority of cases of domestic FBI. These principles also formed the basis for much of the Safer Food Better Business food safety management pack that was launched in 2005. Since 2007, the FSA has led the UK-wide Food Safety Week Initiative to promote these same themes.[19]

4.6.2   CHANGES IN FOOD MANUFACTURING, RETAIL, FOOD DISTRIBUTION, AND STORAGE

Influence of the food industry on consumption patterns and food quality is a major factor in FBD incidence. While the move toward large centralized food processing presents an opportunity to apply better quality control through centralized prevention strategies, it also carries a risk for larger outbreaks.[7,8,9] Food products are now available from diverse sources, and food is often transported over large distances and/or handled many times between its points of production and consumption. Refrigeration is frequently used as a means of preservation. In contrast to the sterility achieved with canning, refrigeration allows the survival of microorganisms. The belief that food would remain safe if kept cold has been challenged with the realization that some pathogens like *Listeria monocytogenes* grow well in some refrigerated foods, and can reach populations as high as 10 cells/g without the product showing adverse signs of spoilage.

The food industry has become one of the largest sectors of Australia's manufacturing industry, employing about 700,000 people with an annual turnover in excess of $80 billion (1000 million). The Australian food industry is expected to be a major source of value-added food in the Asian region by the year 2000 with a target of $7 billion annually in exports. The main emphasis of the food export initiative is the image of Australian food as "clean and green." A failure to produce safe food would have a significant financial effect on the food industry.

4.6.3   EATING PREFERENCES

Changing patterns of food consumption have had a major influence in the increasing incidence of FBD. In the past, food was produced and consumed locally. Traditionally, Australian food was simple and well cooked. Today, the pattern is fewer meals cooked at home and more reliance on ready-to-cook, ready-to-eat foods, and takeaway meals. A trend toward eating fresh unprocessed foods and processed foods without preservatives permits the growth of foodborne pathogens. Minimally processed and extended shelf life food also carry inherent risks to increased contamination. Raw foods of animal origin are increasingly being included into our diet. For example, raw fish, popular in Japanese, Korean, and several other cultures, has become increasingly popular with Australian consumers who are unaware of the different handling and storage techniques used in the traditional preparation of these raw foods or the risks of food-FBD caused by *Vibrio parahaemolyticus*, the principal cause of foodborne outbreaks in Japan.[19]

4.6.4   NEW MODES OF TRANSMISSION

Food can become contaminated by harmful microbes at almost any point on its way from the farm of these microbes in animal or human fecal matter. Meat can come in contact with the fecal matter in animals intestines during slaughter and become contaminated with pathogens such as *E. coli* or *salmonella*. Infected hens and cows can also pass these microbes to consumers through their eggs and milk. Fruits and vegetables can come in contact with fecal matter if they are washed with water that has been contaminated by sewage in the ocean. Another way that FBI spreads is through poor hygiene of people who handle food, for example, not washing their hands properly. The spread of FBI can also occur during processing or preparation if food touches other food, knives, cutting boards and utensils that are contaminated. FBI occurs when a group of people consume the same contaminated food and two or more of them come down with the same illness.[12] It may be a group that ate a meal together somewhere, or it may be a group of people who do not know each other at all, but who all happened to buy and eat the same contaminated item from a grocery store or restaurant. For an outbreak to occur, something must have happened to contaminate a batch of food that was eaten by the group of people. Often, a combination of events contributes to the outbreak. A contaminated food may be left out a room temperature for many hours, allowing the bacteria to multiply to high numbers, and then be insufficiently cooked to kill the bacteria

4.7   EFFECTS OF FOODBORNE PATHOGENS

FBI cause numerous physical, economic, and social problems in the United States; however, there is disagreement over how serious their impact is abdominal pain, diarrhea, and vomiting are the most common physical symptoms of FBI and usually last only a few days.[5,15]

Sometimes it may cause serious complications and even lifelong health problems including kidney failure, seizures, paralysis, hearing and vision problems, and mental retardation. Foodborne pathogens have much different effect on a body.[14] After a pathogen is swallowed there is an incubation period before illness begins, usually ranging from one hour to three weeks because of this incubation period it is often difficult for a person to know which food caused his or her illness. In some cases, the body may destroy a food pathogen with a person experiencing no symptoms at all. But in other cases, a person becomes ill after the incubation period. When the pathogen enters the stomach and intestines, it causes irritation resulting in the common symptoms of vomiting and diarrhea.[5] Foodborne pathogens harm the body through either infection or intoxication. An infection occurs when a harmful bacteria, viruses, or parasites is ingested and grows in the intestinal tract and causes illness. *Salmonella* and *E. coli* are examples of infection, intoxication or poisoning is caused by ingesting food that is already contaminated with a poisonous toxin. The toxin is what causes illness, not the bacteria that produce it, botulism and shellfish poisoning are examples of intoxication.[4,14,16]

4.7.1   MONITORING CASES OF FBI

Local state and federal agencies all play a role in monitoring cases of FBI, identifying outbreaks and taking action to stop further illness, health and taking action to stop further illness. Health-care providers and laboratories usually report FBI cases to local health agencies who then report to state agencies.[12]

When an FDA- or FSIS-regulated food is involved in an outbreak, the appropriate agency takes action in order to ensure that contaminated food is recalled and the source of the outbreak eliminated.

If a company refuses to issue a recall of unsafe food, the FDA has the authority to mandate one. The FSIS cannot issue a recall, but it has the authority to detain and seize contaminated food. Foodborne pathogens have much different effect on body. After a pathogen is swallowed, here is an incubation period before illness begins, usually ranging from 1 h to 3 weeks. Because of this incubation period, it is often difficult for a person to know which food caused his or her illness. In some cases, the body may destroy a food pathogen with a person experiencing no symptoms at all, but in other cases a person becomes ill after the incubation period. When the pathogen in the stomach and intestines is present, it causes irritation resulting in the common symptoms of vomiting and diarrhea.[5]

Foodborne pathogens harm the body through either infection or intoxication. An infection occurs when a harmful bacteria, virus, or parasitic is ingested and grows in the intestinal tract and causes illness. *Salmonella* and *E. coli* are examples of infection. Intoxication or poisoning is caused by ingesting food that is already contaminated with a poisonous toxin. The toxin is what causes illness, not the bacteria that produce it botulism and shellfish poisoning are examples of intoxication.[4,12,16,18]

4.7.2   SHORT- AND LONG-TERM EFFECTS ON HEALTH

In a small percentage of cases, the pathogens that cause FBI harm essential body organs, causing serious complications and health problems or even death. The FDA estimates that 2 to 3% of all FBI result in long-term health problems such as paralysis, seizures, neurological problems, and kidney failure.[1]

4.7.3   AWAKENING OF OTHER DISEASES

There is evidence that in some cases FBIs trigger the development of other diseases. When a person ingests a foodborne pathogen, the immune system tries to protect the body by attacking that pathogen. However, sometimes the immune system reacts inappropriately, attacking the body's own cells as well. Such an inappropriate response can lead to a chronic health problem. Guillain–Barré syndrome (GBS) can cause permanent paralysis and hospitalization, and it kills approximately 100 people every year in the United States.[20] In addition, the Center for Foodborne Illness Research & Prevention reports that approximately one-third of people with GBS need to be cared for in rehabilitation or long-term care facilities.[9] Some research suggests that FBI may be a cause of other chronic health problems, including irritable bowel syndrome and Cohn's disease—long-term disorders that involve various digestive problems.[1,6]

4.7.4 EXPENSES BURDEN DUE TO THE TREATMENT OF FBDS

Past estimates range greatly, depending on the specific illnesses and health costs that are included. Estimating cost is further complicated by the facts that statistics about FBI are limited and there is limited knowledge about the extent of long-term health problems. One widely accepted estimate of cost is from 1997 by the US Department of Agriculture (USDA), which estimates that the public cost of illness and death from FBI is $35 billion.[12,15]

4.7.5 NEGATIVE EFFECTS ON BUSINESS AND ECONOMY

When a case or outbreak of FBI is attributed to a food supplier, then the business can experience financial harm in numerous ways. If a food item is contaminated, it cannot be sold, meaning lost revenue. In addition, if it must be recalled, there are large costs associated with carrying out that recall, which can involve finding and retrieving thousands or even millions of items. Loss of confidence can also be financially harmful to a business because it reduces future sales, and costly public relations campaigns may be necessary to restore customer confidence.

4.7.6 LOSING FAITH AND CONFIDENCE OF CONSUMERS

In addition to health problems and financial costs, FBI can be harmful to society by causing people to lose confidence in the safety of their food supply. When the food that comes from those producers is the cause of FBI, consumers often feel they cannot rely on them. The Safe Food Coalition complains: "The last few years have been marked by widespread, severe outbreaks of food borne illness caused by common foods contaminated with deadly pathogens".[7,12,15]

4.8 PREVENTION OF FOODBORNE ILLNESS

Foodborne illnesses can be prevented by properly storing, cooking, cleaning, and handling foods. Raw and cooked perishable foods—foods that can spoil—should be refrigerated or frozen promptly. If perishable foods stand at room temperature for more than 2 h, they may not be safe to eat. Refrigerators should be set at 40°C or lower and freezers should be set at 0°C. Foods should be cooked long enough and at a high enough temperature to kill the harmful bacteria that cause illnesses. A meat thermometer should be used to ensure foods are cooked to the appropriate internal temperature, 145° for roasts, steaks, and chops of beef, veal, pork, and lamb, followed by 3 min of rest time after the meat is removed from the heat source; 160° for ground beef, veal, pork, and lamb; 165° for poultry, cold foods should be kept cold and hot foods should be kept hot.[19]

Fruits and vegetables should be washed under running water just before eating, cutting, or cooking. A produce brush can be used under running water to clean fruits and vegetables with firm skin. Raw meat, poultry, seafood, and their juices should be kept away from other foods.

People should wash their hands for at least 20 s with warm, soapy water before and after handling raw meat, poultry, fish, shellfish, produce, or eggs. People should also wash their hands after using the bathroom, changing diapers, or touching animals.

Utensils and surfaces should be washed with hot, soapy water before and after they are used to prepare food. Diluted bleach—1 teaspoon of bleach to 1 quart of hot water—can also be used to sanitize utensils and surfaces.

The CDC has following simple recommendations how to decrease the risk of developing a FBD:

- Cook meat, poultry, and eggs thoroughly.
- Separate cooked and uncooked food. Avoid cross-contamination by not using platters or utensils contaminated by raw foods for cooked foods. Put cooked foods on clean platters, not the ones that held the raw meat.
- Chill leftovers promptly. Do not leave food out for more than 4 h.

- Clean produce. Wash hands before preparing food and immediately after handling raw foods.
- Report suspected FBI to the local health department.

FBIs are preventable, but prevention requires that those in the food industry and in health and regulatory agencies be constantly vigilant to ensure that the hazards are understood, and questionable operating procedures are avoided. Therefore, implementing a hazard analysis critical control point system can provide high assurance of food safety. This system is based upon detecting hazards from epidemiologic data and observations of operations; identifying critical control points establishing control measures, criteria, and critical limits; monitoring the critical control points to ensure that the process is under control; and taking immediate action whenever the criteria are not met.[9,17]

A 2009 survey of 1005 registered voters by Hart research associated and public opinion strategies found that improved food safety is very important to Americans, with 89% supporting new food safety measures and 72% saying they would pay more at the grocery store to have new safety measures for food.[15] Many experts maintain that individual consumers play a large role in food-handling practices and this practice paves an important way to prevent FBI. Specific proposals to reduce the incidence of FBI in the United States include greater use of irradiation more food safety inspections a revised government regulatory structure and a revised government regulatory structure and an improved tracking and recall system.

4.8.1   METHODS TO PREVENT FBI

4.8.1.1   BY PRACTICING SAFE FOOD-HANDLING PRACTICES

People, who prepare food both in restaurants and at home, can help prevent FBI by following safe food-handling practices. The CDC and the FSIS advise that when handling food, people should always remember to clean, separate, cook, and chill. Cleaning items that may have become contaminated and separating potentially contaminated items from others helps prevent the spread of pathogens. Cooking destroys many harmful microbes, and chilling helps prevent microbes from multiplying on food.

These safe food-handling practices should begin during purchase, when buyers should separate raw meat and poultry by wrapping it in plastic bags and keeping it away from other foods in order to avoid the spread of harmful microbes. During food preparation, the preparer should also prevent the spread of harmful microbes by washing his or her hands, utensils, and cutting boards with soap and water before and after contact with raw meat, seafood, poultry, and eggs.[3,15]

4.8.1.2 CONSUMER EDUCATION

Research shows that many consumers are not well educated about how serious FBI can be and are not aware of safe food-handling practices. As a result, some people argue that one way to prevent FBI is to educate consumers. Many consumers think of FBI as an unpleasant but short-term experience and are unaware that there can be some serious long-term consequences. For example, in a 2009 survey of 1000 Americans conducted for manufacturer National Pasteurized Eggs, researchers found that 87% of those surveyed could not identify the possible long-term complications of *Salmonella* such as joint pain and heart damage.[1,12]

4.8.1.3 INCREASED INSPECTION

The FDA is responsible for the safety of approximately 80% of the country's food supply, yet it inspects only a small percentage of that. In a 2007 report, the Subcommittee on Science and Technology argues that the current FDA inspection is not sufficient. It says there is an appallingly low inspection rate; the FDA cannot sufficiently monitor either the tremendous volume of products manufactured domestically or the exponential growth of imported products. During the past 35 years, the decrease in FDA funding has forced FDA to impose a 78% reduction in food inspections. The FDA inspects food manufacturers once in every 10 years. There are similar concerns about the US Food Safety and Inspection Service (FSIS), the agency responsible for inspecting all meat and poultry. Research and advocacy organization OMB Watch says, "The inspectors FSIS does employ are spread too thinly, and Oversight of slaughterhouses and other facilities is insufficient."

4.8.1.4   IRRADIATION

It involves briefly exposing food to radiant energy such as X-rays or gamma rays to kill the microbes that can cause illness. According to the Grocery Manufacturers Association, irradiation has been approved in more than 37 countries. In the United States, it has been approved for use on eggs, uncooked meat and poultry, and a number of other foods, including fresh fruits and vegetables, spices, and wheat flour. Despite its increasing use, irradiation of food is controversial. Proponents maintain that research has proved that food treated with irradiation is generally as nutritious as untreated food and that irradiation does not change taste, texture, or appearance, and does not form toxic compounds or cause food to become radioactive.

4.8.1.5   IMPROVED TRACKING AND RECALL SYSTEM

Improving the speed and efficiency of the tracking and recall system for contaminated food is a goal of public health and safety officials. When a FBI outbreak occurs in the United States, officials need to find the cause and take steps to prevent further illness from occurring. That means finding out where the contaminated food or ingredient comes from, then preventing any more public consumption of it by recalling the remainder of the food. The Food Safety Working Group explains why speed and efficiency are important. If a food is found to be contaminated, there is disagreement over whether or not government agencies should have the power to recall it. Currently, the FDA has that power, but the FSIS does not.

4.8.1.6   REVISED REGULATORY STRUCTURE

A major criticism of America's food-safety system is that regulation is the responsibility of numerous separate federal agencies. While the FDA and FSIS are the primary regulatory agencies, other agencies are also involved in regulation. In a 2007 report, David M. Walker of the Government Accountability Office finds that "the federal oversight of food safety is fragmented, with 15 agencies collectively administering at least 30 laws related to food safety." Walker adds that this system has resulted in "inconsistent oversight, ineffective coordination, and inefficient use of resources."

4.8.1.7   TREATMENT

The commonly occurring symptoms of diarrhea and vomiting often cause weakness and dehydration, so most people simply rest and drink lots of fluids to replace those that are lost in a small number of cases; however, illness is more severe and requires treatment such as intravenous fluid blood transfusions or dialysis. Most people recover from FBI within days recover from FBI within days or sometimes months in more severe cases. [5]

Doctors do prescribe antibiotics in some cases; however, antibiotics are not always effective against FBI in some cases they can actually make the illness worse because they can prevent the growth of normal healthy bacterial in the body and allow harmful bacteria to become better established. Botulism patients are given an antitoxin as soon as possible to stop the toxin from damaging the body any further.[12]

4.8.1.8   GOOD MANUFACTURING PRACTICES RELATED TO PREVENT FBI

4.8.1.8.1   Keep Clean

- Wash your hands before handling food and often during food preparation
- Wash your hands after going to the toilet
- Wash and sanitize all surfaces and equipment used for food preparation
- Protect kitchen areas and food from insects, pests and other animals

4.8.1.8.2   Separate Raw and Cooked Food

- Separate raw meat, poultry, and seafood from other foods
- Use separate utensils such as knives and cutting boards for handling raw foods
- Store food in containers to avoid contact between raw and prepared foods

4.8.1.8.3   Cook Thoroughly

- Cook food thoroughly, especially meat, poultry, eggs and seafood

- Bring foods like soups and stews to boiling to make sure that they have reached 70°C
- Reheat cooked food thoroughly

4.8.1.8.4   Keep Food at Safe Temperatures

- Do not leave cooked food at room temperature for more than 2 h
- Refrigerate promptly all cooked and perishable food (preferably below 5°C)
- Keep cooked food piping hot (more than 60°C) prior to serving
- Do not store food too long even in the refrigerator
- Do not thaw frozen food at room temperature

4.8.1.8.5   Use Safe Water and Raw Materials:

- Use safe water or treat it to make it safe
- Select fresh and wholesome foods
- Choose foods processed for safety, such as pasteurized milk
- Wash fruits and vegetables, especially if eaten raw.

4.9   CONCLUSIONS

FBI is one of the dangerous diseases caused by bacteria, fungi, pests, or viruses or harmful chemicals or toxins. Illness seriously affected the human body causes many of the diseases such as diarrhea, vomiting, abdominal pain, abdominal cramps, giddiness, headache in some severe cases it causes kidney failure and lungs gets affected and some cases organs stop functioning.[12] Commonly FBI is caused by improper handling of utensils and improper maintenance and bad hygienic practices and also when food gets exposed to open atmosphere. The most common microorganisms caused FBIs are clostridium, *E. coli*, *Salmonella*, listeria monocytoges, shrigella, campylobacter, bacillus. FBDs are largely preventable, though there is no simple one-step prevention measure like a vaccine. Instead, measures are needed to prevent or limit contamination all the way from farm to table. A variety of good agricultural and manufacturing practices can reduce the spread of microbes among animals and prevent the contamination of foods. Careful review of the whole food production process can identify the principal hazards, and the control points where contamination can be prevented, limited, or eliminated.[9]

A formal method for evaluating the control of risk in foods exists is called the Hazard Analysis Critical Control Point (HACCP) system. This was first developed by the NASA to make sure that the food eaten by astronauts was safe. The HACCP safety principles are now being applied to an increasing spectrum of foods, including meat, poultry, and seafood. For some particularly risky foods, even the most careful microbe-killing step must be included in the process.[8] For example, early in the century, large botulism outbreaks occurred when canned foods were cooked insufficiently to kill the botulism spores.[7] After research was done to find out exactly how much heat was needed to kill the spores, the canning industry and the government regulators went to great lengths to be sure every can was sufficiently cooked. As a result, botulism related to commercial canned foods has disappeared in this country. Similarly the introduction of careful pasteurization of milk eliminated a large number of milk-borne diseases.[17] This occurred after sanitation in dairies had already reached a high level. In the future, other foods can be made much safer by new pasteurizing technologies, such as in-shell pasteurization of eggs, and irradiation of ground beef. Just as with milk, these new technologies should be implemented in addition to good sanitation, not as a replacement for it as hygiene and sanitation are insufficient to prevent contamination and a definitive one.[11]

4.10 SUMMARY

FBIs, either sporadic or outbreak, can be caused by different pathogenic microorganisms some of which are more predominant than others. The different parameters associated with the predominant factors (pathogenic bacteria and viruses) are also described in this chapter. Control involves the collection and sorting of information, using this information to illuminate those products, places or people that must be controlled in order to minimize the spread of the illness. It has two faces: control of primary spread and control of secondary spread.

Different tools are used to prevent primary and secondary spread. Regulatory techniques concentrate on securing safe spaces and safe marketplaces. Different formal and informal control powers are used. Foodborne pathogens in this group include several Gram-positive spore formers and Gram-negative rods. The cells of spore formers, when consumed through contaminated foods, sporulate in the GI tract and release the toxin. In contrast, after consumption, Gram-negative rods multiply rapidly in the GI tract and die off, releasing toxins. The symptoms, mainly enteric, are associated with the toxins. They are usually required in very high numbers to cause the diseases. Proper sanitation and refrigeration can be used to reduce the incidence.

KEYWORDS

- cause
- CDC
- contamination
- cramps
- criticality
- cross-contamination
- disease
- effect
- food
- food safety
- food-handler
- food-hygiene
- FSSAI
- hazard
- health
- host
- illness
- infection
- intoxication
- pathogen
- prevention
- recoverability
- symptoms
- syndrome
- toxicity
- transmission
- USDA
- WHO

REFERENCES

1. Batz, M. Long-Term Consequences of Foodborne Infections. *Infect. Dis. Clin. North Am.* **2013,** *27,* 599–616.
2. Bibek, R. *Fundamentals Food Microbiology;* CRC Press: Boca Raton, FL, 2005; pp 323–391.

3. Burgess, C. M.; Rivas, L.; McDonnell, M. J.; Duffy G. Biocontrol of Pathogens in the Meat Chain. Chapter 12, In *Meat Biotechnology*; Toldra, F., Ed.; Springer: New York, 2008; pp 253–288.
4. Center for Disease Control and Prevention. Incidence and Trends of Infection with Pathogens Transmitted Commonly Through Food. *Foodborne Diseases Active Surveillance Network (10 U.S. Sites, 1996–2012 Morbidity and Mortality Weekly Report)*, 2013. *62*(15), 283–287.
5. Cohen, M. B. Etiology and Mechanisms of Acute Infectious Diarrhea in Infants in the United States. *J. Pediatr.* **1991**, *118*, S34–S39.
6. Cremon, C. *Salmonellosis* Gastroenteritis During Childhood is a Risk Factor for Irritable Bowel Syndrome in Adulthood. *Gastroenterology* **2014**, *147*, 69–77.
7. EFSA. The European Union Summary Report on Trends and Sources of Zoonoses, Zoonotic Agents and Foodborne Outbreaks in 2011. *EFSA J.* **2013**, *11*(4), 3129.
8. Gerba, C. P. Sensitive Populations: Who is at the Greatest Risk? *Int. J. Food Microbiol.* **1996**, *30*, 113–123.
9. Lund, B. M.; O'Brien, Sarah J. The Occurrence and Prevention of Foodborne Disease in Vulnerable People. *Foodborne Pathog. Dis.* **2011**, *8*, 961–973.
10. Mols, M.; Pier, I.; Zwietering, M. H.; Abee, T. The Impact of Oxygen Availability on Stress Survival and Radical Formation of *Bacillus cereus. Int. J. Food Microbiol.* **2009**, *135*(3), 303–311.
11. Montville, T. J.; Matthews, K. R. Eds. *Food Microbiology: An Introduction;* ASM Press: Washington D.C., 2005; pp 232.
12. Nakaya, A. C. *Food Borne Illness Compact Research Series*. Popular works; Reference PointPress: San Diego, CA, 2011; pp 20–73.
13. Payne, D. C. Norovirus and Medically Attended Gastroenteritis in U.S. Children. *N. Engl. J. Med.* **2013**, *368*, 1121–1130.
14. Qiong, H.; Xiaoyan, W.; Lingling, L.; Xiaoling, D.; Zihui, C. The Effectiveness of Foodborne Diseases Training Among Clinicians in Guangdong, China. *Food Control* **2013**, *33*(1), 268–227.
15. Rajkowski, K. T.; Bennett, R. W. *Bacillus cereus*. Chapter 3, In *International Handbook of Foodborne Pathogens;* Miliotis, M. D., Bier, J. W., Eds.; Marcel Dekker: New York, 2003; pp 27–39.
16. Richard, H. *Regulating Food Borne Illness*. Hart publishing, 2015; pp 10–22.
17. Schoeni, J. L.; Wong, A. C. L. *Bacillus cereus* Food Poisoning and its Toxins. *J. Food Prot.* **2005**, *68*(3), 636–648.
18. Todd, E. C. D.; Bartle, S.; Charles, A.; Guzewich, J. J.; Lee M.; Tan, A. Procedure to Investigate Food Borne Illness. *Interact. Food Prot. Assoc.* **2011**, *2*(1), 55–70.
19. Torgerson, P. R.; de Silva, N. R.; Eric, M. F.; Fumiko, K.; Mohammad, B. R. The Global Burden of Food Borne Parasitic Diseases: An Update. *Trends Parasitol.* **2014**, *30*(1), 20–26.
20. Toyofuku, H. Foodborne Diseases Prevalence of Foodborne Diseases in Western Pacific Region. *Ref. Module Food Sci. Encycl. Food Saf.*, **2014**, *1*, 312–322.

PART II

Specific Fruits, Spices, and Dairy-Based Functional Foods for Human Health

CHAPTER 5

FUNCTIONAL AND MEDICINAL VALUES OF FENUGREEK— A REVIEW

MURLIDHAR MEGHWAL[1,*] AND IRENA MLADENOVA[2]

[1]*Assistant Professor, Department of Food Science and Technology, National Institute of Food Technology Entrepreneurship & Management, Kundli - 131028, Sonepat, Haryana, India; Mobile: +91 9739204027*

[2]*Department of Hygiene, Epidemiology and Infectious Diseases, Trakia University, Medical Faculty, Stara Zagora, 11 Armeiska Str., Bulgaria, Tel.: +35-9897324472, E-mail: imladenova@yahoo.com*

*Corresponding author. E-mail: murli.murthi@gmail.com

CONTENTS

5.1   INTRODUCTION

Fenugreek (in Hindi: *Methi*) is one of the most used spices in the world and is consumed worldwide. Fenugreek is a semiarid-type crop and is cultivated worldwide. Generally, its seeds and green leaves are used in food preparations as well as in medicinal applications. Fenugreek seeds and leaves are good source of natural food fiber and other nutrients required in our body.[36] It has a specific flavor and aroma due to which it is more liked and consumed. "Cemen" is a popular food in Turkey and is a paste prepared from ground fenugreek seeds.[13] Fenugreek seed or powder or ball made from it is used in ghee (clarified butter) production and preservation. India is the largest producer as well as consumer of fenugreek and in the world. Fenugreek has the ability to modify the food texture. There are several medicinal uses and applications of fenugreek, for example:

- Functional food applications
- Used as nutraceutical
- Physiological utilization: has antibacterial, anticarcinogenic, antiulcer, anthelmintic, antioxidant, diabetes management, hypoglycemic, and hypocholesterolemic properties
- To improve the digestion

This chapter reviews the major functional food-related uses and application of seed and leaves of fenugreek.

5.2   IMPORTANCE OF FENUGREEK LEAVES AND SEEDS

The green fenugreek leaves are used as herb, for curry preparations, and other food items. Fenugreek green leaves have β-carotene, ascorbic acid, fiber, calcium, and zinc.[36] Fenugreek leaves help to manage the glycemia. [19,32,37]

Fenugreek seeds are golden-yellow in color, hard, small in size, and have four-faced stone-like structure (Fig. 5.1).[7] These are viscous, gummy, fibrous, and sticky. Saponin and alkaloid are compounds of fenugreek, and are antinutritional factors.[19] Volatile oil and fixed oil contents in the fenugreek seed are in small quantities.[34] The seeds of fenugreek are most important and useful products. Unroasted and uncooked fenugreek seeds have maple flavor and bitter taste, but their bitterness can be reduced and flavor can be enhanced by roasting. Seeds are used as food seasoning and condiments. They have multiple uses and applications, such as ground powder is used in pickles, vegetables dishes, and as a spice.

FIGURE 5.1   Fenugreek: seeds (left); leaves (right).

5.2.1 FENUGREEK GUM

The endosperm of fenugreek seeds has gum, which consists of galactose and manose. Fenugreek gum makes high-viscosity solutions in water. The fenugreek is used for thickening, stabilizing, and emulsifying of the food agents.[5,13] Fenugreek gum has highest water solubility due to more galactose content. Fenugreek gum is a type of soluble dietary fiber and it can be incorporated in bread-making. The application of fenugreek gum in breadmaking showed volumes and texture comparable with the control samples of bread.[30] Bread-making by substituting bread flour with fenugreek gum dough showed more water absorption compared to the dough which is without fenugreek gum.

5.2.2 DIETARY FIBER

Fenugreek seeds are rich source of soluble dietary fiber.[32] Raju et al.[25] reported that the fiber content of fenugreek extract can moderate metabolism of glucose in the digestive tract. Water absorption on the outer surface makes seeds coat soft and mucilaginous. The advantages of fenugreek fiber are: it binds to toxins and protects the colon mucus membrane from cancer-causing toxins; the amino-acid, 4-hydroxy isoleucine, present in the fenugreek seed has facilitator action on insulin secretion; it lowers the rate of glucose absorption in the intestines and controls blood sugar levels; and enhances for glucose level tolerance. Fenugreek seeds are rich in mucilaginous fiber which is composed of galactomannans. Galactomannan is a major soluble fiber of the fenugreek seeds. It decreases the bile salts uptake in intestine and also reduces the digestion and absorption of starch in body.[17]

5.2.3   PROTEIN

Fenugreek endosperm is rich in proteins such as globulin, histidine, albumin, and lecithin[18] and endosperm is found to contain highest protein.[20] Isikli et al.[13] reported that it has protein content ranging from 20 to 30% as well as amino acid, in particular 4-hydroxyisoleucine, which has high potential for insulin-stimulating activity. The fenugreek protein fraction is lysine-rich and comparable in quality to that of soybean.[8] The residual proteins play an important role in decreasing the surface tension at the oil–water interface. The molecular weight of fenugreek gum was increased with removing the attached proteins and viscosity of fenugreek gum was increased with increase in gum concentration or with reduction of residual protein.

5.2.4   VITAMINS

Fenugreek is rich in choline. Vitamins A, B_1, B_2, C, niacin, and nicotinic acid are reported in fenugreek seed; germinating fenugreek seeds contain pyridoxine, cyanocobalamin (vitamin B_{12}), calcium pantothenate, biotin, and vitamin C.[15] Fenugreek consumption can potentially improve body health, in particular body fat percentage, and vitamin C of fenugreek seed has important role according to Poole et al.[23]

5.3   HUMAN HEALTH AND FENUGREEK

Fenugreek possesses pharmacological properties such as antimicrobial, anticholesterolemic, carminative, emollient, febrifuge, laxative, restorative, uterine tonic, expectoral, galactogogue, anticarcinogenic, anti-inflammatory, antiviral, antioxidant, demulcent, and hypotensive properties. In addition, it regulates several enzymatic activities, relieves fever, reduces body pain and fat, alleviates swelling, augments appetite, and promotes lactation and sex hormones. Compounds isolated from fenugreek have remarkable biological activities including protection against cancer, malaria, allergies, bacteria, and viruses.[20] Fenugreek, in particular, is abundant in polyphenolics that inhibit peroxidation and remarkably reduce oxidative hemolysis in human erythrocytes. Moreover, their optimal consumption may lower triglycerides and cholesterol concentrations in the blood, prevent cancer,[26] and control diabetes mellitus. The oral intake of ethyl acetate extract from green seeds has been tested to reduce triglycerides and low-density lipoprotein cholesterol (LDL-C), while high-density lipoprotein cholesterol (HDL-C) is increased, and hence it had noteworthy antioxidant and hypocholesterolemic effects.[108]

5.3.1   FENUGREEK AS AN ANTICANCER AGENT

Fenugreek is a useful medicinal spice and herb for cancer prevention. Fenugreek can change and modify apoptosis and free-radical lipid peroxidation.[1,3] A steroid sapogenin diosgenin ($C_{27}H_{42}O_3$) is useful as a beginning material for steroid hormone synthesis (cortisone and progesterone). Alsemari[2] found that fenugreek extract has a very selective cytotoxicity against cancer cell lines such as T-cell lymphoma (TCP), B-cell lymphomas, thyroid papillary carcinoma, and breast cancer (MCF7). This clearly indicates that fenugreek has selective cytotoxic effects against cancer cells. This is the first human case in which established malignant central nervous system (CNS) cancer showed regression and disappearing of the cancer lesion with daily use of fenugreek extract. Fenugreek may serve as a potential therapeutic in the treatment of lymphoid malignancy and other cancers. Gas chromatography–mass spectrometry (GC–MS) analysis of fenugreek extract indicated the presence of several compounds with anticancer properties, including gingerol, cedrene, zingerone, vanillin, and eugenol. The whole crude extract has shown to provide more selective potential than the individual compounds used in isolation. Further, it must be considered that the biologically active agents in fenugreek may vary on the basis of geographical environments.[109]

The results show that functional capacity evaluation (FCE) exhibits anticancer effects by blocking the proliferation of MCF-7 cells and inducing apoptosis in part by modulating expression levels of caspase-3, 8, 9, p53, FAS, FADD, Bax, and Bak. The induction of apoptosis by FCE is affected by its ability to regulate the expression of pro-apoptotic genes such as caspase-3, 8, 9, p53, FAS, FADD, Bax, and Bak. Considering this, it is most likely that FCE induced, at least in part, p53, FAS, FADD, Bax, Bak, and caspases-mediated apoptosis in MCF-7 cells. This study demonstrated that FCE significantly inhibits the growth of MCF-7 human breast cancer cells in vitro, and it provides the underlying mechanism for the anticancer activity. These studies support the use of FCE for breast cancer chemoprevention due to its influence in suppressing growth of immortalized breast cells without significant toxicity.[110]

5.3.2   HYPOCHOLESTEROLEMIC EFFECTS

Hypocholesterolemia refers to the deficiency of cholesterol in the blood, and fenugreek can help to manage the effects of hypocholesterolemiac.[33] Fenugreek seeds have lowered serum cholesterol, triglycerides, and LDL in patients suffering from hypercholesterolemia and experimental models. Fenugreek consumption could reduce triglyceride accumulation in the liver but did not interfere with the plasma insulin or glucose levels obesity suffering rats.[27]

5.3.3   HYPOGLYCEMIC PROPERTIES OF FENUGREEK

Hypoglycemia is state of human body where there is an abnormal decrease of sugar level in blood. Fenugreek seeds were found to have hypoglycemic and hypocholesterolemic effects. Fenugreek is helpful in utilization of glucose, and in increase of glucose tolerance. Injection of extracts of fenugreek seed was found to improve plasma glucose and insulin response and it also reduced urinary concentrations.[32] The consumption of germinated seeds in comparison with the boiled fenugreek seeds could reduce blood glucose levels.[93] Germinated fenugreek seeds are observed to be more beneficial than dried seeds because germinated fenugreek seed increases the bioavailability of different constituents of fenugreek.

5.3.4   ANTIOXIDANT PROPERTIES OF FENUGREEK

Fenugreek has powerful antioxidant property that has beneficial effect on liver and pancreas.[3] The presence of phenolic and flavonoid compounds in fenugreek enhances its antioxidant capacity.[20] The antioxidant activity of the germinated fenugreek seeds is due to the presence of flavonoids and polyphenols.[6] Grover et al.[11] reported that fenugreek seeds have hypoglycemic and antihyperglycemic effects due to the presence of antioxidant carotenoids.

5.3.5   IMMUNOMODULATORY PROPERTIES OF FENUGREEK

Fenugreek and its food items are immunomodulatory in nature. Based on body weight, relative thymus weight, cellularity of lymphoid organs, delayed type of hypersensitivity response, plaque forming cell assay, hemagglutination titer, quantitative hemolysis assay, phagocytosis, lymph proliferation, and a significant increase in phagocytic index and phagocytic capacity of macrophages, observations showed stimulatory immunomodulatory effects of fenugreek. More research studies need to be conducted so as to confirm these effects.

5.3.6   REDUCTION IN INFERTILITY PROBLEMS WITH FENUGREEK

Fenugreek is suggested to have vigor-strengthening effect in consumers. It improves the digestive system and bioavailability to have healthy and strong sperms in males. It has been also found that fenugreek seeds help to manage the menstruation cycle-related complexities in females.

5.3.7   CONTROL OF DIABETES WITH FENUGREEK

Fenugreek has a lot of dietary fibers and it is good for diabetic patients. Fenugreek and its high-fiber content help to reduce the glycemic index and therefore they are recommended to include in the healthy diets.[24] According to Yadav et al.[38] the aqueous extract of fenugreek seeds has higher hypoglycemic and antihyperglycemic potential and may be used as complementary medicine to treat the diabetic population.

For a diabetic person, fenugreek can be a good source for proteins also.[8] Fenugreek was found to reduce type I and II diabetes. Sharma and Raghuram[31] reported decrease in fasting food glucose levels in the glucose tolerance test, and urinary glucose excretion reduction, serum total cholesterol decrease, LDL and very low-density lipoprotein cholesterol and triglyceride levels decrease without alteration in the HDL-C fraction. Fenugreek seeds have been found to improve glycemic control and decrease insulin resistance in diabetic patients. Fenugreek significantly reduced blood sugar in fasting and postprandial subjects but it did not affect platelet aggregation, fibrinolytic activity, and fibrinogen.[4]

5.3.8 ANTIULCER (ANTI-HELICOBACTER PYLORI-INFECTION) PROPERTIES

The aqueous extract and a gel fraction isolated from the seeds showed significant ulcer-protective effects. It has soothing effect on gastritis and gastric ulcer.[8,35] *Helicobacter pylori* is an etiological agent of chronic gastritis, peptic ulcer disease, gastric cancer, and mucosa-associated lymphoid tissue (MALT) lymphoma, and over 50% of the world population is infected with *H. pylori*.

Treatment of *H. pylori* infection with antibiotics is the best way to eradicate this bacterium, but the antibiotic resistance is increasing in many countries, and retreatment after eradication failure is difficult. Search for alternative therapies against *H. pylori* is needed. Randhir and Shetty[28,29] have studied the antimicrobial efficacy of fenugreek sprout extracts on the growth of *H. pylori*. A direct correlation between the extract amount and inhibition diameter has been observed. Early-stage extracts with high inhibition have been associated with higher total phenolic content and associated antioxidant activity. The authors have proposed that the mechanism of antimicrobial activity by the phenol-containing extracts is by creating an acidic environment that causes the bacterial cell membrane to disrupt. The antimicrobial activity observed in fenugreek could also be due to the presence of scopoletin, a coumarin derivative of coumaric acid. Scopoletin is a lactonized phenol which has the potential to inhibit the electron transport chain in prokaryotes. Other possibility is that the phenolics-enriched fenugreek extracts alter the urease enzyme activity of *H. pylori*, which provides the alkaline microenvironment in the stomach for its survival.

Several studies have assayed the effect of plant extracts on *H. pylori*. Some of these studies have been validated in animals and confirmed the potential benefit of using plants as the source of antimicrobial agents against *H. pylori*. O'Mahony et al.[21] have investigated both the bactericidal and antiadhesive properties of 25 plants against *H. pylori*. Sixteen of them have never been tested before against *H. pylori* and are plants that are frequently used in cooking as well as in medicine in many countries in Asia. In that study, fenugreek appeared inactive but the seeds have been examined rather than sprouts, which may be the reason for such results.

5.3.9   FENUGREEK AS DIGESTION AID

Spices consumed in diet positively influenced the pancreatic digestive enzymes. Patel and Srinivasan[35] reported dietary fenugreek-enhanced pancreatic lipase activity. Non-starch polysaccharides of the fenugreek increase the bulk of the food and help to have proper bowel movements. It also helps in smooth digestion whereas high fiber of fenugreek helps in relieving constipation ailments. For intestinal microbes, fenugreek acts as a good prebiotics.

5.3.10   FENUGREEK APPLICATION IN FOOD INDUSTRY

Fenugreek is a good agent to stabilize and emulsify the food constituents. Fenugreek seeds have very good solubility and emulsifying properties. If fenugreek gum is mixed with soy isolate, then its emulsifying properties increase many times. The solubility and emulsifying properties of soy protein isolate with fenugreek gum dispersions were stable over a wide range of pH, ion strength, and temperature.[12] Fenugreek is used as prebiotics in functional food.[14] Garti et al.[10] demonstrated that fenugreek gum shows an emulsifying capability for stabilizing oil-in-water emulsions and they further concluded that critical coverage gum/oil ratio for stable non-coalesced emulsion is smaller than the ratio obtained for guar or other gums, indicating its emulsification properties are superior to those of other galactomannans. Fenugreek gum has very good application in making soups because it modifies the rheological properties and interaction of starch and other soup ingredients. Losso et al.[16] incorporated fenugreek in bread-making and demonstrated that fenugreek in food help to reduce blood sugar but the use of fenugreek is a barrier due to its bitter taste and strong odd flavor. Fenugreek can be incorporated in baked products in acceptable limits, which will reduce insulin resistance and treat diabetes patients.

5.3.11 MISCELLANEOUS APPLICATIONS

Fenugreek can be used for anthelmintic applications to remove the parasitic intestinal worms. If someone is having health issues related to hepatic lipids, fenugreek can be very useful to manage this problem.[25] Fenugreek is a good source of fiber which can help to control the cholesterol in human body which can avoid heart-related issues. Fenugreek is diuretic emmenagogue and emollient in nature, therefore it helps to manage the menstruation flow problems in woman.

5.4   PRECAUTIONS IN FENUGREEK USE

Food safety and food quality should be taken care to avoid any food toxicity or food-related health hazard. Some people can have allergies to fenugreek consumption; therefore, such consumers should avoid taking leaves, seeds, or other parts of the fenugreek. Due to high fiber content, it should be used in minute quantities only. Specific people such as diabetic patients should take extra precautions and should not take fenugreek along with therapeutic medication because fenugreek could interfere with the absorption of those therapies that control blood sugar. People face another problem of bad odor due to sweat after consuming fenugreek, and hence it should be used in limited quantities.[9] Some of the applications in ingestion, inhalation, and external application of fenugreek seed powder may cause rhinorrhea, wheezing, and fainting.[22] Defatted fenugreek seeds are not bitter in taste and can be easily consumed. Masking helps to overcome the bitterness of fenugreek.[24]

5.5 SUMMARY

The major beneficial properties of fenugreek have been discussed in this chapter. Fenugreek and its products have many health benefiting properties such as antidiabetic, antioxidant, anticarcinogenic, anthelmintic, antiulcer, antifertility, immune-enhancing, enzymatic pathway modifier, hypoglycemic, and hypocholesterolemic properties. Such medicinal properties have been investigated by various researchers. Fenugreek seeds have good dietary fiber content, protein, minerals, vitamins, and other nutrients. Fenugreek is gummy in nature. Various modern and latest findings show the potential applications of fenugreek for food and health. In certain specific cases, there is a need to take extra care, such as fenugreek consumption by pregnant women, patient suffering from chronic asthmatic diseases, people who have allergy to fenugreek should avoid or minimize fenugreek consumption. One should consult a physician regarding its dose and form of use.

KEYWORDS

- allergy
- anthelmintic
- anticarcinogenic
- antidiabetic
- antifertility
- antioxidant
- antiulcer
- asthmatic
- diabetes
- digestion
- disease
- enzymatic
- fenugreek
- fiber
- food
- health
- hypocholesterolemic
- hypoglycemic
- immunomodulatory
- leaves
- nutrition
- patient
- quality
- safety
- seeds

REFERENCES

1. Aggarwal, B. B.; Shishodia, S. Molecular Targets of Dietary Agents for Prevention and Therapy of Cancer. *Biochem. Pharmacol.* **2006,** *71*(10), 1397–421.
2. Alsemari, A. The Selective Cytotoxic Anti-Cancer Properties and Proteomic Analysis of *Trigonella Foenum-Graecum. J. BMC Complementary Altern. Med.* **2014,** *14*(114), 1–9.
3. Bhatia, K.; Kaur, M.; Atif, F.; Ali, M.; Rehman, H.; Rahman, S.; Raisuddin, S. Aqueous Extract of *Trigonellafoenum-graecum* L. Ameliorates Additive Urotoxicity of Buthionine Sulfoximine and Cyclophosphamide in Mice. *Food Chem. Toxicol.* **2006,** *44*, 1744–1750.
4. Bordia, A.; Verma, S. K.; Srivastava, K. C. Effect of Ginger (*Zingiber officinale Rosc.*) and Fenugreek (*Trigonella foenum graecum* L.) on Blood Lipids, Blood Sugar and Platelet Aggregation in Patients with Coronary Artery Disease. *Prostaglandins Leukot. Essent. Fatty Acids.* **1997,** *56*(5), 379–384.
5. Brummer, Y.; Cui, W.; Wang, Q. Extraction, Purification and Physicochemical Characterization of Fenugreek Gum. *Food hydrocolloids* **2003,** *17*, 229–236.
6. Dixit, P.; Ghaskadbi, S.; Mohan, H.; Devasagayam, T. P. A. Antioxidant Properties of Germinated Fenugreek Seeds. *Phytother. Res.* 977–983.
7. Ebubekir, A.; Engin, O.; Faruk, T. Some Physical Properties of Fenugreek (*Trigonella-foenum-graceum* L.) Seeds. *J. Food Eng.* **2005,** *71*, 37–43.
8. Faeste, C. K.; Namork, E.; Lindvik, H. J. Allergenicity and Antigenicity of Fenugreek (*Trigonella-foenum-graecum*) Proteins in Foods. *Allergy Clin. Immunol.* **2009,** *123*(1), 187–194.
9. Faeste, C. K.; Christians, U.; Eliann E.; Jonscher, K. R. Characterization of Potential Allergens in Fenugreek (*Trigonella-foenum-graecum*) Using Patient Sera and MS-based Proteomic Analysis. *J. Proteomics* **2010,** *73*, 1321–1333.
10. Garti, N.; Madar, Z.; Aserin, A.; Sternheim, B. Fenugreek Galactomannans as Food Emulsifiers. *LWT.* **1997,** *30*, 305–311.
11. Grover, J. K.; Yadav, S.; Vats, V. Medicinal Plants of India with Anti-Diabetic Potential. *J. Ethnopharmacol.* **2002,** *81*, 81–100.
12. Hefnawy, H. T. M.; Ramadan, M. F. Physicochemical Characteristics of Soy Protein Isolate and Fenugreek Gum Dispersed Systems. *J. Food Sci. Technol.* **2011,** *48*(3), 371–377.
13. Isikli, N. D.; Karababa, E. Rheological Characterization of Fenugreek Paste (Cemen). *J. Food Eng.* **2005,** *69*, 185–190.
14. Lee, E. E. L. Genotype-x Environment Impact on Selected Bioactive Compound Content of Fenugreek (*Trigonella foenum-graecum*). Department of biological sciences: University of Lethbridge, Canada, 2006; pp 1–150.
15. Leela, N. K.; Shafeekh, K. M. Fenugreek. In *Chemistry of Spices;* CAB International: Oxford, UK, 2008; pp 242–259.
16. Losso, J. N.; Holliday, D. L.; Finley, J. W.; Martin, R. J.; Rood, J. C.; Yu, Y.; Greenway, F. L. Fenugreek Bread: A Treatment for Diabetes Mellitus. *J. Med. Food* **2009,** *12*(5), 1046–1049.

17. Madar, Z.; Shomer, I. J. Polysaccharide Composition of a Gel Fraction Derived from Fenugreek and its Effect on Starch Digestion and Bile Acid Absorption in Rats. *J. Agric. Food Chem.* **1990**, *38*, 1535–1539.

18. Mathur, P.; Choudhry, M. Consumption Pattern of Fenugreek Seeds in Rajasthani Families. *J. Hum. Ecol.* **2009**, *25*(1), 9–12.

19. Muralidhara, L.; Narasimhamurthy, K.; Viswanatha, S.; Ramesh, B. S. Acute and Subchronic Toxicity Assessment of Debitterized Fenugreek Powder in the Mouse and Rat. *Food Chem. Toxicol.* **1999**, *37*, 831–838.

20. Naidu, M. M.; Shyamala, B. N.; Naik, Sulochanamma, G.; Srinivas, P. Chemical Composition and Antioxidant Activity of the Husk and Endosperm of Fenugreek Seeds. *LWT—Food Sci. Technol.* **2011**, *44*, 451–456.

21. O'Mahony, R.; Al-Khtheeri, H.; Weerasekera, D.; Fernando, N; Vaira, D; Holton, J.; Basset, C. Bactericidal and Anti-Adhesive Properties of Culinary and Medicinal Plants Against *Helicobacter pylori*. *World J. Gastroenterol.* **2005**, *11*(47), 7499–7507.

22. Patil, S. P.; Niphadkar, P. V; Bapat, M. M. Allergy to Fenugreek (*Trigonella-foenum-graecum*). *Ann. Allergy Asthma Immunol.* **1997**, *78*(3), 297–300.

23. Poole, C.; Bushey, B.; Foster, C.; Campbell, B.; Willoughby, D.; Kreider, R.; Taylor, L.; Wilborn, C. The Effects of a Commercially Available Botanical Supplement on Strength, Body Composition, Power Output, and Hormonal Profiles in Resistance-Trained Males. *J. Int. Soc. Sports Nutr.* **2010**, *27*, 34.

24. Raghuram, T. C.; Sharma, R. D.; Pasricha, S.; Menon, K. K.; Radhaiah, G. Glycaemic Index of Fenugreek Recipes and its Relation to Dietary Fiber. *Int. J. Diabetes Dev. Countries* **1992**, *12*, 1–4.

25. Raju, J.; Bird, R. P. Alleviation of Hepatic Steatosis Accompanied by Modulation of Plasma Andliver Tnf-Alpha Levels by *Trigonellafoenum-graecum* (fenugreek) Seeds in Zucker Obese (fa/fa) Rats. *Int. J. Obes.* **2006**, *30*(8), 1298–1307.

26. Raju, J.; Patlolla, J. M.; Swamy, M. V.; Rao, C. V. Diosgenin, a Steroid Saponin of *Trigonella-foenum-graecum* (Fenugreek), Inhibits Azoxymethane-Induced Aberrant Crypt Foci Formation in F344 Rats and Induces Apoptosis in HT-29 Human Colon Cancer Cells. *Cancer Epidemiol Biomarkers Prev.* **2004**, *13*(8), 1392–1398.

27. Raju, J; Gupta, D.; Rao, A. R.; Yadava, P. K.; Baquer, N. Z. *Trigonella-foenum-graecum* (Fenugreek) Seed Powder Improves Glucose Homeostasis in Alloxan Diabetic Rat Tissues by Reversing the Altered Glycolytic, Gluconeogenic and Lipogenic Enzymes. *Mol. Cell. Biochem.* **2001**, *224*, 45–51.

28. Randhir, R.; Lin, Y. T.; Shetty, K. Phenolics, their Antioxidant and Antimicrobial Activity in Dark Germinated Fenugreek Sprouts in Response to Peptide and Phytochemical Elicitors. *Asia Pac. J. Clin. Nutr.* **2004**, *13*, 295–307.

29. Randhir, R.; Shetty, K. Improved α-amylase and *Helicobacter pylori* Inhibition by Fenugreek Extracts Derived via Solid-State Bioconversion Using *Rhizopus oligosporus*. *Asia Pac. J. Clin. Nutr.* **2007**, *16*(3), 382–392.

30. Roberts, K. T.; Cui, S. W.; Chang, Y. H.; Ng, P. K. W.; Graham, T. The Influence of Fenugreek Gum and Extrusion Modified Fenugreek Gum on Bread. *Food Hydrocolloids.* **2012**, *26*, 350–358.

31. Sharma, R. D. Effect of Fenugreek Seeds and Leaves on Blood Glucose and Serum Insulin Responses in Human Subjects. *Nutr. Res.* **1986**, *6*(12), 1353–1384.

32. Sharma, R. D.; Raghuram, T. C.; Rao, N. S. Effect of Fenugreek Seeds on Blood Glucose and Serum Lipids in Type I Diabetes. *Eur J. Clin. Nutr.* **1990,** *44*(4), 301–306.

33. Singhal, P. C.; Gupta, R. K.; Joshi, L. D. Hypocholesterolmic Effect of Fenugreek Seeds. *Curr. Sci.* **1982,** *51*, 136–137.

34. Sowmya, P.; Rajyalakshmi, P. Hypo-Cholesterolemic Effect of Germinated Fenugreek Seeds in Human Subjects. *Plant Foods Hum. Nutr.* **1999,** *53*, 359–365.

35. Srinivasan, K. Fenugreek (*Trigonellafoenum-graecum*): A Review of Health Beneficial Physiological Effects. *Food Rev. Int.* **2006,** *22*(2), 203–224.

36. Thomas, J. E.; Bandara, M.; Lee, E. L.; Driedger, D.; Acharya, S. Biochemical Monitoring in Fenugreek to Develop Functional Food and Medicinal Plant Variants. *New Biotechnol.* **2011,** *28*, 122–120.

37. Yadav, S.; Sehgal, S. Effect of Home Processing and Storage on Ascorbic Acid and, β-Carotene Content of Bathua (*Chenopodium album*) and Fenugreek (*Trigonella-foenum-graecum*) Leaves. *Plant Foods Hum. Nutr.* **1997,** *50*, 239–247.

38. Yadav, U. C. S.; Moorthy, K.; Baquer, N. Z. Effects of Sodium Orthovanadate and *Trigonella-foenum-graecum* Seeds on Hepatic and Renal Lipogenic Enzymes and Lipid Profile during Alloxandiabetes. *J. Biosci.* **2004,** *29*, 81–91.

CHAPTER 6

FRUITS AS A FUNCTIONAL FOOD

P. P. JOY[1,*], R. ANJANA[2], T. A. RASHIDA-RAJUVA[3],
AND RATHEESH ANJANA[4]

[1]*Pineapple Research Station, Kerala Agricultural University,
PO Vazhakulam, Ernakulam 686670, Kerala, India,
Tel.: +91-9446010905*

[2]*Pineapple Research Station, Kerala Agricultural University,
PO Vazhakulam, Ernakulam 686670, Kerala, India,
Tel.: +91-9946619746, E-mail: anjanasumathi@gmail.com*

[3]*Pineapple Research Station, Kerala Agricultural University,
PO Vazhakulam, Ernakulam 686670, Kerala, India,
Tel.: +91-9349903270, E-mail: rashh.ta@gmail.com*

[4]*Pineapple Research Station, Kerala Agricultural University (KAU),
Vazhakulam, Ernakulam 686670, Kerala, India, Tel.: +919497277406,
E-mail: anjanaratheesh@gmail.com*

Corresponding author. E-mail: joy.pp@kau.in, joyppkau@gmail.com

CONTENTS

6.1   INTRODUCTION

Food is any substance that we eat or drink, which provides nutrition in order to maintain growth and sustain life. In an era of expedite lifestyle, people look for things to be achieved easily, even in the case of foods. Instead of having anything to eat and following the age-old routine diet, it is better to take food, which provides us both flavor and health. As the father of medicine, Hippocrates, said more than 2000 years ago, "Let food be thy medicine and medicine be thy food."

6.1.1   FUNCTIONAL FOOD

The term "functional food" was introduced first in Japan in the mid-1980s.[97] A food can be regarded as "functional" if it is beneficial for smooth performance of one or more target functions in the body, beyond adequate nutrition, a factor that improves health and well-being or reduces the risk of disease.[30] It simply means that those food items that have relevant functional ingredients can be called as functional foods. It refers to food containing significant levels of naturally occurring, biologically active components that impart health benefits beyond the basic essential nutrients. These components play a vital role in disease prevention and health promotion.[45] Functional foods [182] can be broadly classified as conventional foods,[182] modified foods, and medical foods (Table. 6.1).

This chapter entrusts the functionalities of major fruits and other relevant fruit crops of the world, thus defining the different antioxidants present in them and how they can be a gift to human health.

TABLE 6.1   Classification of Functional Foods. (*Source:* Modified from.[182])

Conventional foods	Modified foods	Medical foods
Whole foods	Fortified	Foods: special dietary use.
	Enriched foods	
	Enhanced foods	

6.2 FRUITS

Fruits are mostly sweet and fleshy part of a tree or other plants that may contain seed and can be wholly eaten as food. It can be included in the conventional foods category and its processed products in the others.

Fruit-producing crops are classified based on their climatic adaptability (Table 6.2). They are tropical, subtropical, temperate, and arctic fruits. Tropical fruits are those which grow in the region between the Tropic of Cancer (23°27′ N latitude) and Tropic of Capricorn (23°27′ S latitude). They require a moist warm climate, can withstand dry weather, and are evergreen. Subtropical fruits grow between temperate and tropical climatic conditions. They are either evergreen or deciduous. They are adapted to low temperature but not frost. Temperate fruit crops are those, which are extreme cold-loving plants. They are deciduous and become dormant in winter. Arctic region is an all-time cold temperature region. Only a few crops thrive in those extreme frost conditions.

TABLE 6.2   Classification of Fruits.

Fruit types	Examples
Tropical	Mango, banana, plantain, papaya, guava, pineapple, passion fruit, mangosteen, jackfruit, rambutan
Subtropical	Citrus fruit (sweet orange, mandarin, tangerine), grapes, pomegranate
Temperate	Apple, peach, pear, nectarine
Arctic	Almond and trifoliate orange

6.2.1 FUNCTIONAL PROPERTIES OF FRUITS

A fruit as functional food has its own advantages, for example, apples are good for overall growth and the development of the human body. It is a reservoir of many antioxidants and vitamins. Banana is another fruit rich in simple sugars, which provide instant energy to athletes and undernourished children. The richest source of phenolic compounds is grapes. Regular intake of grape juice can lower the free radicals in the body. Guava is an underutilized crop which has been suggested for diarrhea and diabetes treatment. Jackfruit is another energy-rich crop, which replenishes body energy loss. Mangos are excellent summer food and are medically used for curing hiccups, sore throat, diarrhea, and dysentery. Oranges are carotenoid-rich fruits and hence are important as an anticancer fruit; these also help in lowering body weight just like pomegranate. Papaya is another rarely used crop which has been used in traditional medicines. Passion fruit is even prescribed by doctors for several contagious diseases and flu. Peaches help in antiaging and curing various diseases. Pineapple is often recommended for digestion problems.[20,25]

6.2.2 FRUIT COMPOSITION AND CONSTITUENTS

Fruits are composed of several macro and micronutrients. Carbohydrates, proteins, and fats come under the category of macronutrients that are required in larger amounts by the body. Micronutrients are only needed in smaller quantities. Common micronutrients are vitamins and minerals. Their constituents and composition in different fruits are mentioned in Tables 6.3–6.6 that were prepared by authors using information from Table 6.3,[4,19,32,36,66,68,69,75,101,138–153,155–157] Table 6.4,[138–153,155–157] Table 6.5,[138–153,155–157] and Table 6.6.[138–153,155–157] The overall nutrient possibilities from a fruit are schematically represented in Figure 6.1.

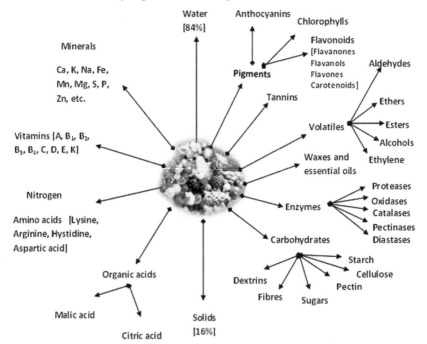

FIGURE 6.1 Schematic representations of chemical constituents of fruits.

TABLE 6.3 Chemical Composition of Fruits—Proximate (values per 100 g edible portion).

Fruits	Water (g)	Energy (kcal)	Protein (g)	Total lipid (fat) (g)	Ash (g)	Carbohydrate, by difference (g)	Fiber, total dietary (g)	Sugars, total (g)
Apple	85.56	52	0.26	0.17	0.2	13.81	2.4	10.39
Banana	74.91	89	1.09	0.33	1.1	22.57	2.6	12.23
Grapes	84.29	57	0.81	0.47	0.5	13.93	3.9	17.00
Guava	80.80	68	2.55	0.95	0.51	15.19	5.4	8.92
Jackfruit	73.46	95	1.72	0.64	1.10	23.08	1.5	19.08
Mango	83.46	60	0.82	0.38	0.55	14.79	1.6	13.66
Mangosteen	80.94	73	0.41	0.58	0.16	17.91	18	–
Nectarine	87.59	44	1.06	0.32	3.90	7.13	1.7	7.89
Orange	86.75	47	0.94	0.12	0.4	11.79	2.4	9.35
Papaya	88.06	43	0.47	0.26	0.42	10.79	1.7	7.82
Passion fruit—purple	85.62	51	0.39	0.05	0.80	13.14	0.2	13.40
Passion fruit—yellow	84.21	60	0.67	0.18	0.50	14.44	0.2	14.25
Peach	88.87	39	0.91	0.25	0.506	9.464	1.5	8.39
Pear	83.96	57	0.36	0.14	0.20	15.34	3.1	9.75
Pineapple	86.00	50	0.54	0.12	0.22	13.12	1.4	9.85
Plantain	65.28	122	1.30	0.37	2.68	30.37	2.3	15.00
Pomegranate	77.93	83	1.67	1.17	0.85	18.38	4.0	13.67
Rambutan	78.04	82	0.65	0.21	0.23	20.87	0.9	–
Tangerine	89.51	38	0.75	0.04	1.07	8.63	1.2	8.25

TABLE 6.4 Chemical Composition of Fruits—Minerals (mg/100 g edible portion).

Fruits	Calcium	Iron	Magnesium	Phosphorus	Potassium	Sodium	Zinc
Apple	6	0.12	5	11	107	1	0.04
Banana	5	0.26	27	22	358	1	0.15
Grapes	37	0.26	14	24	203	1	0.11
Guava	18	0.26	22	40	417	2	0.23
Jackfruit	24	0.23	29	21	448	2	0.13
Mango	11	0.16	10	14	168	1	0.09
Mangosteen	12	0.30	13	8	48	7	0.21
Nectarine	6	0.28	9	26	201	0	0.17
Orange	40	0.10	10	14	181	0	0.07
Papaya	20	0.25	21	10	182	8	0.08
Passion fruit—purple	4	0.24	17	13	278	6	0.05
Passion fruit—yellow	4	0.36	17	25	278	6	0.06
Peach	6	0.25	9	20	190	0	0.17
Pear	9	0.18	7	12	116	1	0.10
Pineapple	13	0.29	12	8	109	1	0.12
Plantain	3	0.60	37	34	499	4	0.14
Pomegranate	10	0.30	12	36	236	3	0.35
Rambutan	22	0.35	7	9	42	11	0.08
Tangerine	12	0.27	11	11	136	5	0.53

TABLE 6.5 Chemical Composition of Fruits—Vitamins (values per 100 g edible portion).

Fruits	Vitamin A (IU)	Vitamin B₁ (mg)	Vitamin B₂ (mg)	Vitamin B₃ (mg)	Vitamin B₆ (mg)	Vitamin B₉ (µg)	Vitamin C (mg)	Vitamin D (IU)	Vitamin E (mg)	Vitamin K (µg)
Apple	54	0.017	0.026	0.091	0.041	3	4.6	0	0.18	2.2
Banana	64	0.031	0.073	0.665	0.367	20	8.7	0	0.1	0.5
Grapes	67	1.5	–	0.000068	0.000016	0.00009	6.5	–	–	–
Guava	624	0.067	0.04	1.084	0.11	49	228.3	0	0.73	2.6
Jackfruit	110	0.105	0.055	0.92	0.329	24	13.7	–	0.34	–
Mango	1082	0.028	0.038	0.669	0.119	43	36.4	0	0.9	4.2
Mangosteen	35	0.054	0.054	0.286	0.018	31	2.9	0	–	–
Nectarine	332	0.034	0.027	1.125	0.025	5	5.4	0	0.77	2.2
Orange	225	0.087	0.04	0.282	0.06	30	53.2	0	0.18	0
Papaya	950	0.023	0.027	0.357	0.038	37	60.9	0	0.3	2.6
Passion fruit—purple	717	0	1.131	1.46	0.05	7	29.8	0	0.01	0.4
Passion fruit—Yellow	943	0	0.101	2.24	0.06	8	18.2	0	0.01	0.4
Peach	326	0.024	0.031	0.806	0.025	4	6.6	0	0.73	2.6
Pear	25	0.012	0.026	0.161	0.029	7	4.3	0	0.12	4.4
Pineapple	58	0.079	0.032	0.5	0.112	18	47.8	0	0.02	0.7
Plantain	1127	0.052	0.054	0.686	0.299	22	18.4	0.7	0.14	0.7
Pomegranate	0	0.067	0.053	0.293	0.075	38	10.2	0	0.6	16.4
Rambutan	3	0.013	0.022	1.352	0.02	8	4.9	–	–	–
Tangerine	1312	0.088	0.031	0.445	0.043	5	33.9	0	0.15	0

TABLE 6.6   Chemical Composition of Fruits—Fats (per 100 g edible portion).

Fruits	Fatty acids, total saturated (g)	Fatty acids, total monounsaturated (g)	Fatty acids, total polyunsaturated (g)
Apple	0.028	0.007	0.051
Banana	0.112	0.032	0.073
Guava	0.272	0.087	0.401
Jackfruit	0.195	0.155	0.094
Mango	0.092	0.140	0.071
Nectarine	0.025	0.088	0.113
Orange	0.015	0.023	0.025
Papaya	0.081	0.072	0.058
Passion fruit—Purple	0.004	0.006	0.029
Passion fruit—Yellow	0.015	0.022	0.106
Peach	0.019	0.067	0.086
Pear	0.022	0.084	0.094
Pineapple	0.009	0.013	0.040
Plantain	0.143	0.032	0.069
Pomegranate	0.120	0.093	0.079
Tangerine	0.004	0.007	0.007

6.2.3   FUNCTIONAL INGREDIENTS AND THEIR PROPERTIES

6.2.3.1   POLYPHENOLS

Polyphenols are a group of dietary antioxidants that are found naturally in fruits and vegetables. They primarily consist of flavonoids including flavanols, flavones, isoflavones, flavonols, flavonones, and anthocyanins, as well as non-flavonoid polyphenols including phenolic acids, lignans, and stilbenes. The mechanisms of antioxidant activity of polyphenols are characterized by direct scavenging or quenching of oxygen-free radicals and inhibition of oxidative enzymes that generate reactive oxygen species.

- **Flavonoids:** Flavonoids neutralize free radicals which may damage cells and bolster cellular antioxidant defenses.

○ Isoflavones are structurally similar to estrogen but are not steroids. They can bind to steroid receptors and can be called as phytoestrogens.

○ Flavanones are flavonoids which are glycosylated at the seventh position to give flavanones. A variety of flavanones are present in fruits. They are discussed below.

- Hesperetin: Hesperidin's aglycone form is hesperetin. It regenerates vitamin C. It slows down the proliferation of cancer cells. It also slows down the replication of viruses such as polio, herpes, and flu.[20] They have chemopreventive effects. They are often used for blood vessel conditions such as hemorrhoids, varicose veins, and poor circulation (venous stasis). It is also used to treat lymphedema, a condition involving fluid retention that can be a complication of breast cancer surgery.[123]

- Naringenin has potential prophylactic properties. It acts as an antioxidant, anti-inflammatory, antiallergic, hypolipidemic, and vasoprotective agent.[28]

- Narirutin is another flavanone beneficial for the treatment of bronchial asthma.[27]

- Neohesperidin is found in citrus fruits.

• **Flavonols:** They occur in unglycosylated forms in fruits. It includes monomers such as catechins and polymers like proanthocyanidins.

○ Catechins (Flavan-3-ols) act as anticarcinogen in lungs, stomach, esophagus, duodenum, liver, pancreas, mammary gland and also prevent chronic inflammation associated with carcinogenesis, inhibits oxidation of low-density lipoprotein (LDL), reduce nitrite production preventing nitrosation and also prevent cardiovascular disease (CVD).[6,17,85]

○ Quercetin was most effective in protecting LDL from oxidation, followed by myricetin and kaempferol.[6,9,17] It inhibits oxidation of LDL, thus reducing atherosclerosis and CVD and inhibits colon cancer too.

- ○ Kaempferol has anti-inflammatory and antioxidant properties.
- ○ Proanthocyanidins are oligomers of catechin and epicatechin and their gallic acid esters. They form tannins.
- **Anthocyanins:** They are water-soluble glycosides and acyl-glycosides of anthocyanidins. They are responsible for the color of fruits.
  - ○ The main classes are cyanidin, delphinidin, petunidin, peonidin, and malvidin.[61] These have antioxidant properties. Its main role is in immunosuppressive mechanisms like antiallergic, anti-inflammatory, antimicrobial, and anticancer.[6,9,17] It has proved to be effective against uterine carcinoma and colon adenocarcinoma.[53]
  - ○ Flavones: The important edible flavones are apigenin, luteolin, and tangeretin. Flavones are not so common in fruits.
  - ○ Apigenin has the medicinal use in the treatment of HIV, inflammatory bowel disease, and skin conditions.[10,21] It also has the potential for treatment of prostate cancer and cervical cancer.[34]
  - ○ Luteolin exhibits antimutagenic, antitumerogenic, antioxidants, and anti-inflammatory properties.[14,49,81,99]
  - ○ Tangeretin is a polymethoxylated flavone. It is 36 times stronger than hesperetin in stopping cancer cell proliferation.[20] It has an important role in the treatment of breast cancer.[59] Other health benefits include cholesterol lowering and neuroprotection. Tangeretin increased the levels of dopamine and has potential neuroprotective activity.[122]
- **Carotenoids:** They inhibit cancer cell growth of the human endometrium, mammary gland, and lungs with greater potency. β-Carotene neutralizes free radicals which may damage cells. β-Carotene and lutein are orange and yellow carotenoids, respectively. They are water insoluble. They can be converted to vitamin A. They benefit against liver cancer and lung cancer.[12,76] β-Carotene is a weak antioxidant but boosts the activity of natural killer immune cells. It gives cornea protection against UV light. Zeaxanthin may contribute to maintenance of healthy vision. Lutein may protect against colon

cancer.[20] Lycopene may contribute to the maintenance of prostate health.

- **Phenolic acids:** They are phenol carboxylic acids separated into two classes—hydroxybenzoic acids and hydroxycinnamic acids. They are rarely seen in fruits and hence limited studies were available.
- Hydroxybenzoic acid: Gallic acid and p-hydroxybenzoic acid are the common types.
- Hydroxycinnamic acid: Phenols and hydroxycinnamic acid inhibit the formation of carcinogen metabolites.
  - ○ Chlorogenic acid: It alleviates colon cancer. It also inhibited lipid peroxidation in rat liver induced by liver carcinogen in nectarines exert inhibitory action against the oxidation of LDL cholesterol, prevents hardening of arteries, and encourages proper circulation of human platelets. This helps in the maintenance of cardiac health. This natural compound has multiple health benefits including the prevention of various tumors and cancer. It also possesses chemopreventive properties which inhibit the proliferation of colon carcinogenesis.[131]
  - ○ Ferulic acid: It inhibits carcinogen metabolites.
  - ○ Stilbenoids are glycosylated forms of stilbenes. Resveratrol in grapes are the major stilbenoids present in fruits.
- **Resveratrol:** It acts as an antioxidant thus reducing the oxidative damage in the DNA of neuronal cells, offering an anti-aging action.[62] It has anticancer factor against colon and prostate cancers. It is beneficial in coronary heart disease (CHD) by preventing vasodilation and platelet aggregation. It inhibits cholesterol synthesis.[52] It alleviates the risk of stroke. It could be effective as a therapeutic or chemopreventive agent against melanoma.[63] It is an antioxidant that is involved in the mechanism inhibition of intracellular signaling transduction pathways.[25] It imparts protection against age-related macular degeneration (AMD).[50] It has role in preventing Alzheimer's disease and viral/fungal infections.
- **Dietary fibers:** insoluble fibers may contribute to the maintenance of a healthy digestive tract. Soluble fiber may reduce the risk of CHD and some types of cancers. Cellulose, hemicellulose, and pectin are some fibers present in fruits.

- **Proteins** in the form of enzymes play a significant role in free radical scavenging activities. Polyphenol oxidases, peroxidises, phenolase, phosphatase, proteases, pectin methyl esterase (PME), polygalacturonase and sucrose, which are present in the skin, are like them.[17,18,85] There are certain other proteins, which are important in the overall functioning of fruits like bromelain in pineapple.
- **Bromelain** is a complex mixture of substances that can be extracted from the stem and core fruit of the pineapple. It has demonstrated to have significant anti-inflammatory effects in conditions such as acute sinusitis, sore throat, arthritis, and gout and speeding recovery from injuries and surgery. Pineapple enzymes have been used successfully to treat rheumatoid arthritis (RA) and to speed tissue repair as a result of injuries, diabetic ulcers, and general surgery. It reduces blood clotting and helps remove plaque from arterial walls. Studies suggest that pineapple enzymes may improve circulation in those patients with narrowed arteries such as those suffering from angina. It is used to cure bronchitis and throat infections. It is efficient in the treatment of anemia. Pineapple is an excellent cerebral toner; it combats loss of memory, sadness, and melancholy. Excessive inflammation, excessive coagulation of the blood, and certain types of tumor growth may all be reduced by therapeutic doses of bromelain when taken as a dietary supplement.
- **Prebiotics** are nondigestible carbohydrates which cannot be broken down by the body. They are food sources for probiotic organisms.
- **Vitamins** contribute to the maintenance of healthy vision, immune function, bone health, cell integrity and help to regulate calcium and phosphorus transport.[42] Fruits are rich in vitamins C, A, and E. Vitamin C is the body's primary water-soluble antioxidant which works against free radicals that attack and damage normal cells. Free radicals have been shown to promote the artery plaque buildup of atherosclerosis and diabetic heart disease, cause the airway spasm that leads to asthma attacks, damage the cells of the colon so that they become colon cancer cells, and contribute to the joint pain and disability seen in osteoarthritis and RA. This would explain why diets rich in vitamin C have been shown to be useful for preventing or reducing the severity of all of these conditions. In addition, vitamin C is vital for the proper functioning of the immune system, making it a nutrient to turn to for the prevention of

recurrent ear infections, cold, and flu. It helps prevent gum diseases and also prevents the formation of plaque, thus keeping the oral cavity healthy. It also contributes in the proper growth and development of the muscles, teeth, and blood vessels of a growing baby. It contributes in collagen synthesis, helps in keeping the tissues toned up, and maintains the youthfulness of the skin.

- **Minerals and electrolytes:** Minerals may reduce the risk of high blood pressure and stroke. Copper and manganese are cofactors of the antioxidant enzyme, superoxide dismutase. Copper is required for the production of red blood cells (RBCs).

  ○ Potassium assists in preventing muscle cramps and keeps up the energy levels by building up the proteins. It is essential for proper nerve and cellular function. It is an important component of cell and body fluids that help regulate heart rate and blood pressure. It helps in protein synthesis, functioning of the muscles, maintains electrolyte balance, and optimally utilizes carbohydrates and supports metabolic processes. Deficit of potassium in the body may lead to disorders like hypokalemia which may degrade the muscular health and may cause cardiac arrhythmia. Thus, it helps to prevent hypokalemia.

  ○ Iron is required for RBC formation. Fluoride is a component of bones and teeth and is essential for prevention of dental caries. It majorly contributes in the general growth and development.

  - Phosphorus, an essential mineral, plays an important role in the normal functioning of cells and tissues. It helps in the formation of strong teeth and bones and assists in the metabolism of carbohydrates and fats. It extends its support in the cellular maintenance and repairs and also assists in renal functioning, muscular contraction, normal heart rate, and nerve communication.[131]

The major functional ingredients and their fruit sources are mentioned in Table 6.7 that was prepared using inform articles.[33,67,95]

TABLE 6.7   Functional Ingredients: Their Fruit Sources and Role.

Functional ingredients		Fruit sources	Role
Flavonoids			
Flavanones	Hesperetin Hesperidin	Tangerines, oranges	Regenerate vitamin C, anticancerous, slow down the replication of viruses like polio, herpes and flu, chemo-preventive, against lymphedema, hemor-rhoids, varicose veins, and poor circulation
	Naringenin		Anti-inflammatory, anti-allergic, hypolip-idemic, vasoprotective, and anticarcinogenic against bronchial asthma
	Naringin	Oranges	For the treatment of obesity, diabetes, hypertension, and metabolic syndrome
	Narirutin		May be effective in the treatment of bronchial asthma
	Eriocitrin	Citrus fruits	Lipid-lowering, capil-lary permeability, antioxidant
	Neoeriocitrin		Protect cartilage tissue

TABLE 6.7   *(Continued)*

Functional ingredients		Fruit sources	Role
Flavonols	Kaempferol	Grapes, plantain, apple	anti-inflammatory and antioxidant
	Myricetin	Grapes, apple, plantain, pineapple	Reduce atherosclerosis
	Quercetin	Grapes, plantain, apple, peaches	Anticarcinogenic, prevents diarrhea, allergies, atherosclerosis, asthma, Hay fever, hypertension, interstitial cystitis prostatitis diabetes, RA athletic endurance
	Isorhamnetin	Plantain, grapes, apple	Reduce the risk of cancer, improve heart health and ease diabetes complications
	Laricitrin	Red grapes, apple	Reduces CVD, improves endothelial function, anticarcinogenic, reduces platelet activity
	Syringetin	Red grapes	Induces human osteoblast differentiation
Flavan-3-ols	Catechin	Grapes, pineapple, peaches	Anticarcinogen in lungs, stomach, esophagus, duodenum, liver, pancreas, mammary gland and also prevent chronic inflammation associated with carcinogenesis
	Epicatechin	Grapes, pineapple, apple, peaches	Attenuation of diabetes, heart health
	Gallocatechin epigallocatechin	Grapes, banana, apple, pear	Improves brain function, fat loss, lowers risk of cancer

TABLE 6.7 *(Continued)*

Functional ingredients		Fruit sources	Role
Anthocyanidins	Malvidin	Guava, grapes, oranges	Anti-inflammatory and anticarcinogenic activity, cardiovascular disease prevention, obesity control, and diabetes alleviation
	Cyanidin		
	Delphidin		
	Peonidin		
	Peltunidin		
Leucoanthocyanidins	Leucocyanidin	Plantain	Protection against ulcer
Flavones	Apigenin	Plantain	Treatment of HIV, inflammatory bowel disease and skin conditions, prostate cancer, and cervical cancer
	Luteolin		Antimutagenic, antitumerogenic, antioxidant and has anti-inflammatory properties
	Tangeretin	Tangerines	Against breast cancer, cholesterol lowering and in neuroprotection
Carotenoids	β-carotene	Tangerines, pineapple, peaches, pear, guava, banana, apple, grapes, orange, passion fruit, jackfruit	Inhibit cancer in endometrium, mammary gland and lungs, liver and colon; cornea protection against UV-induced erythema
	Lycopene		
	β-cryptoxanthin		
	lutein		
	α-carotene		
	Zeaxanthin		
	Astaxanthin		
Phenolic acids			
Hydroxybenzoic acid	Gallic acid	Grapes, guava, pineapple	Reduce hypertension, atherosclerosis and dyslipidemia
	p-hydroxybenzoic acid	Plantain	
	Gentisic acid	Grapes	

TABLE 6.7 *(Continued)*

Functional ingredients		Fruit sources	Role
Hydroxycinnamic acid	Caffeic acid	Plantain, grapes	Chemo protective agent in oral cancer, helps in cardiac health and antihyperglycemic
	Chlorogenic acid	Pineapple, peaches, grapes	Reduce colon cancer, prevents hardening of arteries
	Ferulic acid	Pineapple, grapes, apple, orange	Protect against cancer, bone degeneration, menopausal symptoms
Stilbenoids	Resveratrol	Grapes	Anti-aging, anticancer factor against colon and prostate cancers, against coronary heart disease, alleviates the risk of stroke, chemopreventive agent against melanoma, preventing Alzheimer's disease and viral/fungal infections
Phytoestrogens	Glycetin	Plantain	relief from menopausal symptoms and lower risk of osteoporosis, heart disease and breast cancer
Tannins	Proanthocyanidins	Grapes	Fight against tooth cavities, diarrhea, protect heart diseases and cancer
	Procyanidin B$_2$	Apple	
Dietary fibers	Cellulose	Plantain, pineapple, grapes, mangos, tangerines, jackfruit	Maintain bowel health, lowers cholesterol levels, helps to control blood sugar levels
	Hemicelluloses		
	Galacto-oligosaccharides		
	Pectin		
	Lignin		

TABLE 6.7 *(Continued)*

Functional ingredients		Fruit sources	Role
Prebiotics	Fructo-oligosaccharides	Banana	Resists gastric acidity, selectively stimulates the growth and/or activity of intestinal bacteria
Other functionally rich compounds	Oleoresins	Plantain	Decrease blood cholesterol concentration
	Shogaol		
	Gingerol		Suppresses collagen induced arthritis
	Astilbin		

6.3 SELECTED FRUITS AND THEIR FUNCTIONAL PROPERTIES

6.3.1 BANANA

Banana is one of the high-calorie tropical fruits. Banana and plantains are grown in about 120 countries.[183,184] The leading producers are Brazil, Ecuador, China, the Philippines, Indonesia, Costa Rica, Mexico, Thailand, and Colombia. In India, banana ranks first in production and third in area among fruit crops. The major banana-producing states are Karnataka, Gujarat, Andhra Pradesh, and Assam.

It contains good amount of antioxidants, minerals, and vitamins beneficial to health. Banana is composed of soft, easily digestible flesh made up of simple sugars like fructose and sucrose that, upon consumption, instantly replenish energy and revitaliz the body. Thus for these qualities, bananas are used by athletes to get instant energy and as supplement food in the treatment plan for underweight children. Fresh banana is a very rich source of potassium. Potassium is an important component of cell and body fluids that helps control heart rate and blood pressure, countering bad effects of sodium.

Structurally, it has a protective outer skin layer and delicious, sweet and tart, creamy-white edible flesh inside.

6.3.1.1   IMPORTANT FUNCTIONAL INGREDIENTS

Hundred grams of banana carry 89 kcal. Besides, it contains good amount of health-benefiting antioxidants, minerals, and vitamins. Banana fruit is composed of soft, easily digestible flesh made up of simple sugars like fructose and sucrose that upon consumption instantly replenishes energy and revitalizes the body. Thus, for these qualities, bananas are used by athletes to get instant energy and as supplement food in the treatment plan for underweight children. The fruit holds a good amount of soluble dietary fiber (2.6 g/100 g) that helps normal bowel movements and 23% carbohydrate.

6.3.1.1.1   Antioxidants

It contains health-promoting flavonoid poly-phenolic antioxidants such as *lutein, zeaxanthin*, and *α-carotenes* in small amounts. These compounds act as protective scavengers against oxygen-derived free radicals and reactive oxygen species that play role in aging and various disease processes.

6.3.1.1.2   Vitamins and Minerals

Banana is a good source of vitamin B_6 (pyridoxine) and provides about 28% of daily recommended allowance. Pyridoxine is an important B-complex vitamin that has beneficial role for the treatment of neuritis and anemia. Further, it helps decrease homocysteine (one of the triggering factors in coronary artery disease (CAD)) levels within the human body.[88]

The fruit is also a moderate source of vitamin C (about 8.7 mg/100 g). Consumption of foods rich in vitamin C helps the body develop resistance against infectious agents and scavenge harmful oxygen-free radicals.[98]

Fresh bananas provide adequate levels of minerals such as copper, magnesium, and manganese. Magnesium is essential for bone strengthening and has a cardiac-protective role as well. Manganese is utilized as a cofactor for the antioxidant enzyme, superoxide dismutase. Copper is required in the production of RBCs. Fresh banana is a very rich source of potassium. Hundred gram fruit provides 358 mg potassium. Potassium is an important component of cell and body fluids that helps control heart rate and blood pressure, countering bad effects of sodium.

6.3.2 PLANTAIN

Plantain is a tropical fruit crop, native to Southeast Asian islands.[111,112,115] The Swedish botanist, Linnaeus, named the banana and plantain family as *Musa* and the plantain especially as *paradisiaca* because he said that it was "the tree of paradise." The largest producers of plantains are African countries such as Uganda, Rwanda, Ghana, and Nigeria, where plantains and bananas provide more than 25% of food energy requirements for about 70 million people.[116] It is the most important trading fruit after citrus. It is cultivated both in home gardens and also as commercial crop. The important groups in plantain are French plantain and Horn plantain. Nendran belongs to the French plantain group. It has a very good shelf life.[77] They are much starchier than banana. The beer brewed from plantain has low alcohol content but has high vitamin content.[82]

Musa paradisiaca is a herbaceous plant (up to 9-m long), and fruits are oblong, fleshy, 5–7-cm long in wild form and longer in the cultivated varieties.[39]

6.3.2.1 IMPORTANT FUNCTIONAL INGREDIENTS

The main nutritional ingredients present in plantain are starch, phenolic compounds, and minerals. It contains carbohydrates (32%), protein (1%), fat (0.4%), water (65%), and some vitamins and mineral elements. The fruit is extremely low in fat and protein, high in fiber and starch. It is a good source of vitamin A, B_6, and C which help to maintain vision, good skin and build immunity against diseases. It is also rich in potassium, magnesium, and phosphate when cooked green. The consumption of unripe plantain products could render a multifaceted action that will terminate the instant generation of free radicals thereby setting the human system free from accumulation of radicals.[86] Saraswathi and Gnanam[83] reported that *M. paradisiaca* inhibits cholesterol crystallization in vitro which may have an effect on atherosclerosis plaque and gallstones in vivo.

Parmar and Kar[72] tested the peel extract of *M. paradisiaca* in rats with diet-induced atherosclerosis. This study reports the protective role of the extract in atherosclerosis and thyroid dysfunction.

6.3.2.1.1 Antioxidants

- *Phytoestrogen:* Phytoestrogens stimulate osteoblast differentiation[100] and increase osteoclast apoptosis.[96] Plant extracts and compounds, primarily isoflavones that mimic or modulate endogenous estrogens, usually by binding to estrogenic receptors, prevent atherosclerotic cardiovascular diseases.[70]
- *Flavonoids:* Several flavonoids and related compounds such as leucocyanidin were isolated from the unripe pulp of plantain.[55,56,78] Lewis et al.[56] reported that natural flavonoid, leucocyanidin, from the unripe banana (*Musa sapientum* var. paradisiaca) pulp protects the gastric mucosa from erosions. Leucocyanidin and the synthetic analogues, hydroxyethylated leucocyanidin and tetraallyl leucocyanidin were found to protect the gastric mucosa in aspirin-induced erosions in rat by increasing gastric mucous thickness.[55] Goel et al. [31] reported that banana pulp powder (*M. sapientum* var. paradisiaca) showed significant antiulcerogenic activity in aspirin-, indomethacin-, phenylbutazone-, prednisolone-induced gastric ulcers and cysteamine- and histamine-induced duodenal ulcers in rats and guinea-pigs, respectively. Flavonoids isolated from unripe fruits showed hypolipidemic activity evidenced by decrease in cholesterol, triglycerides (TGs), free fatty acids, and phospholipid levels in serum, liver, kidney, and brain of rats. The cholesterol lowering effect was attributed to a higher degradation rate of cholesterol than synthesis.[93] It was demonstrated that plantain fruit in its unripe state has active antiulcerogenic flavonoid, leucocyanidin.[57]
- *Flavonols*: Several flavonoids and related compounds (quercetin and its 3-Ogalactoside, 3-O-glucoside, and 3-O-rhamnosyl glucoside) were isolated from the unripe pulp of plantain.[55,56,78,86]
- *Capsaicinoid:* It has anti-inflammatory property.[48] It is used as a topical cream for neuropathical ailments.
- *Phenolic acids*: Caffeic acid is a hydroxyl cinnamic acid. It also acts as a chemoprotective agent in oral cancer. Antihyperglycemic

action of caffeic acid[117] is a noted feature, whereas p-hydroxybenzoic acid is a popular antioxidant with less toxicity.

- *Oleoresin*: Shogaol acts as chemopreventive role in the development of hepatocellular carcinoma in cirrhotic patients. Gingerol
  plays the role of an anti-inflammatory agent and a promising antioxidant.[86]

6.3.2.1.2   Dietary Fiber

Cellulose and hemicelluloses have been isolated from pulp and peel of
M. paradisiaca.[24,47] Fibers from *M. paradisiaca* fruit increased glycogen-
esis in the liver and lowered fasting blood glucose.[92] Hemicellulose and
other neutral-detergent fibers from the unripe *M. paradisiaca* fruit showed
low absorption of glucose and cholesterol and low serum and tissue levels
of cholesterol and TGs.[91] Peels are rich in total dietary fiber.[3]

- *Carbohydrates:* Starch, iron, crystallizable and noncrystallizable
  sugars have been found in the fruit pulp of *M. paradisiaca* and *M.
  sapientum*.[29] Starch content of ripe fruit is 5–10%. Total sugars are
  15%.[3]
- *Vitamins:* Vitamin C and B vitamins have been found in the fruit
  pulp of *M. paradisiaca*.[29] The plantain is richer in ascorbic acid
  and higher in carotene content than the bananas. Folic acid content
  is 22 µg/100 g.[3]
- *Minerals*: Mineral salts have been found in the fruit pulp of *M. paradisiaca*.[29] The green fruit of *M. paradisiaca* has been reported to
  have hypoglycemic effect due to the stimulation of insulin production and glucose utilization.[65] Its high potassium and sodium
  content has been correlated with the glycemic effect.[79] Potassium
  content is 499 mg/100 g.[3]

6.3.3   APPLE

Out of all the deciduous fruits, apple is the most important in terms of
production and extent. Apple was introduced to India by the British in the
Kullu Valley as far back as 1865, while the colored delicious cultivars of

apple were introduced to Shimla hills of the same state in 1917. Apples are a very significant source of flavonoids in the diet in the United States and in Europe.[94]

Apple fruit features oval or pear shape.[159–161] Its outer peel comes in different hues and colors depending upon the cultivar type. Internally its crispy, juicy pulp is off-white to cream in color, and has a mix of mild sweet and tart flavor. Its seeds are bitter in taste, and therefore, inedible.

6.3.3.1   IMPORTANT FUNCTIONAL INGREDIENTS

Delicious apple fruit is notable for its impressive list of phytonutrients and antioxidants. These components are essential for optimal growth, development, and overall wellness. Apples are rich in antioxidant phytonutrients flavonoids and polyphenolic compounds.[22]

6.3.3.1.1   Flavonoids

Some of the important flavonoids in apples are *quercetin, epicatechin*, and *procyanidin B₂*. Apples are also good in tartaric acid that gives tart flavor to them. Altogether, these compounds help the body protect from deleterious effects of free radicals.[88]

6.3.3.1.2   Dietary Fiber

Apples are low in calories; 100 g of fresh fruit slices provide just 52 kcal. They contain no saturated fats or cholesterol. The fruit is rich in dietary fiber, which helps prevent absorption of dietary-LDL or bad cholesterol in the gut. The fiber also saves the colon mucous membrane from exposure to toxic substances by binding to cancer-causing chemicals inside the colon.

6.3.3.1.3   Vitamins and Minerals

Apple fruit contains good quantities of vitamin C and β-carotene. Vitamin C is a powerful natural antioxidant. Consumption of foods rich in vitamin C helps the body develop resistance against infectious agents and scavenge harmful, pro-inflammatory-free radicals from the body.[54] The fruit is a good source of B-complex vitamins such as riboflavin, thiamine, and pyridoxine. Together, these vitamins help as cofactors for enzymes in metabolism as well as in various synthetic functions inside the human body. Apples also carry a small amount of minerals such as potassium, phosphorus, and calcium. Potassium is an important component of cell and body fluids that helps controlling heart rate and blood pressure, thus countering the bad influences of sodium.

6.3.4   GRAPES

Grapes are deciduous fruit crop. They are ancient fruit crops whose history dates back to even the prima phase of cultivation by human beings. It is a native of North America and Europe. In India, commercial cultivation commenced only during the 20th century.[71,77] It was introduced to India from Iran and Afghanistan. Wild grapes were used to make wine during earlier times in Himachal Pradesh. Now it is grown in the states of Maharashtra, Karnataka, Tamil Nadu, Andhra Pradesh, Punjab, Haryana, Delhi, Uttar Pradesh, Rajasthan, and Madhya Pradesh.[15]

Grapes are of mainly two types as dark blue and green.[109,113,185] Of these, dark blue ones are more functionally significant. Grapes are berries; each berry has skin, pulp, and seeds. The skin of grape is a thin layer. Inside the skin there is pulp and seeds.[110] The pulp is the fleshy part of the fruit that can be squeezed to get juice. The seeds are normally four in number that are inside the pulp. The seeds are rich in tannins.

6.3.4.1   IMPORTANT FUNCTIONAL INGREDIENTS

The fruits are used as fresh, raisins, canned, for making juice and wine. Almost 65% of world productions of grapes are used for consumption as wine and juice whereas 20% as fresh fruits. It has a total antioxidant concentration 2.42 mmol/100 g.[35] The main nutritional ingredients present in grapes are polyphenolic compounds called resveratrol and anthocyanins.

6.3.4.1.1   Polyphenols

- *Resveratrol*: It is a distinctive phytoestrogen present in the skin of grapes.

- *Anthocyanins*: They are flavonoids abundantly present in the red grapes.
- *Flavone*: Guercetrin is the flavone present which is the yellow pigment seen in both red and green grapes.[77]
- *Flavanonols*: Engeletin and astilbin are the flavanonols present in grapes. They are faint yellow compounds in green grapes. They are mainly 3-glycosides or 3-glucuronides.[85,17]
- *Flavan-3-ols*: Catechin, gallocatechin, epicatechin, epigallocatechin are flavan 3-ols present. They are colorless polymers present in grape skin and seeds. Catechins are a class of flavonoid tannin group antioxidants found in both red and green varieties.
- *Flavanols*: The flavanols present in grapes are kaempferol, quercetin, myricetin, and isorhamnetin. Quercitron is a glycoside of guercetrin which gives yellow pigment to both red and green grapes.[77]

6.3.4.1.2 Dietary Fiber

Grapes are not rich in dietary fibers. They have several polymeric carbohydrates such as cellulose, hemicellulose, and pectin.[37]

6.3.4.1.3 Carbohydrates

Grapes are high in carbohydrates. The sugar predominant is fructose followed by glucose and sucrose.

6.3.4.1.4 Proteins

Grapes are sources of several oxidizing enzymes such as polyphenol oxidases, peroxidises, phenolase, phosphatase, proteases, PME, polygalacturonases, and sucrase which are present in the skin.[17,18,85]

6.3.4.1.5 Vitamins

Grapes are a good source of vitamins C, A, B-complex, and K. Vitamin C accounts to 4–7 mg/L while vitamin A is 67 IU, thiamine (B_1) ranges 160–170 µg/L, riboflavin (B_2) 3–60 µg/L, nicotinamide (B_3) 0.68–2.6 µg/L, pantothenic acid (B_5) 0.5–1.4 µg/L, pyridoxine (B_6) 0.16–0.5 µg/L, biotin

(B_8) 1.5–4.2 µg/L, folic acid (B_9) 0–1.8 µg/L, cyanocobalamin (B_{12}) 0–0.2 µg/L. Vitamin K in trace amounts is also present in grapes.[37]

6.3.4.1.6 Minerals and Electrolytes

Grapes are rich source of micronutrients such as copper, iron, and manganese. Good amounts of calcium and magnesium are also present in grapes. Iron is mostly found in raisins. Grapes are also a good source of the electrolytes such as sodium and potassium. Potassium amounts to 190–203 mg/100 g fresh fruit.

6.3.5 MANGO

Mango is one of the most popular fruit also known as "The king of the fruits." It is nutritionally rich fruit with unique flavor, fragrance, taste, and health-promoting qualities that make it a functional food, often labeled as "super fruit." Botanically, this exotic fruit belongs within the family of *Anacardiaceae*, a family that also includes numerous species of tropical-fruiting trees in the flowering plants. The tree is believed to have origi-nated in the sub-Himalayan plains of the Indian subcontinent. Mangos were introduced to California (Santa Barbara) in 1880.[173] India occu-pies the top position among mango-growing countries of the world and produces 40.48% of the total world mango production. China, Thailand, Pakistan, Mexico, and Indonesia are the major mango-growing countries.

Each mango fruit measures 5–15 cm in length and about 4–10 cm in width and has typical "mango" shape, or sometimes oval or round.[124,165] Its weight ranges from 150 to 750 g. Outer skin (pericarp) is smooth and is green in unripe mangos but in ripe fruits, it turns to golden yellow, crimson red, yellow, or orange-red depending upon the cultivar type. Mango comes in different shapes and sizes depending upon cultivar types. Internally, its flesh (mesocarp) is juicy, orange-yellow in color with numerous soft fibrils radiating from its centrally placed flat, oval-shaped stone (enveloping a single large kidney-shaped seed). Its flavor is pleasant and rich, and tastes sweet with mild tartness. A high-quality mango fruit should feature no or very less content and minimal tartness. Mango seed (stone) may either have a single embryo, or sometimes polyembryonic.[164]

6.3.5.1 IMPORTANT FUNCTIONAL INGREDIENTS

Mango fruit is rich in prebiotic dietary fiber, vitamins, minerals, and polyphenolic flavonoid antioxidant compounds.

6.3.5.1.1 Vitamins and Minerals

Mango fruit is an excellent source of vitamin A, and 100 g of fresh fruit provides 765–1082 IU or 25% of recommended daily levels of vitamin A. Consumption of natural fruits rich in carotenes is known to protect from lung and oral cavity cancers. Fresh mango is a good source of potassium. A 100-g fruit provides 156–168 mg of potassium while just 2–4 mg of sodium. Potassium is an important component of cell and body fluids that helps controlling the heart rate and blood pressure.

It is a very good source of vitamin B_6 (pyridoxine), vitamins C and E. Vitamin B_6 is required for GABA (γ-amino butyric acid) hormone production within the brain. It controls homocysteine levels within the blood, which may otherwise be harmful to blood vessels resulting in CAD and stroke.

Further, it is composed of moderate amounts of copper. Copper is a co-factor for many vital enzymes, including cytochrome c-oxidase and superoxide dismutase. (Other minerals that function as co-factors for this enzyme are manganese and zinc.) Copper is also required for the production of RBCs.

Additionally, mango peel is also rich in phytonutrients such as the pigment antioxidants like carotenoids and polyphenols.[120]

6.3.5.1.2 Dietary Fiber

Pectin is the most important dietary fiber present in mangos (7%)-healthy probiotic fiber.[112]

6.3.5.1.3 Flavonoids

Antioxidants such as quercetin and astragalin are the two kinds of antioxidants found in mango fruit.[168]

6.3.5.1.4   Carotenoids

Major carotenoids found are β-carotene, α-carotene, and β-cryptoxanthin. According to a new research study, mango fruit has been found to protect against colon, breast, leukemia, and prostate cancers. Several trial studies suggest that polyphenolic antioxidant compounds in mango are known to offer protection against breast and colon cancers.[120]

6.3.6   ORANGE

Orange is a tropical to semitropical evergreen, small flowering tree, growing to about 5–8-m tall and bears seasonal fruits that weigh about 100–150 g.[181,186,187] The orange is unknown in the wild state, and is assumed to have originated in southern China, northeastern India, and perhaps southeastern Asia. It was carried to the Mediterranean area possibly by Italian traders after 1450 or by Portuguese navigators around 1500. Citrus fruits have been valued for their wholesome nutritious and antioxidant properties. It is a scientifically established fact that citrus fruits especially oranges by virtue of their abundance in vitamins, antioxidants, and minerals can benefit in many ways. Moreover, it also has the other biologically active, nonnutrient compounds such as phytochemical antioxidants, soluble and insoluble dietary fibers that help in cutting down cancer risk, chronic diseases like arthritis, obesity, and CHDs.

6.3.6.1   IMPORTANT FUNCTIONAL INGREDIENTS

Delicious and juicy orange fruit contains an impressive list of essential nutrients, vitamins, minerals for normal growth and development and the overall well-being. Orange peel contains many volatile oil glands in pits. Interior flesh is composed of segments, called carpels, made up of numerous fluid-filled vesicles.

6.3.6.1.1   Vitamins and Minerals

Oranges are an excellent source of vitamin C. It also contains very good levels of vitamin A. It is again a very good source of B-complex vitamins such as thiamine, pyridoxine, and folates. These vitamins are essential in the sense that the human body requires them from external sources to replenish.[13]

Orange fruit also contains a very good amount of minerals like potassium and calcium. Nutrients in oranges are plentiful and diverse. The fruit is low in calories, contains no saturated fats or cholesterol, but is rich in dietary fiber, pectin.

6.3.6.1.2 Pectin

Pectin, by its virtue as a bulk laxative, helps protect the mucous membrane of colon by decreasing its exposure time to toxic substances as well as by binding to cancer-causing chemicals in the colon. Pectin has also been shown to reduce blood cholesterol levels by decreasing its reabsorption in the colon by binding to bile acids in the colon.

6.3.6.1.3 Antioxidants

Orange fruit contains a variety of phytochemicals. *Hesperetin, naringin,* and *naringenin* are flavonoids found in citrus fruits. *Naringenin* is found to have a bioactive effect on human health as antioxidant, free radical scavenger, anti-inflammatory, and immune system modulator. Oranges also contain very good levels of flavonoid antioxidants such as *α- and β-carotenes, β-cryptoxanthin, zeaxanthin* and *lutein.* Consumption of natural fruits rich in flavonoids help human body protects from lung and oral cavity cancer.

6.3.7 TANGERINES

Tangerine is a fruit crop native to tropical and subtropical Southeast Asia .[118,121] These plants are among the oldest fruit crops domesticated. The term "tangerine" originally was used in the 19th century to designate only this Mediterranean type, with the word based on the city of Tangier [46]. The types of tangerines are Honey tangerine, originally called a murcott, is very sweet and other popular types include the Sunburst and Fairchild tangerines.

The tangerine fruit is smaller than most oranges, and the skin of some varieties tends to be loose and peels off more easily. Good-quality tangerines will be firm to slightly hard, heavy for their size, and pebbly-skinned with no deep grooves. They have a darker reddish-orange skin when compared to orange. They are generally distinguished from oranges by their smaller size, loose, easily peelable skin (pericarp) and sweeter juicy flesh (arils). Being loose-skinned, the fruit is hard to pick without damage and is the highest priced fruit.[41] The number of seeds in each segment (carpel) varies significantly (up to 59).

6.3.7.1   IMPORTANT FUNCTIONAL INGREDIENTS

Tangerines are a good source of vitamin C, folate, and β-carotene. They also contain some potassium, magnesium and vitamins B_1, B_2, and B_3. A medium-sized tangerine (70 g, weighed without the peel) provides 1.2–1.5 g of dietary fiber and supplies 25–38 kcal energy.[7] Tangerine essential oils are good for ailments.

6.3.7.1.1   Flavanones

Hesperidin is mostly found in pulp and peel. Tangerines have the highest hesperidin at 19.26 mg aglycone/100 g edible fruit. *Narirutin, Eriocitrin, Naringin, Narirutin, Neo eriocitrin, Neohesperidin,* and *Poncirin* are also present.[8]

6.3.7.1.2   Flavanoids

Didymin is a novel chemotherapeutic agent for the treatment of nonsmall cell lung cancer.[38]

6.3.7.1.3   Flavones

Tangeretin is the major flavone present.[119]

6.3.8   PEAR

Sweet, delicious, and rich-flavored pears offer crunchiness of apples yet juicy as peach and nectarine. They are widely popular, particularly in the whole of the northern hemisphere, for their unique nutrient qualities. They bear medium-sized fruits that characteristically have several small seeds at their center encased in tough coat.[102,163,166] The area under pear is steadily increasing in North India.

In structure, pear fruit features bell or pyriform shape, weighing about 200 g. Fresh fruit is firm in texture with mild apple flavor. Externally its skin is very thin and depending upon the cultivar type, it can be green, red-orange, or yellow-orange in color. Inside, it is off white color flesh which is soft and juicy. In case of completely ripe fruits, its flesh may turn to grainy texture with gritty sensation while cutting with a knife. The center of the fruit is more or less similar to apple in appearance with centrally located tiny inedible seeds.

6.3.8.1   IMPORTANT FUNCTIONAL INGREDIENTS

Pears are a good source of dietary fiber. Regular eating of this fruit may offer protection against colon cancer. Most of the fiber in them is nonsoluble polysaccharide, which functions as a good bulk laxative in the gut. Additionally, its gritty content binds to cancer causing toxins and chemicals in the colon, protecting its mucous membrane from contact with these compounds. In addition, pear fruit is one of the very low-calorie fruits; it provides only 57–58 calories per 100 g. A low-calorie but good diet may help to bring significant reduction in body weight and blood LDL cholesterol levels.

6.3.8.1.1   Vitamins and Minerals

They contain good quantities of vitamin C. Fresh fruits provide about 7% of recommended dietary allowance (RDA) per 100 g. The fruit is a good source of minerals such as copper, iron, potassium, manganese, and magnesium as well as B complex vitamins such as folates, riboflavin, and pyridoxine (vitamin B_6). Pears have been suggested in various traditional medicines in the treatment of colitis, chronic gallbladder disorders, arthritis, and gout. Ash content varies from 0.20 to 0.34%.

6.3.9   PINEAPPLE

Pineapple is one of the important tropical fruit crops of the world. This fruit is native to Southern Brazil.[179,189]

The early reports indicate that domesticated pineapple was already very widely distributed in the Americas (Orinoco, Amazon and coastal Brazil around Rio de Janeiro) and the Caribbean prior to the arrival of Columbus. In India, it is grown in Karnataka, Meghalaya, West Bengal, Kerala, Assam, Manipur, Tripura, Arunachal Pradesh, Mizoram, and Nagaland. Mauritius and Kew varieties are commercially cultivated in India. Vazhakulam Pineapple (Mauritius variety), exclusively grown in the Vazhakulam area of Ernakulam district in Kerala, has geographical indication because of its exquisite taste and flavor. Pineapple is a fruit with outer peel, fleshy pulp, and inner core.[126]

6.3.9.1   IMPORTANT FUNCTIONAL INGREDIENTS

This tropical fruit has exceptional juiciness, vibrant flavor, and immense health benefits. It is a reservoir of vitamins, antioxidants, and minerals. A group of sulfur-containing proteolytic (protein digesting) enzymes (bromelain) in pineapple help in digestion. Pineapple is a digestive aid and a natural anti-inflammatory fruit. It contains large amount of phenolic substances. Antioxidant activity of those phenols was studied.[1] Pineapple is known to be very effective in curing constipation and irregular bowel movement. This is because it is rich in fiber, which makes bowel movements regular and easy. For any kind of morning sickness, motion sickness or nausea, drinking pineapple juice works effectively. It is loaded with essential nutrients and vitamins that are needed by the body for the overall growth and development. Juice from fresh pineapple can be used to relieve bronchitis, diphtheria, and chest congestion.[126]

6.3.9.1.1   Polyphenols

- Flavonoid: Myricetin was the major polyphenol identified in pineapple fiber.[43]
- Flavan-3-ols: They are hydrophenolic compounds in pineapple peel, while in pineapple pulp and core, only a little epicatechin was detected.[125] Catechin (58.51 mg/100 g) while epicatechin (50 mg/100 g) was observed in higher amounts in the dried fruit extracts.

6.3.9.1.2 Phenolic Acids

Gallic acid (31.76 mg/100 g dry extracts) was found to be the main polyphenol in pineapple peels.

6.3.9.1.3 Proteins

Fresh pineapples are rich in bromelain.

6.3.9.1.4 Vitamins

Pineapple is also a good source of vitamins C, B_1, and B_6.

6.3.9.1.5 Minerals

Pineapple is an excellent source of the trace mineral manganese, which is an essential cofactor in a number of enzymes important in energy production and antioxidant defenses. For example, the key oxidative enzyme superoxide dismutase, which disarms free radicals produced within the mitochondria (the energy production factories within our cells), requires manganese. Pineapple contains considerable calcium and potassium. Manganese is required for the growth of healthy bones and tissues.[126]

6.3.10 PEACHES AND NECTARINES

Peaches and nectarines are deciduous crops mostly grown in and are native to China. They are also cultivated in the United States and Europe. They are called "stone fruits" (since the seed is large and stony). Peaches differ from nectarines in having fuzzy-like appearance on the surface while nectarines are smooth-skinned.[127,128,130] Both have the same systematic position and functional characteristics. The skin of the nectarine fruit has a higher concentration of antioxidant components as compared to the pulp, so having the fruit unpeeled is recommended.

6.3.10.1   IMPORTANT FUNCTIONAL INGREDIENTS

Research studies have shown that nectarines contain bioactive compounds such as anthocyanins, catechins, chlorogenic acids and quercetin derivatives which have the potential to combat obesity-related medical conditions such as diabetes and cardiac disorders.[37] Among phytochemicals, polyphenols deserve a special mention due to their free-radical-scavenging activities and in vivo biological activities.

They are a moderate source of antioxidants, vitamins C and A, and ß-carotene. They are rich in many vital minerals such as potassium, fluoride, and iron. Peaches contain health-promoting flavonoid polyphenolic antioxidants such as lutein, zeaxanthin and ß-cryptoxanthin. Hydroxycinnamic acid derivatives and flavan-3-ols are also present. The flavonols and anthocyanins are mainly located in the peel.[60]

6.3.10.1.1   Flavonoids

They are found in nectarines that may help in preventing rise in the aggregation of platelets and reduce the risk of development of atherosclerosis.[20]

6.3.10.1.2   Phenolic Acids

Chlorogenic acid and anthocyanins are present.

6.3.10.1.3   Carotenoids

Lutein, ß-carotene, zeaxanthin and β–cryptoxanthin are present.

6.3.10.1.4   Vitamins

The high antioxidant capacity of peaches and nectarines are due to its highvitamin C content. The total ascorbic acid (vitamin C) content ranged from 48 to 132 mg/kg. Pregnant women are recommended nectarines for the folate content present which helps in reducing the risk of neural tube defects such as *spina bifida* and also contributes in the overall health of the mother and the baby. Vitamin wealth of nectarines also include vitamin A, vitamin B_1 (thiamine), vitamin B_2 (riboflavin), vitamin B_3 (niacin), pantothenic acid, vitamin B_6, folate, vitamin E (alpha-tocopherol), and vitamin K (phylloquinone).

6.3.10.1.5 Dietary Fiber

Fiber keeps up the digestive health. It also helps in preventing the binding of toxins to the colon walls and promotes detoxification by eliminating the toxins out of the body.[129,131]

6.4 OTHER FUNCTIONALLY SIGNIFICANT FRUITS

6.4.1 GUAVA

Guava (*Psidium guajava* L.) is a tropical fruit and belongs to the family *Myrtaceae*.[132] It is a native of Central America.[87] It is consumed as fresh[2] and is popularly known as the poor man's apple. It is produced mostly in India, Brazil, Egypt, South Africa, Columbia and the United States. It is a simple crop which needs less care and attention for production with amazing nutritive value. In India, the best-quality guava is produced in Allahabad, Uttar Pradesh.[15] Also, in India, it is found in the states of Madhya Pradesh, Gujarat, Rajasthan, Maharashtra, Andhra Pradesh, Tamil Nadu, West Bengal, Punjab, Assam, Orissa, Karnataka, and Kerala.

6.4.1.1 FUNCTIONAL SIGNIFICANCE

Guava occurs in mainly two forms as pink[133,134] and white.[135] The most functionally rich one is pink type. The parts of the fruit both peel and pulp are a source of natural antioxidants. It is used to obtain antioxidant dietary fiber (AODF), a new item which combines in a single natural product the properties of dietary fiber and antioxidant compounds.[2] They are rich sources of polyphenols like flavonoids.[58] The estimated antioxidant levels in the fruit[84] consist of: total phenolics 2473 ± 45 µg gallic acid/g fresh weight; flavonoids 209 ± 10 µg; proanthocyanidins 263 ± 31 µg; vitamin C 1426 ± 26 µg/g; lycopene 1150 µg/100 g wet wt. edible portion; and β-carotene 984 µg/100 g wet wt. edible portion.

Lycopene source of guava fruit is twice greater than that of tomatoes with 5204 µg/100 g of fruit. The fruit is a rich source of soluble dietary fiber (5.4 g/100 g fruit). It is a good laxative. It is a rich source of pectin. Pectin content increases during fruit development but decreases in over-ripe fruits.[37] Vitamin C content varies distinctly with varieties, pink flesh with highest 228 mg/100 g fruit.[51] It is more than thrice the amount required in the daily intake. Outer rind is the major source than the inner pulp. The guava fruit is a moderate source of pantothenic acid (B5), niacin (B3), pyridoxine (B6), and vitamins E and K.

They are also sources of magnesium, copper, iron, and manganese. It is a good source of minerals such as phosphorus (23–40 mg/100 g), calcium (14–30 mg/100 g), iron 0.2–1.4 mg/100 g.[37] Iron is present in its seeds.

The fruit is also a rich source of potassium. It provides more of it than other fruits like banana. Softening enzymes such as polygalacturonase, pectin esterase, β-galactosidase, cellulose increase with ripening.[23] Fruit shelf-life is about 10 days.[11] The guava fruits are low in calories (68 kcal) and are free of cholesterol.

6.4.2   PASSION FRUIT

Passion fruit (*Passiflora edulis* Sims.) is a delicious subtropical fruit of the family Passifloraceae.[103] It is a native of Brazil which is the largest exporter of passion fruit. In India, it is grown in Nilgiris, Wayanad, Coorg, Himachal Pradesh, Nagaland, and Mizoram. They come in both yellow and purple forms.

6.4.2.1   FUNCTIONAL SIGNIFICANCE

Passion fruit juice is nutritious and known for its blending capabilities.[15] It is rich in antioxidants, minerals, vitamins, and fibers. The dried passion flowers contain an alkaloid passiflorin which is used for relieving pain and inducing sleep.[77] The major pigments in passion fruit are carotenoids, β- and γ-carotene and phytofluene in purple variety and trace quantities of flavones but no anthocyanins.[16] β-Cryptoxanthin is also present. The fruit is a good source of dietary fiber. Hundred grams of fruit provides 0.2–10.40 g fiber. Vitamin C content is 18–30 mg/100 g of fruit. Vitamin A content is 700–1274 IU/100 g passion fruit.

It is a rich source of minerals such as iron, copper, sulfur, magnesium, and phosphorus. Calcium is 4–12 mg/100 g. Iron is 0.2–1.6 mg/100 g. The fresh fruit is rich in high amount of potassium (278–348 mg/100 g fruit). Sodium is 6–28 mg/100 g. Chlorides are also present. They have no cholesterol and energy-rich fruit (84–97 kcal).

6.4.3 JACKFRUIT

Jackfruit (*Artocarpus heterophyllus* Lam. of the family Moraceae) is the native fruit of India and is the largest fruit in the world, which is grown as a homestead crop.[104–107] It is widely grown in the eastern and southern parts of India.

6.4.3.1 FUNCTIONAL SIGNIFICANCE

The ripened fruit flakes are a good nutrient source. It has sugars like fructose and sucrose making the body revitalize instantly. It is rich in carotene, thiamine, pectin, iron, phosphorus, and calcium. It is used as a vegetable when unripe. The seeds are rich sources of starch.[77] β-carotene and lutein are present as carotenoids. Xanthin and β-cryptoxanthin are the xanthophylls from the fruit. Apart from this, flavonoids are also present in it.

The rind is a rich source of dietary fibers like pectin. The flakes, seeds, sterile flowers, skin, and core contain calcium pectate.[15] The dietary fiber is about 1.5 mg/100 g fruit. Jackfruit is a good source of antioxidant vitamin C. It is 13.7 mg/100 g fruit. The fruit is also rich in vitamin A with 110 IU/100 g fruit. It is one of the rare fruits which are rich in B-complex vitamins.

The fresh fruit is a rich source of magnesium, manganese, and iron. The potassium level provided by the fruits is 303–448 mg/100 g fruit. Lectin is the natural protein present in the fruit. Jacaline is an extract obtained from jackfruit. The ripened fruit is a good energy source with 95 kcal and is cholesterol free.

Lectin is used in the cancer treatment. Jacaline inhibited the growth of HIV infection in vitro.[77]

6.4.4 POMEGRANATE

Pomegranate (*Punica granatum*) is one of the most popular, nutritionally rich fruit of the family Punicaceae, with unique flavor, taste, and health-promoting characteristics.[158,180,188] Together with subarctic-pigmented berries and some tropical exotics such as mango, it too has novel qualities of functional foods, often called as super fruits. Botanically, it is a small-sized fruit bearing deciduous tree. The tree is thought to have originated in Persia and the Sub-Himalayan foothills of Northern India. Pomegranate tree grows to about 5–8-m tall. It is cultivated at a commercial scale in vast regions across the Indian subcontinent, Iran, Caucasus, and Mediterranean regions for its fruits. Pomegranate is cultivated commercially only in Maharashtra. Small-scale plantations are also seen in Gujarat, Rajasthan, Karnataka, Tamil Nadu, Andhra Pradesh, Uttar Pradesh, Punjab, and Haryana.

Completely established tree bears numerous spherical, bright red, purple, or orange-yellow-colored fruits depending on the cultivar types. On an average, each fruit measures about 6–10 cm in diameter and weighs about 200 g. Its tough outer skin (rind) features a leathery texture. Interior structure of the fruit is separated by white, thin, spongy, membranous, bitter tissue into discrete compartments. Such sections are packed as sacs, filled with tiny edible sweet, juicy, pink pulp encasing around a single, angular, soft or hard (in case of over mature fruits) seed.

6.4.4.1   FUNCTIONAL SIGNIFICANCE

The fruit is moderate in calories, holds about 83 kcal/100 g, slightly more than that of in the apples. It contains no cholesterol or saturated fats. It is a good source of soluble and insoluble dietary fibers, providing about 4 g/100 g (about 12% of RDA). Nutritionists recommend pomegranate in the diet for weight reduction and cholesterol-controlling programs. Regular inclusion of fruits in the diet boosts immunity, improves circulation, and offers protection from cancers.

Certain ellagitannin compounds such as granatin B and punicalagin are found abundantly in the pomegranate juice. Studies suggest that punicalagin and tannins can be effective in reducing heart-disease risk factors by scavenging harmful free radicals from the human body. Further, it is an also good source of many vital B-complex vitamins such as pantothenic acid (vitamin B_5), folates, pyridoxine and vitamin K, and minerals such as calcium, copper, potassium, and manganese.[89]

The fruit is also a good source of antioxidant vitamin C, provides about 10–17% per 100 g of daily requirement. Consumption of fruits rich in vitamin C helps the body develop resistance against infectious agents by boosting immunity. Regular consumption of pomegranate has also been found to be effective against prostate cancer, benign prostatic hyperplasia (BPH), diabetes, and lymphoma.

6.4.5 MANGOSTEEN

Mangosteen (*Garcinia mangostana*) of the family *Clusiaceae* (*Guttiferae*) is an evergreen, erect tree.[169–171] The location of mangosteen's origin is not known with certainty. It is commonly found in the tropical rainforests of Indonesia, Malaysia, Thailand, and the Philippines as well as in some cultivated orchards in Sri Lanka and India, where annual precipitation and relative humidity are favorable for its growth. Fresh purple fruits are available in the markets from June to October. Mangosteen is grown in four areas of India, as all of them are tropical, have high humidity and decent rainfall; It is grown in Nilgiri hills, the southern districts of Tirunelveli, Kanyakumari in Tamil Nadu and Kerala. Trees fruit prolifically, especially those aged 20 years and above; however, the fruits must be picked from the tree.

Unique for its appearance and flavor, mangosteen is often revered as the queen of tropical fruits, particularly in the South-East Asian regions. This exotic, round, purple color fruit is quite popular for its snow-white, juicy, delicious arils all over the Asian countries, and in recent years by the European and American fruit lovers as well.

Each tree bears several deep purple, round-shaped fruits capped with light-green calyx at the stem end. Completely matured fruit measures about 3–7 cm in diameter. Its outer tough rind is about 7–12-mm thick which contains bitter yellow latex that stains clothes black. Internally, the fruit features 4–10 juicy, snow-white, soft, fleshy, triangular segments as in oranges. Each segment may carry 1–4 off-white-colored seeds. Seeds are inedible and bitter in taste. The flavor of the fruit can be described as sweet, mild tangy, fragrant, and delicious.

6.4.5.1 FUNCTIONAL SIGNIFICANCE

Delicious and juicy, mangosteen is one of the popular tropical fruits. It comprises an impressive list of essential nutrients which are required for normal growth and development and overall nutritional well-being. It is moderately low in calories and contains no saturated fats or cholesterol. It is rich in dietary fiber (100 g provides about 13% of RDA).

Mangosteen is a good source of vitamin C and provides about 12% of RDA/100 g. Vitamin C is a powerful water-soluble antioxidant. Consumption of fruits rich in vitamin C helps human body develop resistance against viral flu and help scavenge harmful, pro-inflammatory free radicals.[44]

Fresh fruit is a moderate source of B-complex vitamins such as thiamine, niacin, and folates. These vitamins are acting as cofactors and help the body to metabolize carbohydrates, protein, and fats. Further, it also contains a very good amount of minerals like potassium, copper, manganese, and magnesium.

6.4.6   PAPAYA

Papaya fruit (*Carica papaya* of the family Caricaceae) is another gift of Mexicans to this world.[172,174] This exotic fruit, also popular as pawpaw, is packed with numerous health-benefiting nutrients. It is one of the favorites of fruit lovers for its nutritional, digestive, and medicinal properties. Papaya plant is grown extensively all over the tropical regions under cultivated farms for its fruits as well as for latex, papain, and an enzyme that found wide applications in the food industry.

Papaya tree bears many spherical or pear-shaped fruits clumped near its top end of the trunk. Inside, the fruit features numerous black pepper corn-like seeds, encased in a mucin coat, at its hollow central cavity as in melons. The flesh is orange in color with either yellow or pink hues, soft in consistency and has deliciously sweet, musky taste with rich flavor. Papaya seeds have been proven natural remedy for many ailments in the traditional medicines. The seeds can be found its application as anti-inflammatory, anti-parasitic, analgesic, and used to treat stomach ache and ringworm infections.

6.4.6.1   FUNCTIONAL SIGNIFICANCE

The papaya fruit is very low in calories (just 39–43 kcal/100 g) and contains no cholesterol; it is a rich source of phytonutrients, minerals, and vitamins. Papayas contain soft, easily digestible flesh with a good amount of soluble dietary fiber that helps to have normal bowel movements, thereby reducing constipation problems. Fresh, ripe papaya is one of the fruits with the highest vitamin C content (provides 60–61.8 mg or about 103% of dietary reference intakes (DRI), more than that of in oranges, or lemons). It is also an excellent source of vitamin A (provides 950–1094 IU/100 g). Papaya fruit is also rich in many essential B-complex vitamins such as folic acid, pyridoxine, riboflavin, and thiamine. These vitamins are essential in the sense that body requires them from external sources to replenish and play a vital role in metabolism. Fresh papaya also contains a good amount of potassium (182–257 mg/100 g) and calcium. Flavonoids like ß-carotene, lutein, zeaxanthin and cryptoxanthin are present in papaya.

6.4.7 RAMBUTAN

The Rambutan (*Nephelium lappaceum* of Sapindaceae family) is a native to Malay-Indonesian region, other regions of tropical Southeast Asia.[175,176,177] It has spread from there to various parts of Asia, Africa, Oceania, and Central America. The widest variety of cultivars, wild and cultivated, is found in Malaysia. Rambutan originates in Malaysia, where it also got its name "Rambut" in Malay means hair. Today, Thailand's Surat Thani province produces most of the world's rambutans, with Indonesia as another top grower. Many diverse regions grow rambutan including parts of India, Sri Lanka, Hawaii, South America, Tanzania, Australia, Zanzibar, Central America, countries of the Caribbean and Pacific Asian countries like the Philippines. The common types are red and yellow.

6.4.7.1 FUNCTIONAL SIGNIFICANCE

A 100 g serving of fresh rambutan pulp provides about 4.9–40 mg of vitamin C, which corresponds to 66% of the daily value (DV) for vitamin C. Vitamin C is not only best known as a dietary remedy for the common cold and flu but also provides a number of other health benefits. Consumption of rambutan increases body's natural ability to flush out heavy metals and other toxins as well as to deal with stress.[178]

Compared with most other fruits, rambutans are also a good source of copper. A copper deficiency may lead, to anemia, ruptures in blood vessels, bone and joint problems, elevated cholesterol levels, frequent infections and chronic fatigue. Copper is also crucial for healthy hair growth and foods rich in copper, such as rambutans, may help prevent hair loss, intensify hair color and prevent premature greying of hair.[74]

6.5   ADVANCED PROCESSING AND PRESERVATION TECHNIQUES

During the last few years, ultrasound-assisted processing has attracted an increasing interest in the fields of food science and technology. The application of ultrasound in food technology can be divided into two different approaches, low-energy diagnostic ultrasound in the MHz range used for nondestructive testing and high-power ultrasound in the kHz range applied for material alteration. Process–structure–function interactions are the basis for several processes in food technology. Consequently, the potential of ultrasound in assisting and influencing these processes has been studied widely.[26,40]

Advanced food preservation includes high-pressure processing (HPP) and the use of various electric methods such as microwave, pulsed electric fields, and between electric fields, ohmic processing.[5,154]

HPP is effective against microorganisms because it results in the rupture of microbial membranes. Recent studies on high-pressure sterilization, achieved through the use of high initial temperatures, have further advanced this technology. In both ohmic and pulsed electric field processing, the electric current is uniformly applied to the entire food product, which creates local heating and also causes rupture of microbial and plant cells.

6.6   EFFECT OF PRESERVATION AND PROCESSING ON FUNCTIONAL PROPERTIES

6.6.1   HIGH-PRESSURE PROCESSING

6.6.1.1   VITAMINS AND TOTAL CAROTENOIDS

Total carotenoids found in fruits are relatively stable to preservation by HPP. Most authors found that the total carotenoid content of fruits and vegetables was either unaffected or increased by preservation using high pressure.[64,90] Researchers have found that the B-vitamins are stable to HPP at room temperature.

Vitamin C content in purees of guava was found to be fairly stable following HPP alone or HPP plus mild heat treatments, but not in HPP plus high heat application. While vitamin C content of orange juice generally decreases during storage time, HPP-treated juice declined less than thermally pasteurized orange juice, at all storage temperatures studied.[5]

6.6.1.2   TOTAL PHENOLICS

Studies on high-pressure preservation effects on total phenolics determined that these compounds were either unaffected or actually increased in concentration and extractability following treatment with high pressure. The catechins, hydroxyl cinnamic acids, and procyanidins were significantly higher in the juice after HPP. These authors stated that polyphenol oxidase was activated in the range of 200–300 MPa and this increased catechins oxidation.

6.6.2   MICROWAVE PRESERVATION

6.6.2.1   VITAMINS AND TOTAL CAROTENOIDS

Most authors reported that the application of microwave preservation treatments resulted in loss of vitamin A or carotenoids. Total carotenoid content of papaya puree was determined after application of various levels of microwave power (285–850 W) and losses were reported to be as high as 57%. A limited number of reports on specific carotenoids, such as β-carotene and lycopene, generally reported losses following microwave treatments. Vitamin C content in fruits and vegetables varied from losses as high as 57% to increases of 10–26% as a result of microwaving. Total vitamin C content in apple puree was similar before and after the microwave process; however, ascorbic acid content decreased (43% retention) and dehydro ascorbic acid increased (57%).[5]

6.6.3   MICROWAVE VACUUM PRESERVATION

6.6.3.1   VITAMINS AND CAROTENOIDS

Researchers have investigated microwave vacuum drying of banana slices[137] and of whole grapes for which vitamins A, B, and C contents were compared to that in fresh and sun-dried grapes. In all cases, the removal of water resulted in a concentration of the initial nutrient content, ranging from 50% to over 700% higher values in the microwave-dried product. However, these authors neglected to report nutrient content on a dry-weight comparison, which would have been of greater merit. The microwave vacuum-dried grapes had a moisture content of 2.8%, whereas the fresh grape moisture content was 73.3%, representing over a 30-fold difference. In most cases, vitamin C content was reduced by microwave vacuum preservation. Microwave processing of apple puree at 652 W (75°C) for 35 s resulted in a loss in the reduced form of vitamin C (43% retention) and an increase in the oxidized form (57% increase).[73]

6.6.3.2   TOTAL PHENOLICS

Total phenolics were retained at higher levels in microwaved vacuum-dried fruits and vegetables than in those that were air-dried; however, at high levels of microwave power (> 500 W), this difference was not significant. When compared to air drying and freeze drying, microwave vacuum drying of fruit had intermediate levels of total phenolics retention. Microwave vacuum drying typically resulted in a loss of these compounds, but quercetin 3-O-glycoside and catechin content actually increased as a result of the drying procedure.[5]

6.6.3.3   MINERALS AND CRUDE FIBER

Microwave vacuum-preserved grapes had higher levels of minerals and crude fiber, most likely due to concentration in the dried product.[5]

6.6.4   RADIATION TECHNOLOGY

Irradiation of food involves the exposure of food to short-wave energy of ionizing radiations, through the radiation field, allowing the food to absorb the desired amounts of radiation energy to achieve specific objectives such as extension of shelf-life, insect disinfestations, and elimination of food-borne parasites. The food never comes in direct contact with radiations and hence they are not radioactive. The application of this technology is so diverse that it has lots of future prospects in preservation sector.[80]

6.7   CONCLUSION

Fruits thus play a vital role in maintaining a sound and healthy life. The antioxidant-rich fruits help to fight against cancer by clearing out reactive oxygen species. Many of them are iron-rich sources which help in replenishing blood-related deficiencies. Fruits like banana and plantain are widely cultivated and their processed products are always in demand. Seasonal crops such as passion fruit, apples, and oranges need to be preserved for year-round usage. Guava, jackfruit, and other underexploited fruits available in homesteads need to be more explored and utilized especially for their antioxidant properties. Many nutraceuticals (nutrient-rich modified foods) are also made from fruits for specific purposes. Food technological aspects must be improved to preserve these nutrients without much loss. The chapter also throws light to the need of orchards at home for making sure of entrapping all the possible nutrients.

6.8   SUMMARY

Functional foods play a vital role in disease prevention and health promotion. Fruits are involved in the whole foods category and promoted as functional foods. Fruits are a rich source of several antioxidants and other nutrients. Polyphenols are a group of dietary antioxidants found naturally in fruits and vegetables. They primarily consist of flavonoids including flavanols, flavones, isoflavones, flavonols, flavonones and anthocyanins, and non-flavonoid polyphenolics including phenolic acids, lignans and stilbenes. Soluble fibers present in fruits may reduce risk of CHD. Proteins in the form of enzymes play a significant role in free radical scavenging activities. Vitamins may contribute to the maintenance of healthy vision, immune function, bone health; cell integrity helps regulate calcium and phosphorus. Fruits are rich in antioxidant vitamins like C, A, and E. Minerals play an important role in the normal functioning of cells and tissues. Functional significance of highly produced fruit crops and some other functionally rich fruits are described and discussed in this chapter. Banana fruit is composed of soft, easily digestible flesh made up of simple sugars like fructose and sucrose that upon consumption instantly replenishes energy and revitalizes the body. It is rich in potassium. Plantain is extremely low in fat and protein, high in fiber, starch, antioxidants, and minerals. Apples are rich in antioxidant phytonutrients flavonoids and polyphenolic compounds. Grapes contain nutritional ingredients such as polyphenolic compound called resveratrol and anthocyanins. Oranges and tangerines are an excellent source of vitamin C and carotenes. Pears are rich in dietary fibers. Pineapple is a source of vitamins, fibers and digestive enzyme bromelain. Peaches and nectarines contain a range of polyphenols. Guava is a great source of antioxidants and pectin. Passion fruit is an antioxidant-rich fruit providing great immunity. Jackfruit is good source of pectin and carotenes. Pomegranate and Mangosteen are rich sources of antioxidants and vitamins. Papaya is an immune booster and has high amount of potassium. Rambutan is an immense source of vitamin C. Functional properties of fruits can be preserved through HPP, microwave vacuum processing, and through radiation technology.

KEYWORDS

- Anacardiaceae
- anthocyanidins
- anthocyanins
- apigenin
- apple
- Arctic
- *Artocarpus heterophyllus* Lam.
- ash
- astragalin
- banana
- Batsch
- bromelain
- calcium
- capsaicin
- carbohydrate
- *Carica papaya*
- carotenoids
- catechins
- chlorogenic acid
- *Citrus sinensis*
- *Citrus tangerina* Tanaka
- coronary artery disease
- didymin
- dietary fibers
- energy
- eriocitrin
- fatty acids, total monounsaturated
- fatty acids, total polyunsaturated
- fatty acids, total saturated
- ferulic acid
- fiber
- flavan-3-ols
- flavanones

- prebiotics
- proanthocyanidins
- procyanidin B$_2$
- proteins
- *Prunus persica* (L.)
- *Psidium guajava* L.
- *Punica granatum*
- *Pyrus communis*
- radiation technology
- rambutan
- resveratrol
- shogaol
- sodium
- stilbenoids
- subtropical
- sugars
- super fruit
- tangeretin
- tangerines
- tannins
- temperate
- total lipid
- tropical
- Vazhakulam pineapple
- vitamin A
- vitamin B$_1$
- vitamin B$_2$
- vitamin B$_3$
- vitamin B$_6$
- vitamin B$_9$
- vitamin C
- vitamin D
- vitamin E
- vitamin K
- vitamins
- *Vitis vinifera* L.
- Zeaxanthin
- Zinc

REFERENCES

1. Adhikarimayum, H.; Kshctriimyum, G.; Mabam, D. Evaluation of Antioxidant Properties of Phenolics Extracted from *Ananas comosus. Not. Sei Biol.* **2010,** *2*(2), 68–71.
2. Antonio, J. E.; Mariela, R.; Raquel, P.; Fulgencio, S. C. Guava Fruit (*Psidium guajava* L.) as a New Source of Antioxidant Dietary Fiber. *J. Agric. Food Chem.* **2001,** *49*(11), 5489–5493.
3. Arun, K. B.; Persia, F.; Aswathy, P. S.; Chandran, J.; Sajeev, M. S.; Jayamurthy, P.; Nisha, P. Plantain Peel—A Potential Source of Antioxidant Dietary Fiber for Developing Functional Cookies. *J. Food Sci. Technol.* 2015, *52*(10), 6355–6364.
4. Ashraf, C. M.; Iqbal, S.; Ahmed, D. Nutritional and Physico-Chemical Studies on Fruit Pulp, Seed and Shell of Indigenous *Prunus persica. J. Med. Plants Res.* **2011,** *5*(16), 3917–3921.
5. Barrett, D. M.; Lloyd, B. Advanced Preservation Methods and Nutrient Retention in Fruits and Vegetables. *J. Sci. Food Agric.* **2012,** *92*, 7–22.

6. Belitz, H. D.; Grosch, W. *Food Chemistry*, 2nd ed.; Springer: Wurzburg, Germany, 1999; pp 301.
7. Bender, D. A.; Bender, A. E. *A Dictionary of Food and Nutrition;* Oxford University Press: New York, 2005; pp 583.
8. Berhow, M.; Tisserat, B.; Kanes, K.; Vandercook, C. Survey of Phenolic Compounds Produced in Citrus. *USDA ARS Tech. Bull.* 1998.1–154.
9. Bombardelli, E.; Morazzoni, P. *Vitis vinifera* L. *Fitoterapia* **1995**, *66*(4), 291–317.
10. Brit, D. F.; Shull, J. S.; Yaktrine, N. L. Chemoprevention of Cancer. Chapter 81, In *Modern Health and Diseases*, 9th ed.; Shils, M. E., Olson, J. A., Shicke, M., Ross, A. C., Eds.; Williams and Wilkins: U.S., 1998; pp 1263–1269.
11. Brown, B. I.; Wills, R. B. H. Post-Harvest Changes in Guava Fruits of Different Maturity. *Sci. Hortic.* **1983**, *19*, 237–243.
12. Cabanis, J. C. Acidos organicos, sustancias minerales, vitaminas y lipidos. In: *Flanzy C, coordonateur. Enologia: Fundamentos Cientificos y Technologicos.* Madrid: Ediciones Mundi-Prensa & A. Madrid Vicente, 2000, pp 43–65.
13. Carr, A.; Frei, B. The Role of Natural Antioxidants in Preserving the Biological Activity of Endothelium-Derived Nitric Oxide. *Free Radical Biol. Med.* **2000**, *28*, 1806–1814.
14. Casagrande, F.; Darbon, J-F. Effects of Structurally Related Flavonoids on Cell Cycle on Human Melanoma Cells: Regulation of Cyclin-Dependent Kinases CDK2 and CDK1. *Biochem. Pharmacol.* **2001**, *61*, 1205–1215.
15. Chadha, K. L. *Hand Book of Horticulture;* Indian Council of Agricultural Research: New Delhi, 2001; p 1031.
16. Chan, H. T. Jr. Passion Fruit. In *Tropical and Subtropical Fruits: Composition, Nutritive Values, Properties and Uses;* Shaw, P. E., Chan, H. T. Jr., Nagy, S., Eds.; CT: AVI Publishing Co.: Westport, 1980; pp 568.
17. Cheynier, V.; Moutounet, M.; Sarni-Manchado, P. Phenolic Compounds. In *Enology: Scientific Fundametals and Technologies;* Flanzy C, Ed.; Ediciones Mundi-Prensa & A Madrid Vicente: Madrid, 2000; pp 114–136.
18. Crouzet, J.; Flanzy, C. Eds. *Enology: Scientific Fundamentals and Technologies.* Ediciones Mundi-Prensa & A. Madrid Vicente: Madrid, 2000; pp 245–273.
19. Delian, E.; Chira, L.; Dumitru, L.; Bădulescu, L.; Chira, A.; Petcuci, A. Mineral Content of Nectarines Fruits in Relation to Some Fertilization Practices. *Sci. Pap., Series B, Hortic.* **2012**, *6*, 73–80.
20. Divya; Pandey, V. *Natural Antioxidants and Phytochemicals in Plant Foods;* Satish Serial Publishing House: Delhi, 2014; p 366.
21. Duthie, G.; Crozier, A. Plant-Derived Antioxidants. *Curr. Opin. Clin. Nutr. Metab. Care* **2000**, *3*, 447–451.
22. Eberhardt, M.; Lee, C.; Liu, R. H. Antioxidant Activity of Fresh Apples. *Nature* **2000**, *405*, 903–904.
23. El-Buluk, R. E.; Babiker, E. E.; Al-Tinay, A. H. Biochemical and Physical Changes in Fruits of Four Guava Cultivars During Growth and Development. *Food Chem.* **1995**, *54*, 279–282.
24. Emaga, T. H.; Andrianaivo, R. H.; Wathelet, B.; Tchango, J. T.; Paquot, M. Effects of the Stage of Maturation and Varieties on the Chemical Composition of Banana and Plantain Peels. *Food Chem.* **2007**, *103*, 590–600.

25. Eskin, N. A. M.; Tamir, S. *Dictionary of Nutraceuticals and Functional Foods;* CRC Press, Taylor and Francis Group: London, 2006; pp 768.

26. Fernandes, F. A. N.; Gallao, M. I.; Rodrigues, S. Effect of Osmosis and Ultrasound on Pineapple Cell Tissue Structure During Dehydration. *J. Food Eng.* **2009**, *90*(2), 186–190.

27. Funaguchi, N.; Ohno, Y.; La, B. L.; Asai, T.; Yuhgetsu, H.; Sawada, M.; Takemura, G.; Minatoguchi, S.; Fujiwara, T.; Fujiwara. Narirutin Inhibits Airway Inflammation in an Allergic Mouse Model. *Clin. Exp. Pharmacol. Physiol.* **2007**, *34*(8), 766–770.

28. Gardana, C.; Guarnieri, S.; Riso, P.; Simonetti, P.; Porrini, M. Flavanone Plasma Pharmacokinetics from Blood Orange Juice in Human Subjects. *Brit. J. Nutr.* **2007**, *98*(1), 165–172.

29. Ghani, A. *Medicinal Plants of Bangladesh: Chemical Constituents and Uses,* 2nd ed.; The Asiatic Society of Bangladesh: Dhaka, Bangladesh, 2003; pp 315.

30. Gibson, G. R.; Williams, C. M. *Functional Foods—Concept to Product*; Wood head Publishing Limited: Abington Hall, CRC Press: Washington DC, 2000; p 374.

31. Goel, R. K.; Gupta, S.; Shankar, R.; Sanyal, A. K. Anti-Ulcerogenic Effect of Banana Powder (*Musa sapientum* var. *paradisiaca*) and its Effect on Mucosal Resistance. *J. Ethnopharmacol.* **1986**, *18*, 33–44.

32. Goswami, C.; Hossain, M. A.; Kaderand R. Islam, H. A. Assessment of Physico-chemical Properties of Jackfruit (*Artocarpus heterophyllus* Lam) pulps. *J. Hortic. For. Biotechnol.* **2011**, *15*(3), 26–31.

33. Guine, R.; Lima, M. J.; Barocca, M. J. Role and Health Benefits of Different Functional Food Components. *J. Educ. Technol. Health,* **2010**, *37*(14), 114–120.

34. Gupta, S.; Afaq, F.; Mukhtar, H. Involvement of nuclear factor—kB, Bax and Bcl-2 in induction of cell cycle arrest and apoptosis by apigenin in human prostate carcinoma cells, *Oncogene,* **2002**, *21*, 3727–3778.

35. Halvorsen, B.L.; Holte, K.; Myhrstad, M.C.W.; Barikmo, I.; Hvattum, E.; Remberg, S. F.; Wold, A. B.; Haffner, K.; Baugerod, H.; Andersen, L. F.; Moskaug, J. O.; Jacobs, D. R.; Blomhoff, R. A systematic screening of total antioxidants in dietary plants. *J. Nutr.* **2002**, *132*, 461–471.

36. Hossen, S.; Kabir, M.S.; Uddin, M. B.; Rahman, A.K.M.L.; Mamun, M.R.A. Effect of Different Extractions of Juice on Quality and Acceptability of Guava Jelly. *J. Innov. Dev. Strategy* **2009**, *3*(4), 27–35.

37. Hui, Y. H. *Handbook of Fruits and Fruit Processing.* Wiley India Pvt. Ltd.: New Delhi, 2006; pp 697.

38. Hung, J. Y.; Hsu, Y. L.; Ko, Y. C.; Tsai, Y. M.; Yang, C. J.; Huang, M. S.; Kuo, P. L. Didymin, a Dietary Flavonoid Glycoside from Citrus Fruits, Induces FAS-Mediated Apoptotic Pathway in Human Non-Small-Cell Lung Cancer Cells in vitro and in vivo. *Lung Cancer* **2010**, *68*(3), 366–374.

39. Imam, M. Z.; Akter, S. *Musa paradisiaca* L. and *Musa sapientum* L.: A Phytochemical and Pharmacological Review. *J. Appl. Pharm. Sci.* **2011**, *1* (5), 14–20.

40. Jambrak, A. R.; Mason T. J.; Paniwnyk, L.; Lelas, V. Ultrasonic Effect on pH, Electric Conductivity and Tissue Surface Of Button Mushrooms, Brussels Sprouts and Cauliflower. *Czech J. Food Sci.* **2007**, *25*(2), 90–100.

41. Janick, J. *Lecture 32: Citrus.* Purdue University. 2005, Retrieved November 14, 2007.

42. Jitendra, K.; Amit, P. An Overview of Prospective Study on Functional Food. *Int. J. Recent Sci. Res.* **2015**, *6*(7), 5497–5500.
43. Jose, A. L.; Rupérez, P.; Calixto, F. S. Pineapple Shell as a Source of Dietary Fiber with Associated Polyphenols. *J. Agric. Food Chem.* **1997**, *45*(10), 4028–4031.
44. Jung, H.; Keller, W.; Mehta, R.; Kinghorn, A. Antioxidant Xanthons from Pericarp of *Garcinia mangostana* (mangosteen). *J. Agric. Food Chem.* **2006**, *54*, 2077–2082.
45. Kathleen, K. *Phytochemicals and Functional Foods: Super Foods for Optimal Health;* Rutgers cooperative Research and Extension, NJAES, Rutgers, The state University of New Jersey, 1999; p 1.
46. Katz, S. H.; Weaver, W. W. *Encyclopedia of Food and Culture;* Schribner: New York, 2003; p 2004 (ISBN 0684805685).
47. Ketiku, A. O. Chemical Composition of Unripe (Green) and Ripe Plantain (*Musa paradisiaca*). *J. Sci. Food Agric.* **1973**, *24*(6), 703–707.
48. Kim, C. S.; Kawada, T.; Kim, B. S.; Han, I. S.; Chroe, S. Y.; Krata, T.; Yu, R. Capsaicin Exhibits Anti-Inflammatory Property by Inhibiting IkB-a Degradation in LPS-Stimulated Petiotoneal Macrophages. *Cell. Sig.* **2003**, *15*, 299–306.
49. Kim, H. K.; Cheon, B. S.; Kim, Y. H.; Kim, S. Y.; Kim, H. P. Effects of Naturally Occurring Flavanoids on Nitric Oxide Production in the Macrophage Cell Line RAW 264.7 and Their Structure–Activity Relationships. *Biochem. Pharmacol.* **1999**, *58*, 759–765.
50. King, R. E.; Kent, K. D.; Bomser, J. A. Resveratrol Reduces Oxidation and Proliferation of Human Retinal Pigment Epithelial Cells via Extracellular Signal-Regulated Kinase Inhibition. *Chemico-Biol. Interact.* **2005**, *151*, 142–149.
51. Kumar, R.; Hoda, N. Fixation of Maturity Standards for Guava (*Psidium guajava* L.). *Indian J. Hortic.* **1974**, *31*, 140–142.
52. Laden, B. P.; Porter, T. D. Resveratrol Inhibits Human Squalene Monoxygenase. *Nutr. Res.* **2001**, *21*, 747–753.
53. Lazze, M. C.; Savio, M.; Pizzala, R.; Cazzalini, O.; Perucca, P.; Scovassi, A. I.; Stivala, L. A.; Bianchi, L. Anthocyanins Induce Cell Cycle Perturbations and Apoptosis in Different Cell Lines. *Carcinogenesis* **2004**, *25*, 1427–1433.
54. Lee, K.; Kim, Y.; Kim, D.; Lee, H.; Lee, C. Major Phenolics in Apple and Their Contribution to the Total Antioxidant Capacity. *J. Agric. Food Chem.* **2003**, *51*, 6516–6520.
55. Lewis, D. A.; Shaw, G. P. A Natural Flavonoid and Synthetic Analogues Protect the Gastric Mucosa from Aspirin-Induced Erosions. *J. Nutr. Biochem.* **2001**, *12*, 95–100.
56. Lewis, D. A.; Field, W. D.; Shaw, G. P. Natural flavonoid Present in Unripe Plantain Banana Pulp (*Musa sapientum* L. var. *paradisiaca*) Protects the Gastric Mucosa from Aspirin-Induced Erosions. *J. Ethno pharmacol.* **1999**, *65*, 283–288.
57. Loganayaki, N.; Rajendrakumaran, D.; Manian, S. Antioxidant Capacity and Phenolic Content of Different Solvent Extracts from Banana (*Musa paradisica*) and Mustai (*Rivea hypocrateriformis*). *Food Sci. Biotechnol.* **2010**, *19*(5), 1251–1258.
58. Luximon-Ramma, A.; Bahorun, T.; Crozier, A. Antioxidants Actions and Phenolic and Vitamin C Contents of Common Mauritian Exotic Fruits. *J. Sci. Food Agric.* **2003**, (5), 496–502.
59. Marc, E. B.; Herman, T. D.; Tom, B.; Veerle, L.; Marck, V.; Krist'l, M. V.; Vanluchene, E.; Margareta, N.; Rudolphe, S.; Marc, M. M. Influence of Tangeretin on Tamoxifen's Therapeutic Benefit in Mammary Cancer. *J. Natl. Cancer Inst.* **1999**, *91*(4), 354–359.

60. Mariäa, I.G.; Francisco, A.; Toma, B. N.; Betty, H. P.; Adel, A. K. Antioxidant Capacities, Phenolic Compounds, Carotenoids, and Vitamin c Contents of Nectarine, Peach, and Plum Cultivars from California. *J. Agric. Food Chem.* **2002**, *50*, 4976–4982.

61. Mazza, G.; Miniati, E. *Anthocyanins in Fruits, Vegetables and Grains;* CRC Press: Boca Raton, Florida, 1993; p 362.

62. Nazar, L.; Csiszar, A.; Veress, G.; Stef, G.; Pacer, P.; Oroszi, G.; Wu, J.; Ungvari, Z. Vascular Dysfunction in Aging: Potential Effects of Resveratrol and Anti-Inflammatory Phytoestrogen. *Curr. Med. Chem.* **2006**, *13* (9), 989–996.

63. Niles, R. M.; McFarland, M.; Weimer, M. B.; Redkar, A.; Fu, Y.-M.; Meadows, G. G. Resveratrol is a Potent Inducer of Apoptosis in Human Melkanoma Cells. *Cancer Lett.* **2003**, *190*, 157–163.

64. Oey, I.; Plancken, I.; van der Loey, A.; Hendrickx, M. Does High Pressure Processing Influence Nutritional Aspects of Plant Based Food Systems? *Trends Food Sci. Technol.* **2008**, *19*, 300–308.

65. Ojewole, J. A.; Adewunmi, C.O. Hypoglycemic Effect Of Methanolic Extract of *Musa paradisiaca* (Musaceae) Green Fruits in Normal and Diabetic Mice. Methods Find. *Exp. Clin. Pharmacol.* **2003**, *25*(6), 453.

66. Olusegun, A. M.; Passy, O. G.; Terwase, D. S. Effects of Waxing Materials, Storage Conditions on Protein, Sugar and Ash Contents of Citrus Fruits Stored at Room and Refrigerated Temperatures. *J. Asian Sci. Res.* **2012**, *2*(12), 913–926.

67. Ottaway, P. B. *Food Fortification and Supplementation: Technological, Safety and Regulatory Aspects,* 1st ed.; CRC Press, Boca Raton – FL; 2008; p 282.

68. Othman, O. C.; Mbogo, G. P. Physicochemical Characteristics of Storage-Ripened Mango *(Mangifera indica L.)* Fruits Varieties of Eastern Tanzania. *Tanz. J. Sci.* **2009**, *35*, 57–65.

69. Othman, O. C. Physical and Chemical Composition of Storage-Ripened Papaya *(Carica Papaya* L.) Fruits of Eastern Tanzania. *Tanz. J. Sci.* **2009**, *35*, 47–55.

70. Pan, W.; Ikeda, K.; Takebe, M.; Yamori, Y. Genistein, Daidzein and Glycitein Inhibit Growth and DNA Syntheses of Aortic Smooth Muscle Cells from Stroke-Prone Spontaneously Hypertensive Rats. *J. Nutr.* **2001**, *131*(4), 1154–1158.

71. Pandey, R. M.; Pandey, S. N. *The grape in India*; Indian Council of Agricultural Research Station: New Delhi, 1990; p 115.

72. Parmar H. S.; Kar, A. Protective role of *Citrus sinensis, Musa paradisiaca*, and *Punica granatum* Peels Against Diet-Induced Atherosclerosis and thyroid Dysfunctions in Rats. *Nutr. Res.* **2007**, *27*, 710–718.

73. Picouet, P. A.; Landl, A.; Abadias, M.; Castellari, M.; Vinas, I. Minimal Processing of a Granny Smith Apple Puree by Microwave Heating. *Innov. Food Sci. Emerg. Technol.* **2009**, *10*, 545–550.

74. Poerwanto, R. Rambutan and Longan Production in Indonesia. *Acta Hortic.* **2005**, *665*, 81–86.

75. Priyanka, P.; Sayed, H. M.; Joshi, A. A.; Jadhav, B. A. Comparative Evaluation of Physico-Chemical Properties of Two Varieties of Pomegranate Fruits: Ganesh and Arakta. *Afr. J. Food Sci.* **2013**, *7*(11), 428–430.

76. Pueyo, E.; Polo, M. C. Composition of Grape Lipids and Wine. *Foods, Equipment Technol.* **1992**, *2*, 77–81.

77. Radha, T.; Mathew, L. Fruit Crops. In *Horticulture Science Series 3*; Peter, K. V. Ed.; New India Publishing Agency: New Delhi, 4, 2007; p 29.

78. Ragasa, C. Y.; Martinez, A.; Chua, J. E. Y.; Rideout, J. A. A Triterpene from Musa errans. *Philippine J. Sci.* **2007,** *136*(2), 167–171.

79. Rai, P. K.; Jaiswal, D.; Rai, N. K.; Pandhija, S.; Rai, A. K.; Watal, G. Role of Glycemic Elements of Cynodondactylon and *Musa paradisiaca* in Diabetes Management. *Lasers Med. Sci.* **2009,** *24*(5), 761–768.

80. Rajaratnam, S.; Ramkete, R. S. *Advances in Preservation and Processing Technologies of Fruits and Vegetables;* New India Publishing Agency: New Delhi, 2011; pp 741.

81. Samejima, K.; Kanazawa, K.; Ashida, H.; Danno, G. Luteolin: A Strong Antimutagen Against Dietary Carcinogen Tr-P-2 in Peppermint, Sage and Thyme. *J. Agric. Food Chem.* **1995,** *43*, 410–414.

82. Samson, J. A. *Tropical Fruits;* John Wiley & Sons Inc., New York; 2014; pp 139–146.

83. Saraswathi, N. T.; Gnanam, F. D. Effect of Medicinal Plants on the Crystallization of Cholesterol. *J. Cryst. Growth* **1997,** *179*, 611–617.

84. Setiawan, B.; Sulaeman, A.; Giraud, D. W.; Driskell, J. A. Carotenoid Content of Selected Indonesian Fruits. *J. Food Compos. Anal.* **2001,** *14*(2), 169–176.

85. Shahidi, F.; Naczk, M. *Food Phenolics. Sources, Chemistry, Effects, Applications;* Technomic Publishing: Lancaster, PA (USA), 1995; p 331.

86. Shodehinde, S. A.; Oboh, G. Antioxidant Properties of Aqueous Extracts of Unripe *Musa paradisiaca* on Sodium Nitroprusside Induced Lipid Peroxidation in Rat Pancreas in vitro. *Asian Pac. J. Trop. Biomed.* **2013,** *3*(6), 449–457.

87. Sidgley, M.; Gardner, J. A. International Survey of Underexploited Tropical and Subtropical Perennials. *Acta Hortic.* **1989,** *250*, 2–6.

88. Stanford School of Medicine, Cancer information Page: Nutrition to Reduce Cancer Risk.

89. Still, D. W. Pomegranates: A Botanical Perspective, In *Pomegranates: Ancient Roots to Modern Medicine;* Seeram, N. P., Heber, D. Eds.; CRC Press: Boca Raton, FL, 2006; pp 199–209.

90. Tiwari, B. K.; O'Donnell, C. P.; Cullen, P. J. Effect of Non-Thermal Processing Technologies on the Anthocyanin Content of Fruit Juices. *Trends Food Sci. Technol.* **2009,** *20*, 137–145.

91. Usha, V.; Vijayammal, P. L.; Kurup, P. A. Effect of Dietary Fiber from Banana (*Musa paradisiaca*) on Cholesterol Metabolism. *Indian J. Exp. Biol.* **1984,** *22*(10), 550–554.

92. Usha, V.; Vijayammal, P. L.; Kurup, P. A. Effect of Dietary Fiber from Banana (*Musa paradisiaca*) on Metabolism of Carbohydrates in Rats Fed Cholesterol Free Diet. *Indian J. Exp. Biol.* **1989,** *27*(5), 445–449.

93. Vijayakumar, S.; Presannakumar, G.; Vijayalakshmi, N. R. Investigations on the Effect of Flavonoids from Banana, *Musa paradisiaca* L. on Lipid Metabolism in Rats. *J. Diet. Suppl.* **2009,** *6*(2), 111–123.

94. Vinson, J.; Su, X.; Zubik, L.; Bose, P. Phenol Antioxidant Quantity and Quality in Foods: Fruits. *J. Agric. Food Chem.* **2001,** *49*, 5315–5321.

95. Watson, R. R.; Preedy, V. R. *Bioactive Foods and Extracts: Cancer Treatment and Prevention;* CRC Press, Boca Raton; 2010; p 423.

96. Winzer, M.; Rauner, M.; Pietschmann, P. Glycitein Decreases the Generation of Murine Osteoclasts and Increases Apoptosis. *Weiner Medizinische Wochesnshrift* **2010,** *160*(17–18), 446–451.

97. Wlodzimierz, G.; Anna, O. Probiotics, Prebiotics and Antioxidants as Functional Foods. *Acta Biochim. Pol.* **2005,** *52* (3), 665–671.

98. Wuyts, N.; De Waele, D.; Swennen, R. Extraction and Partial Characterization of Polyphenol Oxidase from Banana (*Musa acuminate* grande naine). *Plant physiol. Biochem.* **2006,** *44*, 308–314.

99. Xagorari, A.; Papapetropolous, A.; Mauromatis; Economou, M.; Fotsis, T.; Roussos, C. Luteolin Inhibits an Endotoxin-Stimulated Phosphorylated Cascade Pro-inflammatory Cytokine Production in Macrophages. *J. Pharmacol. Exp. Ther.* **2001,** *296*, 181–187.

100. Yoshida, H.; Teramoto, T.; Ikeda, K.; Yamori, Y. Glycitein Effect on Suppressing the Proliferation and Stimulating the Differentiation of Osteoblastic MC3T3-E1 Cells. *Biosci., Biotechnol. Biochem.* **2001,** *65*(5), 1211–1213.

101. Zakpaa, H. D.; Mak-Mensah, E. E.; Adubofour, J. Production and Characterization of Flour Produced from Ripe Plantain (*Musa sapientum* L. var. *paradisiaca*) Grown in Ghana. *J. Agric. Biotechnol. Sust. Dev.* **2010,** *2*(6), 92–99.

E-SOURCES

102. http://all4desktop.com/data_images/original/4241140-pear.jpg (accessed March 18, 2016, 2.35 pm).

103. http://plants.usda.gov/core/profile?symbol=PAED (accessed March 11, 2016, 6:52 am).

104. http://www.liveloveraw.com/wp-content/uploads/2014/01/Cutting-out-the-jackfruit-core-the-easy-way-650x487.jpg (accessed March 15, 2016, 3.09 pm).

105. http://plants.usda.gov/core/profile?symbol=arhe2 (accessed March 11, 2016, 6.34 am).

106. https://upload.wikimedia.org/wikipedia/commons/1/11/Jackfruit_Bangladesh_(3). JPG1 (accessed March 14, 2016, 7.15 am).

107. http://www.borongaja.com/data_images/out/13/622969-jackfruit-bread-fruit-tree. jpg2(accessed March 16, 2016, 8.20 am).

108. http://faostat3.fao.org/download/Q/QC/E (accessed Jan 4, 2016, 2.06 pm).

109. http://plants.usda.gov/java/ClassificationServlet?source=display&classid=VITIS (accessed March 8, 2016, 10.19 am).

110. http://www.academicwino.com/wp-content/uploads/2013/07/grape-cross-section_The_Academic_Wino.png (accessed March 8, 2016, 10:43 am).

111. http://dinefresh.in/wp-content/uploads/2015/11/banana-nendran-700x700.jpg (accessed March 22, 2016, 10.01 am).

112. http://www.care2.com/greenliving/10-health-benefits-of-mangos. html#ixzz43iIeI37u (accessed March 23, 2016, 2.20 pm).

113. https://upload.wikimedia.org/wikipedia/commons/5/5e/Wine_grapes03.jpg (accessed March 18, 2016, 11.40 am).

114. http://1.bp.blogspot.com/_GU-vyNNogzo/S9kcoJzpRuI/AAAAAAAADXs/ Ymv8j_EwPcE/s1600/Picture+426.jpg (accessed March 8, 2016, 8.25 pm).

115. http://pattismenu.com/wp/wp-content/uploads/2014/12/plantain-ripeness-IMG_8512.jpg (accessed March 8, 2016, 8.37 pm).

116. http://www.cgiar.org/our-strategy/crop-factsheets/bananas/ (accessed March 8, 2016, 8.55 pm).

117. https://pubchem.ncbi.nlm.nih.gov/compound/caffeic_acid#section=Metabolism-Metabolites (accessed March 9, 2016, 10.28 pm).

118. https://livinglightinternational.files.wordpress.com/2011/02/tangerine-trees.jpg (accessed March 10, 2016, 12.07 pm).

119. http://i0.wp.com/rosemaryandthegoat.com/wp-content/uploads/2011/11/DSC_2687.jpg (accessed March 10, 2016, 9.56 pm).

120. http://www.nutrition-and-you.com/mango-fruit.html (accessed March 22, 2016, 9.36 am).

121. http://www.theplantlist.org/tpl/record/kew-2724391 (accessed March 4, 2016, 10.30 pm).

122. http://www.phytochemicals.info/phytochemicals/tangeretin.php (accessed March 10, 2016, 2.08 pm).

123. http://www.webmd.com/vitamins-supplements/ingredientmono-1033-hesperidin.aspx?activeingredientid=1033&activeingredientname=hesperidin (accessed March 10, 2016, 12.43 pm).

124. http://plants.usda.gov/core/profile?symbol=MAIN3USDAmango (accessed March 22, 2016, 11.32 am).

125. http://en.cnki.com.cn/Article_en/CJFDTOTAL-SPFX200602048.html (accessed March 10, 2016, 9.10 pm).

126. http://prsvkm.kau.in/book/benefits (accessed March 10, 2016, 9.21 pm). i. http://www.ornamental-trees.co.uk/images/products/zoom/1348493288-26391300.jpg (accessed March 10, 2016, 3.05 pm).

127. http://bastet-plants.com/wp-content/uploads/2015/04/nectarine-seedling.jpg (accessed March 10, 2016, 10.13 am).

128. http://www.yokesfreshmarkets.com/sites/default/files/styles/flavortrail_full/public/asset/image/white_nectarine_-_zephyr.jpg?itok=jdbfWxMm (accessed March 10, 2016, 10.15 am).

129. http://plants.usda.gov/core/profile?symbol=PRPE3 (accessed March 10, 2016, 2.43 pm).

130. https://www.organicfacts.net/health-benefits/fruit/nectarines.html (accessed March 2, 2016, 10.34 am).

131. http://plants.usda.gov/java/ClassificationServlet?source=display&classid=PSIDI (accessed March 11, 2016, 5.53 am).

132. http://healthbubbles.com/gb/?p=2535 (accessed March 10, 2016, 8.05 pm).

133. http://s262.photobucket.com/user/7_Heads/media/Fruit%20Trees/Fruit_Trees/Haw_Guava_45.jpg.html (accessed March 10, 2016, 8.40 pm).

134. http://www.vegetexcohcm.com/data/upload/whiteguava.png (accessed March 15, 2016, 2.49 pm).

135. http://plants.usda.gov/core/profile?symbol=MUPA3 plantain(accessed March 2, 2016, 4.08 pm).

136. http://www.21food.com/products/microwave-vacuum-dried-fruit---banana-197353.html (accessed March 23, 2016, 3.24 pm).

137. 137.https://ndb.nal.usda.gov/ndb/foods?format=&count=&max=35&sort=&fgcd =Fruits+and+Fruit+Juices&manu=&lfacet=&qlookup=&offset=35&order=desc (accessed March 7, 2016, 2.05 pm).

138. https://ndb.nal.usda.gov/ndb/foods/show/2122?fgcd=Fruits+and+Fruit+Juices&m anu=&lfacet=&format=&count=&max=35&offset=&sort=&qlookup=   (accessed March 7, 2016, 2.07 pm).

139. https://ndb.nal.usda.gov/ndb/foods/show/2238?fgcd=Fruits+and+Fruit+Juices&ma nu=&lfacet=&format=&count=&max=35&offset=105&sort=&qlookup= (accessed March 7, 2016, 2.08 pm).

140. https://ndb.nal.usda.gov/ndb/foods/show/2284?fgcd=Fruits+and+Fruit+Juices&ma nu=&lfacet=&format=&count=&max=35&offset=140&sort=&qlookup= (accessed March 7, 2016, 2.10 pm).

141. https://ndb.nal.usda.gov/ndb/foods/show/2271?fgcd=Fruits+and+Fruit+Juices&ma nu=&lfacet=&format=&count=&max=35&offset=140&sort=&qlookup= (accessed March 7, 2016, 2.13 pm).

142. https://ndb.nal.usda.gov/ndb/foods/show/2407?fgcd=Fruits+and+Fruit+Juices&ma nu=&lfacet=&format=&count=&max=35&offset=280&sort=&qlookup= (accessed March 7, 2016, 2.14 pm).

143. https://ndb.nal.usda.gov/ndb/foods/show/2326?fgcd=Fruits+and+Fruit+Juices&ma nu=&lfacet=&format=&count=&max=35&offset=175&sort=&qlookup= (accessed March 7, 2016, 2.15 pm).

144. https://ndb.nal.usda.gov/ndb/foods/show/2340?fgcd=Fruits+and+Fruit+Juices&ma nu=&lfacet=&format=&count=&max=35&offset=210&sort=&qlookup= (accessed March 7, 2016, 2.16 pm).

145. https://ndb.nal.usda.gov/ndb/foods/show/2311?fgcd=Fruits+and+Fruit+Juices&ma nu=&lfacet=&format=&count=&max=35&offset=175&sort=&qlookup= (accessed March 7, 2016, 2.17 pm).

146. https://ndb.nal.usda.gov/ndb/foods/show/2279?fgcd=Fruits+and+Fruit+Juices&ma nu=&lfacet=&format=&count=&max=35&offset=140&sort=&qlookup= (accessed March 7, 2016, 2.21 pm).

147. https://ndb.nal.usda.gov/ndb/foods/show/2309?fgcd=Fruits+and+Fruit+Juices&ma nu=&lfacet=&format=&count=&max=35&offset=175&sort=&qlookup= (accessed March 7, 2016, 2.24pm).

148. https://ndb.nal.usda.gov/ndb/foods/show/2310?fgcd=Fruits+and+Fruit+Juices&ma nu=&lfacet=&format=&count=&max=35&offset=175&sort=&qlookup= (accessed March 7, 2016, 2.31 pm).

149. https://ndb.nal.usda.gov/ndb/foods/show/2305?fgcd=Fruits+and+Fruit+Juices&ma nu=&lfacet=&format=&count=&max=35&offset=175&sort=&qlookup= (accessed March 7, 2016, 2.33 pm).

150. https://ndb.nal.usda.gov/ndb/foods/show/2249?fgcd=Fruits+and+Fruit+Juices&ma nu=&lfacet=&format=&count=&max=35&offset=105&sort=&qlookup= (accessed March 7, 2016, 2.34 pm).

151. https://ndb.nal.usda.gov/ndb/foods/show/2246?manu=&fgcd=Fruits%20and%20 Fruit%20Juicesguava (accessed March 7, 2016, 2.50 pm).

152. https://ndb.nal.usda.gov/ndb/foods/show/2359?manu=&fgcd=Fruits%20and%20 Fruit%20Juices (accessed March 7, 2016, 3.10 pm).

153. http://www.hiperbaric.com/en/fruits-vegetables (accessed March 23, 2016, 3.31 pm).

154. https://ndb.nal.usda.gov/ndb/foods/show/2373?fgcd=&manu=&lfacet=&format=F ull&count=&max=35&offset=&sort=&qlookup=09301 (accessed March 11, 2016, 2.50 pm).

155. https://ndb.nal.usda.gov/ndb/foods/show/2272?fgcd=&manu=&lfacet=&format=F ull&count=&max=35&offset=&sort=&qlookup=09177 (accessed March 15, 2016, 4.40 pm).

156. http://www.doa.gov.lk/index.php/en/crop-recommendations/1099 (accessed March 17, 2016, 4.26 pm). http://silverbulletin.utopiasilver.com/pomegranate-can-serve-as-a-backup-ovary-2/ (accessed March 22, 2016, 10.30 am).

157. http://plants.usda.gov/core/profile?symbol=MAPU (accessed March 10, 2016, 8.35 pm).

158. https://pixabay.com/en/apple-fruit-delicious-vitamins-1112047/) (accessed March 23, 2016, 11.35 am).

159. http://assets.inhabitat.com/wp-content/blogs.dir/1/files/2012/09/apples-growing. jpeg (accessed March 22, 2016, 10.00 am).

160. http://plants.usda.gov/java/ClassificationServlet?source=display&classid=MUSA2 (accessed March 12, 2016, 10.34 am).

161. http://www.pearrecipes.co.uk/wp-content/uploads/pear_tree.jpg (accessed March 14, 2016, 2.45 pm).

162. http://portuguese.alibaba.com/product-gs/organic-factory-supply-mango-seed-extract-powder-60227753492.html (accessed March 23, 2016, 12.35 pm).

163. https://www.pinterest.com/pin/253327547764967784/ (accessed March 22, 2016, 2.30 pm).

164. http://plants.usda.gov/core/profile?symbol=PYCO (accessed March 13, 2016, 10.15 am).

165. (http://www.seedman.com/papaya.html) –(accessed on March 21, 2016, 3.45 pm).

166. http://ssu.ac.ir/cms/fileadmin/user_upload/Mtahghighat/tfood/asil-article/Food_ Chemistry2-2013/Total_phenolics__antioxidant_activity__and_functional_proper-ties_of.pdf (accessed March 23, 2016, 11.30 am.

167. http://thewisegardener.com/photos%20and%20artwork/TWG%20Plants/Fruits/ Mangosteen4.jpg (accessed March 14, 2016, 3.32 pm).

168. http://www.mangosteen-natural-remedies.com/images/what-is-mangosteen.jpg (accessed March 14, 2016, 3.33 pm).

169. http://plants.usda.gov/core/profile?symbol=GAMA10 (accessed March 14, 2016, 11.50 am).

170. http://farmersalmanac.com/wp-content/uploads/2011/02/what-the-heck-is-a-papay (accessed March 14, 2016, 3.47 pm).

171. https://www.crfg.org/pubs/ff/mango.html. (accessed March 23, 2016, 9.30 am.

172. http://plants.usda.gov/core/profile?symbol=CAPA23-papaya (accessed March 23, 2016, 9.20 am.

173. http://previews.123rf.com/images/inm_imah/inm_imah1312/inm_ imah131200027/24235759-yellow-rambutan-on-the-tree-Stock-Photo.jpg (accessed March 20, 2016, 10.55 am).

174. https://www.google.co.in/imgres?imgurl=http://fruitionhawaii.com/wp-content/uploads/2013/04/Cassowary-Coast-Rambutan-on-tree.jpg&imgrefurl=http://bigaina.com/portfolio-posts/single-portfolio-wide-image-8/&h=768&w=1024&tbnid=8YOIgzbGSbEHdM:&docid=KR1cbjeFoUfisM&ei=8XbmVpWWBY2WuASt0pm4BQ&tbm=isch&ved=0ahUKEwjV0qDj4b_LAhUNC44KHS1pBlcQMwg-fKAQwBA (accessed March 22, 2016, 10.15 am).

175. http://plants.usda.gov/core/profile?symbol=NELA7 (accessed March 20, 2016, 10.22 am).

176. http://www.healwithfood.org/nutrition-facts/rambutan-fruit-healthbenefitsphp#ixzz42Z1HdJtF (accessed March 23, 2016, 3.55 pm).

177. http://plants.usda.gov/core/profile?symbol=PYCO (accessed March 12, 2016, 10.10 am).

178. http://plants.usda.gov/core/profile?symbol=PUGR2 (accessed March 15, 2016, 9.54 am).

179. http://plants.usda.gov/core/profile?symbol=CISI3 (accessed March 12, 2016, 1.50 pm).

180. https://www.acfchefs.org/download/documents/ccf/nutrition/2010/201006_functional_foods.pdf (accessed March 4, 2016, 2.55 pm).

181. http://www.specialtyproduce.com/produce/sppics/9496.png (accessed March 10, 2016, 3.45 pm).

182. http://www.harvesttotable.com/wp-content/uploads/2009/07/Bananas-on-tree1.jpg (accessed March 10, 2016, 5.40 pm).

183. https://s-media-cache-ak0.pinimg.com/236x/42/df/01/42df01cae754a405df0d41c26be88bbb.jpg (accessed March 10, 2016, 5.45 pm).

184. http://images.all-free-download.com/images/graphiclarge/delicious_fruit_03_hd_pictures_166657.jpg (accessed March 10, 2016, 6.00 pm).

185. http://umad.com/img/2015/8/thumb/orange-tree-wallpaper-3974-4190-hd-wallpapers-thumb.jpg (accessed March 10, 2016, 6.05 pm).

186. https://s-media-cache-ak0.pinimg.com/564x/38/94/16/3894168941bc6f3acddd4a86116b0691.jpg (accessed March 10, 2016, 6.05 pm).

187. http://www.thehindubusinessline.com/multimedia/dynamic/01565/BL29_AGRI_PINEAPPL_1565638g.jpg (accessed March 10, 2016, 6.10 pm).

CHAPTER 7

FUNCTIONAL FERMENTED DAIRY PRODUCTS—REVIEW: CONCEPTS AND CURRENT TRENDS

SHAIK ABDUL HUSSAIN[1], MANJU GAARE[2,*], AND ASHISH KUMAR SINGH[3]

[1]*Dairy Technology Division, ICAR—National Dairy Research Institute, Karnal 132001, Haryana, India, Tel.: +91-9896668983, E-mail: abdulndri@gmail.com*

[2]*Department of Dairy and Food Microbiology, GN Patel College of Dairy and Food Technology, Sardarkrushinagar Dantiwada Agricultural University, Sardarkrushi Nagar, Dantiwada 385506, Banaskantha District, Gujarat, India, Tel.: +91-8971769001*

[3]*Dairy Technology Division, ICAR—National Dairy Research Institute, Karnal 132001, Haryana, India, Tel.: +91-184-2259291, E-mail: akndri@gmail.com*

*Corresponding author. E-mail: manjugdsc@gmail.com

CONTENTS

7.1   INTRODUCTION

Functional foods are foods or food products which provide some health benefits in addition to basic nutrition. With an annual average growth rate of about 8.5%, the global functional food market is expected to exceed $305.4 billion by 2020.[80] Functionality in foods is inherent or it can be incorporated through various means such as fortification, enrichment, alteration, and enhancement. Functional foods can be consumed as part of daily diet, which can improve the general condition of health and/or decrease the disease conditions. Various categories of functional foods found in the world market include fruits, vegetables, energy drinks, fortified juices, breakfast cereals, fresh dairy products, fiber-rich foods, probiotics and synbiotics, and so forth.

Among the various categories of functional foods, fermented milks have occupied a major share in the food market. The beneficial effects of fermented milks can be attributed to a variety of health-promoting components including good bacteria and their metabolites, bioactive peptides, and other functional molecules derived during fermentation. Apart from the inherent functional constituents present in fermented milks, plant-, animal-, and microbial-derived ingredients can be added to improve its health status.

The technique of fermenting the milk dates back to times immemorial and is regarded as the best way to preserve the milk while improving its nutritional status. Lactic acid bacteria (LAB) species belonging to the genera *Leuconostoc, Lactobacilli, Streptococci,* and *Lactococci* are predominantly used for fermenting the milk. In some cases, few yeast and mold species are also used in combination with LAB to ferment the milk. Current tendency is to supplement starter cultures with probiotics to improve the therapeutic potential of the fermented milks.

Health benefits such as immunity enhancement,[49] cholesterol lowering,[119] regulating blood pressure,[90] cancer prevention,[57] antidiabetic effects,[139] alleviation of lactose intolerance,[38] and constipation[115] upon consumption of fermented milks have been reported by several authors. With increasing health awareness among the people, the future for functional fermented foods appears to be more promising.

This chapter discusses nutritional and health status of several indigenous fermented dairy foods and innovative approaches adopted by several authors to induce or enhance the functionality.

7.2   CONCEPT OF FUNCTIONAL FOOD

The concept of functional foods started in Japan in 1984 when its Ministry of Education, Science and Culture, started a project titled "systematic analysis and development of food functions" to know the diet and health relationship as an objective. It was then followed by two more projects on understanding the food functionality. Later in 1991, a policy to legally permit the commercialization of functional foods (Foods for specific health uses (FOSHU)) was defined. However, the ideology of foods of providing medicinal benefits is not new. In fact, it dates back to 2500 years ago, which was conceived by famous Hippocrates, who is regarded as father of Western medicine. Nowadays, the term "functional food" is interchangeably used with medicinal foods, therapeutic foods and foodiceuticals.

The concept of "functionality" in fermented foods was conceived by Metchnikoff,[125] who hypothesized that consumption of fermented milk (yogurt) leads to long and healthy living. Metchnikoff also proposed that yogurt contains beneficial bacteria, which later was named as probiotics, may improve the gut health. Previous to reports of Metchnikoff, Mann and Spoerry[120] indicated that consumption of yogurt fermented with wild strains of *Lactobacillus* spp. resulted in lowering of blood serum cholesterol. Further, Harrison and Peat[69] reported that the addition of *Lactobacillus acidophilus* to infant formula decreased serum cholesterol levels. Later, several studies were carried out and numerous reports on health benefits associated with fermented foods have been published. The majority of the reports indicates that the health benefits of fermented foods are due to the presence of live bacteria or good bacteria, which favorably alters the microbial ecology in the human gastrointestinal tract (GIT).[144] Further studies in fermented milks have led to new incentives in the area of beneficial flora, that is, probiotics. In 1994, the World Health Organization (WHO) considered probiotics as the second most important immune defense system next to antibiotics in immune therapy.[114] Functional foods have been developed in all food categories. From the product perspective, functional foods can be labeled as:[19]

- **Altered foods**—Foods from which a component has been removed, reduced, or replaced by another with beneficial effects, for example, fat replacement or reduction in foods using fiber.
- **Enhanced commodities**—Foods in which one of their components have been enhanced to impart the functionality, For example, milk with increased protein content, eggs with increased omega-3 content, and so on.
- **Enriched foods**—Foods in which functionality was induced by the addition of new components, which are not normally found in a particular food, for example, addition of probiotics or prebiotics, and so on.
- **Fortified foods**—Foods in which additional nutrients have been added, for example, milk fortified with iron and calcium.

7.3 COMMONLY FERMENTED MILK PRODUCTS IN THE WORLD

7.3.1 YOGURT

Yogurt or yoghurt was believed to be initially made by nomads of ancient Turkey. Symbiotic thermophilic starter cultures are *Lactobacillus delbrueckii* ssp. *bulgaricus* and *Streptococcus salivarius* ssp. *thermophilus*. Recent trends are to incorporate probiotic organisms along with yogurt starter to improve nutritional and therapeutic value of yogurt. Yogurt is available in several physical forms, namely set, stirred or fluid (drinking yogurt), frozen and dried yogurt. In some countries, namely the United States and European countries, presence of live yogurt culture ($>10^6$ CFU/ mL) is essential so that the product can be named as yogurt.[212]

The symbiotic blend of yogurt starter cultures helps each other for their growth during yogurt fermentation. *S. thermophilus* grows faster and produces acid and CO_2. The metabolites produced by *S. thermophilus* favor the growth of *L. bulgaricus*. Peptides and amino acids produced by *L. bulgaricus* are utilized by *S. thermophilus* for its growth. Ultimately, the typical yogurt texture and flavor are due to starter culture combination. Acetaldehyde is regarded as the principal flavor compound in yogurt.[202]

7.3.2   CHEESE

Cheese is a most popular fermented milk product (FMP) obtained by the combined action of rennet and starter cultures. According to the historical perspective, cheese was accidentally made in approximately 7000 BC by an Arabian merchant during his long-day's journey as he put his supply of milk into a pouch made of a sheep's stomach containing rennet, which became cheese and whey. Historical records pertaining to the manufacture of cheese in India (6000 to 4000 BC), Egypt (4000 BC), and Babylonia (2000 BC) were also found indicating its widespread popularity.

Cheese is recognized as of high nutritional value food due to its richness in protein, calcium, riboflavin, and vitamins A and D3. Today, more than thousand cheese varieties are produced worldwide. One of the varieties of cheese is manufactured by altering different aspects of cheese manufacturing, such as starter type, fermentation conditions, renneting, cutting the curd, draining of whey, salting, ripening period and conditions, additives (like herbs, spices, and beneficial flora), and so on.[74]

7.3.3   DAHI *AND* LASSI

Dahi and *lassi* (stirred *dahi*) are popular FMPs of the Indian origin, which are similar to plain yogurt and stirred yogurt, respectively, in appearance and consistency. *Dahi* is prepared by fermentation of milk using a mixed culture combination of mesophilic bacteria. Diacetyl is the principal flavor component in *dahi*, which is liked by the Indian palate. *Dahi* is consumed as such or can be utilized in various forms in many Indian culinary preparations. The use of *dahi* was mentioned in various ancient scriptures of India. It is highly regarded for its recuperative effects by Ayurveda, the Indian medicinal literature. Generally, *dahi* is prepared from cow or buffalo milk. In the Himalayan region, it is also prepared from the milk of yak and zomo.

Lassi is prepared from *dahi* by mixing it with sugar in an earthen pot with the help of a wooden stirrer. Salted and spiced *lassi* (called *chhash*) can be made by adding salt and spices like ginger, coriander, and mint in the form of a paste in place of sugar and the desired consistency is attained with the addition of cold water. Fruits and their preparations are generally added to *dahi* and *lassi* to increase the health status. Fruits are rich source of fiber, essential vitamins, and minerals. Due to their characteristic taste and color, fruits also enhance the palatability of the fermented milks.

7.3.4 VILLI

Villi is mesophilic fermented milk having ropy consistency native to Scandinavia. Villi is popularly known as *viilia* in Finland.[26] Villi is characterized by pleasant sharp taste and a good diacetyl aroma linked to a stringy texture and can be cut easily with a spoon. Villi culture consists of a combination of mesophilic LAB and yeast.[207] Yeast provides a unique flavor to villi besides promoting the LAB to produce more exopolysaccharides.[206] Villi is consumed at breakfast and as a snack food. Villi can be best relished together with cereals and fruit. A very traditional way of eating villi is by mixing it with cinnamon and sugar.[112]

7.3.5 KEFIR

Kefir is a fermented beverage originated in the Caucasus Mountains. *Kefir* has been produced from milk of various species including cow, ewe, goat, and buffalo.[5] It is obtained by fermenting milk with kefir grains, which harbor several species of LAB, yeast, and acetic acid bacteria. Traditional method of *kefir* preparation involves pouring of milk in skin bags on daily basis, thus leading to natural fermentation. The finished product has high acidity and varying amounts of alcohol and carbon dioxide.

7.3.6 KUMYS

Kumys or kumiss or koumiss is a FMP made from mare's milk, which is popular in central Asia, Russia, and Eastern Europe. Kumiss has been consumed as food as well as alcoholic drink. Starter cultures that used koumiss include a variety of LAB and yeasts. Like-wise in kefir, both lactic acid and alcohol fermentations occur in koumiss. Koumiss is regarded as health-promoting beverage that improves metabolism and protects the nervous system and kidneys.[206]

7.3.7 AYRAN

Ayran is a popular FMP of Turkey resembling drinking yogurt. Butter milk obtained after the production of butter during churning of yogurt added with water is also known as ayran in Turkey.[213] According to Turkish Food Codex,[197] ayran is defined as "drinkable fermented product prepared by the addition of water to yogurt or by the addition of yogurt culture to stan-dardized milk." In ayran production, table salt is used as an additive, which is added to yogurt after dilution with water at a concentration of 0.5–1.0%. According to this Turkish Food Codex definition, ayran preparation resem-bles that of "majjiga," a fermented milk drink native to southern parts of India, obtained by diluting *dahi* with water followed by addition of salt.

7.4 HEALTH STATUS OF FERMENTED MILK PRODUCTS

The history of fermented milk dates back to more than 10,000 years ago. Renewed interest in fermented milk was created by Élie Metchnikoff in the early 1900s.[125] Since then, the consumption of FMPs has greatly increased. The sale of fermented milk has been greatly boosted during the late 1990s due to increased awareness and knowledge among the consumers about health benefits of yogurt and other FMPs. Further innovations such as the addition of fruits and other nutrients, process modifications to enhance the esthetic quality and ensure the safety, refining of manufacturing procedure for commercial production have led to further improvement in the sale of FMPs. The varieties of FMP, that were developed, depended on the availability of raw materials, environmental conditions, and the flavor preferences of the local people.

Fermentation of milk results in the release of simpler or more digestible compounds from the native milk constituents, which is attributed to the metabolic activity of added starter cultures. Fermentation of milk causes breakdown of proteins (mainly casein, ~1–2%) by bacteria releasing small peptides and amino acids having physiological benefits. Lactose and milk carbohydrate will also be broken down (~20–30%) into simpler compounds (namely lactic acid, etc.) through different pathways. The FMPs will vary according to the type of starter culture that is used for the production of fermented milks. Many bioactive peptides will be released as a result of fomentation of milk. During fermentation, the conversion of lactose into lactic acid is a proven benefit for the lactose intolerant people, who could not able to digest raw milk.[189] Gilliland[59] advocated that reduction of lactose concentration and the presence of large number of starter culture having β-galactosidase enzyme is responsible for the improved digestion of fermented milk compared to normal milk. Also, several studies have reported immunostimulatory function of FMPs. Villena et al.[200] suggested that the immunological action of yogurt is due to the improved digestion of proteins, presence of live starter culture, which may effectively modulate mucosal immunity and also due to the release of bioactive molecules during fermentation. Reported health benefits of various FMPs are given in Table 7.1.

7.5   CURRENT TRENDS IN FUNCTIONAL FERMENTED MILKS

Due to the increased awareness among the consumers regarding the beneficial health effects of fermentation, the demand for fermented milks is ever increasing. Food manufacturing units are trying to bring innovative FMPs into the market in more convenient, attractive and palatable forms to meet the consumer demands. Also, research efforts put to further enhance the safety and functionality have led to the development of new kind of FMPs. Technology of various fermented milks and the recent trends in their manufacture have been discussed in the following sections.

TABLE 7.1   Reported Health Benefits of Fermented Milk Products.

Fermented milk/ milk products	Reported physiological effects	Type of study or experiment	References
Dairy probiotic drink	Reducing overall incidence of illness	Clinical	[124]
FMPt with probiotic	Modulation of brain activity	Clinical	[195]
Functional yogurt NY-YP901	Reduced low-density lipopro-tein (LDL)-cholesterol, body weight, and body mass index (BMI)	Clinical	[23]
Kefir	Eradication of *Helicobacter pylori*	Clinical	[16]
Kefir	Immunomodulatory activity	Cell line study	[88]
Kefir	Arresting the development of cancerous growths in vitro	Animal	[156]
Kefir	Wound-healing properties	Animal	[81]
Kefir	Antiadipogenic effect	In vitro	[76]
Probiotic *dahi*	Antiallergic	Animal	[93]
	Anticarcinogenic	Animal	[130]
	Antiobesity	Animal	[155]
	Alleviation of age-inflicted oxidative stress	Animal	[104]
Probiotic kefir	Improvement in cardiovas-cular function in spontane-ously hypertensive rats	Animal	[50]
Probiotic yogurt	Increased antioxidant status in type 2 diabetic patients	Clinical	[42]
	Significant decrease in atherogenic indices in Type 2 diabetes patients	Clinical	[43]
	Management of nonalcoholic fatty liver disease)	Clinical	[134]
	Alleviation of gastrointestinal symptoms, improvement of productivity, nutritional intake, and tolerance to anti-retroviral treatment in HIV patients	Clinical	[89]
	Cholesterol lowering activity	Clinical	[100]

TABLE 4.1   *(Continued)*

Fermented milk/ milk products	Reported physiological effects	Type of study or experiment	References
Probiotic yogurt and normal yogurt	Decrease in total cholesterol: High-density lipoprotein (HDL)-cholesterol ratio	Clinical	[170]
Yoghurt	Augmenting of natural killer cell activity and reduction in the risk of catching the common cold in elderly individuals	Clinical	[118]

7.5.1   PROBIOTIC FMPs

Probiotics are defined as "live microorganisms that, when administered in adequate amounts, confer a health benefit on the host".[208] This definition is internationally recognized and the viability of probiotics in the final product is important, especially when it has been documented as one of the prerequisites for immune effects.[58] The researchers have recognized that probiotics should be capable of exerting health benefits on the host through their activity in the human body. Although various strains of LAB have been described as probiotic, only few meet the standards of having clinical trial documentation, and many are too sensitive to intense acidity and the presence of bile salts in the human GIT, so they die en route to the gut.[59]

Health benefits attributed to intake of probiotics include alleviation of lactose intolerance, suppression of cancer, reduction in serum cholesterol concentrations, improved gastrointestinal immunity, prevention of urinary tract infections, the management of food allergies and prevention of diabetes, and so on.[24]

The importance of probiotic-containing products (commonly regarded as probiotic foods) in the maintenance of health and well-being is becoming a key factor affecting consumer choice.[121] The majority of probiotic products available in the marketplace contain species of *Lactobacillus* and *Bifidobacterium*, which are main genera of Gram-positive bacteria currently characterized as probiotics. The microorganisms used in probiotic preparations should be generally recognized as safe, and they should be resistant to bile, hydrochloric acid, and pancreatic juice, have anticarcinogenic activity and stimulate immune-system, have reduced intestinal permeability, produce lactic acid, able to survive both acidic conditions of the stomach and alkaline conditions of the duodenum.[201] Both, single and mixed cultures of live microorganisms are used in probiotics preparations.[33] Microorganisms used in probiotic products around the world are:[144,176] (a) **Lactobacilli:** *L. casei*, *L. crispatus*, *L. johnsonii*, *L. plantarum*, *L. reuteri*, *L. rhamnosus*, *L. salvarius*, *and L. acidophilus*; (b) **Bifidobacteria:** *B. breve*, *B. infantis*, *B. bifidum*, *B. longum*, *B. adolescentes*, *B. lactis*, and *B. Animalis*; (c) **Other LAB:** *E. faecalis*, *Streptococcus thermophiles*, *Streptococcus cremoris*, *Pediococcus acidilactici*, *S. salivarius*, *S. diacetylactis*, *S. intermedius*, and *Enterococcus faecium*; (d) **Non-LAB:** *Escherichia coli Nissle 1917*, *Saccharomyces boulardii*, *Clostridium butyricum*, *Propionibacterium freudenreichii*, and *Bacillus cereus var. toyoi*.

Lactobacillus groups are most frequently employed in the production of probiotic products. As the *Lactobacillus* species are of human intestinal origin, it is generally accepted that they are better suited to the physiological needs of the host and can more easily colonize in the human intestine than wild strains or strains that exist in the colon of other animals. Also, an important technological reason for the use of dairy products as carriers of *Lactobacilli* is that many of them have already been optimized to some extent for the survival of live fermentation microorganisms. Thus, the existing technologies can be readily adapted to allow the incorporation of probiotic *Lactobacilli*.[73]

Fermented dairy foods, including milk and yoghurt, are among the most accepted food carriers for delivery of viable probiotic cultures to the human GIT.[85] Other dairy products used for delivery of probiotics include soft, semi-hard, and hard cheeses and frozen fermented dairy desserts. Probiotic dairy products, namely Cheddar cheese,[138] yoghurt,[148] kefir,[29] dahi,[85] lassi,[84] ayran,[108] and so on with acceptable sensory and textural qualities have been manufactured using different strains of probiotics.

Hard varieties of cheeses such as Cheddar, Edam, and Gouda are preferred over yoghurt like products for the delivery of probiotics as they offer advantages like high total solids content, reduced acidity and typical texture, which may protect the probiotics during their passage through GIT. Few studies have demonstrated that *Bifidobacteria* survived well in Cheddar and Gouda cheeses. Several reports suggested that *Bifidobacteria* introduced into hard-pressed cheeses remain viable during storage up to 3 to 4 months and could be metabolically active. Probiotic bacteria are generally added directly to the vat or as freeze-dried cultures along with salt.[41] The growth of probiotics, which are microaerophilic or anaerobic in nature, are favored by the anaerobic conditions prevalent in the cheese matrix.

Stanton et al.[186] studied the survivability of *Lactobacilli*, isolated from upper GIT of human subjects during surgery, in Cheddar cheese. The authors reported that two *L. paracasei* strains grew and sustained in high numbers (10^8 CFU/g cheese) in cheese whereas *L. salivarius* strains did not survive during ripening. The authors also observed that incorporation of these strains had no negative effects on cheese quality, including aroma, flavor, and texture. In pig-feeding trials, it was shown that mature probiotic Cheddar cheese, containing high levels of probiotic bacteria compares favorably with fresh yogurt as an effective delivery system in colonizing the GIT. This study indicated that probiotic Cheddar cheeses containing high levels of *L. paracasei* strains can be manufactured by conventional manufacturing procedures at a relatively low cost. Daigle et al.[34] prepared a hard-pressed Cheddar-like cheese using microfiltered milk and cream enriched with native phosphocaseinate retentate and fermented by *B. infantis*. After preparation, the cheese was packaged in vacuum-sealed bags and kept at 4°C. The *Bifidobacteria* remained viable (above 3×10^6 CFU/g) for 12 weeks. No significant difference was observed between the control cheese and *Bifidobacteria* incorporated cheese in terms of moisture content, pH and salt, and so on. *Bifidobacterium* strain Bo and *L. acidophilus* strain Ki have been used by researchers to produce probiotic cheeses. Gomes et al.[61] manufactured a probiotic cheese from the milk of a native Portuguese goat breed using these organisms. The authors suggested that high rates of inoculum (3.5%) were necessary to meet technological requirements. After 9 weeks of storage period, the number of *Bifidobacteria* and *Lactobacilli* were $6–18 \times 10^8$ CFU/g and $0.2–5 \times 10^7$ CFU/g, respectively.

Gobbetti et al.[60] incorporated *Bifidobacterium* spp. (viz. *B. bifidum*, *B. infantis* and *B. longum*) into Crescenza cheese individually or as multispecies mixture at a concentration of 10^6 CFU/mL of cheese milk. Addition of *Bifidobacteria* did not alter the growth of *S. thermophilus* (used as a starter culture) and the composition of the cheese. However, higher levels of pH 4.6-soluble nitrogen and more pronounced activities of amino-, imino-, di- and tri-peptidase were detected in cheeses added with *Bifidobacteria*. Sabikhi and Mathur[167] prepared probiotic Edam cheese incorporating *B. bifidum* (American Type Culture Collections [ATCC] 15696).

The probiotic cheese had over 7.5 log CFU/g of viable *Bifidobacteria* after three months of ripening. The ingestion of the probiotic Edam cheese resulted in intestinal implantation of *Bifidobacteria* in Albino rats with the concomitant reduction in the fecal coliform count. The anticarcinogenic activity of the *Bifidobacteria* was also illustrated by the reduction in the activity of β-glucuronidase. Probiotic semi-soft cheese was prepared by incorporating *L. fermentum* ME-3 into "Pikantne" cheese.[182] The probiotic strain was found to withstand the processing conditions and exhibited moderate antimicrobial and high antioxidative activity throughout the ripening and storage period. The ripened cheese contained approximately 5×10^7 CFU/g of viable cells.

Consumption of commercial dahi and lassi samples significantly enhanced the immune response in human volunteers.[145] The investigators also reported nonsignificant reduction in total and HDL cholesterol levels in the subjects and decrease in systolic blood pressure in hypertensive patients. Sinha et al. developed probiotic *dahi* containing probiotic *L. acidophilus* National Collection of Dairy Cultures, NCDC14 and *L. casei* NCDC19 and investigated the probiotic potential. They investigated the antidiabetic,[211] antibacterial,[94] immune-modulating,[92] and antiallergenic[93] potential of the product.

A fermented beverage, with high probiotic counts of 10^{11} to 10^{12} of *L. acidophilus* NCDC-13, antidiarrheal, and potential anticarcinogenic properties, was prepared from a combination of whey, skim milk, pearl millet, and barley malt.[52] Fermentation by the probiotic organisms helped to reduce the antinutrient content in the cereals.[53] Some fermented probiotic dairy foods available in the world market[18,116] are: "b-Activ" probiotic curd, "Nesvita" probiotic yoghurt, Acidophilus yeast milk, Acidophilus yoghurt, Actimell, B-active, Biokys, Biomild, Cultra milky, LC-1, Mil-Mil, Mona fysig, Nutrish A/B milk, Probiotic curd, Probiotic ice creams (Amul prolife, Prolite and Amul Sugar free), Progurt Acidophilus milk, Symbalance, Vifit, Yakult, and Yakult (Miru Miru).

7.5.2   CLINICAL STUDIES RELATED TO HEALTH BENEFITS OF PROBIOTIC FERMENTED DAIRY FOODS

Food products especially fermented milk preparations containing different strains of probiotic bacteria have also been reported to provide health benefits in several studies. Moises et al.[131] studied the effect of milk fermented with *L. paracasei* ssp. *paracasei* CRL-431 and *L. acidophilus* on patients having superficial urinary bladder carcinoma. The authors also reported that oral administration of selected probiotics significantly minimized the tumor recurrence, and maintained or diminished the tumoral grade. No adverse side effects, including hepatomegaly, splenomegaly or blood alterations, were observed. A daily dose of (240 mL) fermented milk preparation containing *L. paracasei* ssp. *paracasei* CRL-431 and *Lactobacillus acidophilus* to the lactose intolerance people has resulted in enhanced lactose digestion, influenced orocecal transit time, and diminished intolerance symptoms in lactose intolerant people.[54] Probiotic milk supplemented with lyophilized strains of *L. paracasei* ssp. *paracasei* and *Lactobacillus acidophilus* CERELA was reported to reduce symptoms of persistent diarrhea in children.[55] Rizzardini et al.[160] conducted a randomized, double-blind, placebo-controlled, parallel-group study to investigate the immunomodulatory effects of a dairy drink containing *Lactobacillus paracasei* ssp. *paracasei* 431. The workers reported that the intake of probiotic dairy drink significantly improved immune function by augmenting systemic and mucosal immune response. The addition of probiotics to infant formula has shown to be an efficient way to increase the number of beneficial bacteria in the intestine in order to promote a gut flora resembling that of breast-fed infants. The use of a prebiotic-containing starter formula supplemented with *L. paracasei* ssp. *paracasei* and *B. animalis* ssp. *lactis* in early infancy resulted in healthy growth with no adverse effects on infant behavior.[203]

7.5.3 CHALLENGES RELATED TO THE USE OF PROBIOTICS IN FERMENTED DAIRY PRODUCTS

In order to provide the consumer with most of putative health benefits, it is understandable that a sufficient amount of viable probiotics must reach the intestine. To provide health benefits, the suggested concentration for probiotic bacteria is 10^6 CFU/g or mL of the product.[178] However, several studies have shown low number of probiotics in some market preparations ($<10^6$ CFU/g or mL), thereby diminishing the potential health benefits by these products.[194] Understanding the survival of probiotics and developing methods to maintain and/or to promote their viability throughout the product shelf-life continues to be an important subject of research in this field.

Number of factors have been claimed to affect the viability of probiotic bacteria in fermented foods including acid and hydrogen peroxide produced by bacteria, oxygen content in the product, and oxygen permeation through the package.[178] In fermented products, viability of probiotic bacteria decline over time because of the increased acidity of the product, fluctuations in storage temperature, prolonged storage time, and depletion of nutrients.[36] During cold storage of fermented products, the number of viable probiotic cells often drops far below the minimum therapeutic level. If probiotic bacteria are to be added as adjunct cultures to fermented dairy products (such as yoghurt and cheese), they must be compatible with the main starter culture. The chemical makeup of the food product is also an essential element when considering the metabolic activities of the probiotics.[73]

On the other hand, the proteolytic, lipolytic, and saccharolytic properties of probiotics in food products would be potentially important for further degradation of proteins, lipids, and complex carbohydrates, thereby leading to changes in the flavor of the food product. From a commercial perspective, it is essential that the flavor and texture of the probiotic fermented product remains appealing to the consumer. Hence, the effect of probiotics on the sensory and textural quality of the food product should also be considered before choosing the probiotic organisms for their incorporation into foods.

7.5.4 SUGGESTED APPROACHES TO MAINTAIN THE VIABILITY OF PROBIOTICS IN FERMENTED FOODS

For probiotics to be delivered through foods, additional amount of cells is likely required prior to processing to account for the loss of cells during the processing and/or storage phase. During the shelf-life of the dairy fermented products, maintaining a high level of viable probiotic cell count is not a simple task. It is well known that many strains of probiotic bacteria (mainly *Lactobacillus* and *Bifidobacterium* strains) grow weakly in milk due to their fairly low proteolytic activity and inability to utilize lactose.[166] These bacteria also need certain compounds for their growth, which are missing in milk.[163] In such cases, the level of inoculums can be increased to overcome the shortfall in numbers, or the milk can be fortified with various additives to promote growth of the culture. In order to improve the viability of probiotic bacteria in fermented milks, various substances have been added to milk, such as fructooligosaccahrides, caseinomacropeptides, whey protein concentrate, tryptone, yeast extracts, certain amino acids, nucleotide precursors and iron source.[95,140,179] Additives such as tomato juice,[14] sucrose,[3] papaya pulp,[109] manganese and magnesium ions,[4] simple fermentable sugars,[183] and a combination of casitone and fructose[174] have also been used to promote the growth of *Lactobacilli* in milk.

Additionally, the selection of probiotic strains (acid and bile-resistant strains) and optimization of the manufacturing conditions (both formulation properties and storage conditions) are of utmost importance in the viability of probiotic bacteria in fermented milk.[122,165] Reducing the dissolved oxygen level, the use of oxygen-impermeable containers, two-step fermentation, stress adaptation,[178] using ruptured or microencapsulated cells, adding prebiotics and continuous neutralization of the medium during fermentation to the initial pH (6.5) at periodic intervals (8 h) and overexpression of genes involved in bacterial survival[40] have been found effective means of keeping the number of probiotic bacteria high enough for therapeutic effects.[142] Likewise in case of yoghurt and cheese, when probiotic bacteria are added as adjunct cultures, probiotic strains having good compatibility with the product starter culture should be chosen.[113]

7.6   HERBAL SUPPLEMENTED FERMENTED MILK PRODUCTS

Since ancient times, herbs have played an integral part of the society, valued both for their culinary and medicinal properties. Herbs have been regarded for their role in maintaining human health, improving the quality of human life and besides their use as valuable components of seasoning, beverages, cosmetics, and dyes. The beneficial health effects of herbs may be due to the presence of micronutrients, vitamins, antioxidants, phytochemicals, and fiber in them. Over 80,000 species of plants have been used for medicinal purpose throughout the world.

Herbs are the primary curing agents for more than 80% of the world's population.[45] According to World Health Organization (WHO) estimate, the demand for medicinal plants by the year 2050 would be valued at US$5 trillion annually. This rising trend in the use of herbs to treat diseases is a global phenomenon that has been seen in both developing and developed countries. Milk and milk products, suitable to incorporate wide variety of functional ingredients, are also preferred over other food categories to carry herbs and herbal nutraceuticals. An array of fermented dairy products supplemented with herbs and herbal preparations has been developed to meet the consumer demands.

Based on their use and toxicity, herbs can broadly be categorized into three categories: food herbs, medicinal herbs, and poisonous herbs. Food herbs are generally used for functional food manufacturing. Food herbs are gentle in action, have very low toxicity and are unlikely to cause an adverse response when consumed. Food herbs can be utilized in substantial quantities over long periods of time without any acute or chronic toxicity. However, a clear understanding of benefits of herbs and possible risks or toxicity should be taken into consideration before their use in food products.

7.6.1   COMMON BIOACTIVE INGREDIENTS IN HERBS

Herbs harbor a variety of chemical constituents including vitamins, minerals, and active ingredients that have a variety of medicinal benefits. The active components present in herbs include alkaloids, anthraquinones, bitters, flavonoids, saponins, tannins, and essential oils (EOs).[31]

7.6.2   HEALTH BENEFITS OF HERBS

Herbs and their preparations have claimed to impart various health benefits, which include reducing risk of cancer, heart disease, reducing the risk or occurrence of hypertension, high cholesterol, excessive weight, osteoporosis, diabetes, arthritis, macular degeneration (leading to irreversible blindness), cataracts, menopausal symptoms, insomnia, diminished memory and concentration, digestive upsets and constipation.[83]

7.6.3   FERMENTED DAIRY PRODUCTS SUPPLEMENTED WITH HERBS AND THEIR EXTRACTS

Ayurveda has suggested several ways in which the medicinal benefits of herbs could be conveyed through dairy products. It has also suggested several forms, namely paste, dried, and whole leafs for incorporation into foods.[84] In the recent past, much attention has been paid to extract and incorporate herbal nutraceuticals into food matrix for the preparation of functional foods.

Herby cheeses are produced with different names in many countries worldwide. Ripened varieties herb supplemented cheeses, namely Otlu, Otlu Cacik, and Otlu. Lor are native to Turkey and Syria.[191] These herby cheeses are prepared by incorporating herbs into the curd mass after coagulation of milk.[72] Besides providing characteristic sensory attributes to cheese, owing to their antimicrobial activity herbs, enhance the shelf-life of cheese.[1] Herbal preparations show antimicrobial activity against selected species, that is, they kill pathogenic bacteria while causing no harm to the growth of starter culture. A phenolic compound oleuropein derived from herb has been found to enhance the growth of fungi, while markedly inhibiting the production of aflotoxins.[62] This property could be advantageous in mold-ripened cheeses, where the growth of molds is desirable without the release of mycotoxins.[96] Herbs supplementation may provide additional vitamins and minerals in herby cheeses.[72] About 25 kinds of herbs (especially, *Allium* spp., *Thymus* spp., *Ferula* spp., *Anthriscus nemorosa*, and so on, in single or in combination) were used to prepare herby cheeses with the rate of addition of herbs varying from 0.5–2.0 kg to the curd obtained from 100 L of milk.[191] Sadeghi et al.[168] reported that incorporation of EO of *Cuminum cyminum* @ 0.03 and 0.015% in feta cheese resulted in highest inhibitory effect on the growth of *Staphylococcus aureus* and the combination of EO with probiotic strain *L. acidophilus* further reduced the counts of the pathogen in feta cheese.

Herbal extracts have also been used as coagulants in the preparation of cheeses. The flowers of *Cynara cardunculus* L. and *Cynara humilis* L. have been used as rennet for the manufacture of farm cheeses such as Serra da Estrela, Serpa, Castelo Branco, Azeitao, Evora, and Niza from ovine milk in Portugal and neighboring regions of Spain. Herbal preparations, being slightly more proteolytic than chymosin, cause more hydrolysis of casein leading to lower firmness of the cheese curd.[46] *Cynara cardunculus* L. extract was also used to prepare cheese from ultrafiltered milk but the texture of the resultant cheese was undesirable.[2] However, Ekici et al.[44] reported that the addition of herbs may facilitate histamine formation in herby cheese, which may lead to histamine poisoning. The authors suggested that the herbs used for herby cheese should be examined for histidine-decarboxylating strain of the microorganisms to avoid histamine poisoning. *Aloe vera* has also found to reduce the population of *Staphylococci* spp. in a lactic acid cheese curd. Significant inhibition effect on the growth of *Staphylococci* was observed when *Aloe vera* was applied to the curd just before heat treatment.[185]

EOs derived from herbs have successfully been used for the preservation of cheese. Smith-Palmer et al.[181] reported that bay, clove, cinnamon, and thyme EOs added at concentrations of 0.1%, 0.5%, and 1% (v/w) effectively inhibited the growth of *Listeria monocytogenes* and *Salmnella enteritidis* in low- and full-fat soft cheese. The typical strong flavor of EOs is a limitation for their deliberate use in food products. However, EOs in combination with other preservative mechanisms can be employed to extend the shelf-life of foods to avoid strong flavor perception given by EOs alone. Combined effect of basil EO (added at 0.4%) and modified atmosphere packaging (MAP) extended the shelf-life of fresh Anthotyros (a Greek whey cheese) by approximately 10–12 days under refrigerated storage conditions.[196]

Herbs contain several essential nutrients (ENs), which may enhance the growth of probiotic organisms. The addition of catechins (at 100–2000 mg/kg) to probiotic yoghurt improved the survivability of the *Bifidobacteria* during its storage.[6] *Aloe vera* added to yoghurt improved the growth of *Bifidobacteria*.[151] Tea catechins and ferulic acid added to yoghurt effectively inhibited the growth of pathogenic bacteria (coliforms and *Salmonella*), while unaltering the LAB count.[162] Peng et al.[146] developed a process for the manufacture of de-alcohol yogurt beverage using *Pueraria lobata* and *Hovenia dulcis*. Improved quality yoghurt was prepared by using aqueous extract of garlic (0.1% level).[71] Lee and Yoon[111] prepared yoghurts supplemented with *Aloe vera* powder and *Aloe vera* juice and compared it with yoghurt prepared using dried skim milk. They found that quality retention of *Aloe vera* yoghurt was better than the control yoghurt during 15 days of refrigerated storage. The *Aloe vera* yoghurt had a shelf-life of 18 days when stored at 4°C. Recently, Panesar and Shinde[143] prepared *Aloe vera* juice fortified probiotic yoghurt. The probiotic species under consideration were *L. acidophilus* and *B. bifidum*. Counts of probiotic bacteria remain more than suggested value of more than 10^7 throughout the storage period (28 days). They suggested that *Aloe vera* fortified probiotic yoghurt could be used as an adequate carrier of probiotic bacteria.

Aloe vera juice-supplemented probiotic *dahi*[85] and probiotic *lassi*[84] with acceptable sensory, physicochemical and textural qualities were prepared. The authors also reported that addition of *Aloe vera* had some favorable effects on the survivability of probiotic strain *L. paracasei* ssp. *paracasei* NCDC 627. Animal study of the *Aloe vera*-supplemented probiotic *lassi* revealed that it had better immunoprotective effects compared to control *lassi*. *Shrikhand*, sweetened strained *dahi*, being a sweetish-sour and semi-soft product can easily harbor herbal preparations without undergoing significant changes in sensory quality. Landge et al.[110] successfully prepared herbal *Shrikhand* by incorporating Ashwagandha powder. The authors found that the addition of 0.5% Ashwagandha powder to *Shrikhand* improved the organoleptic quality and the product was remained acceptable up to 52 days under refrigeration. A method to improve the storage stability and organoleptic properties of a carbonated fermented milk beverage by adding a mixture of EOs of coriander, basil, and fennel to a salt solution of whey has been reported by Askerova et al.[11] Tarakci et al.[192] studied the influence of different herbs on physicochemical and organoleptic quality of *Labneh* during storage. They reported that the sensorial scores of the *Labneh* samples were influenced by the variety of herb added and the storage times. Addition of dill and parsley to *Labneh* resulted in the highest sensory scores.

7.7   FERMENTED DAIRY FOODS SUPPLEMENTED WITH FRUITS

Fruits are the best food items provided by nature to human beings. Fruits contain various important phytonutrients, namely vitamins, minerals, antioxidants, and dietary fibers. Scientific studies published on the health benefits of fruit consumption are numerous. Current evidences collectively demonstrate that fruit intake is associated with improved health, reduced risk of various types of cancers, cardiovascular disease (CVD), hypertension and possibly delayed onset of age-related indicators. Supplementation of fruits into fermented dairy foods can improve the health status of fermented milks. Addition of fruit preparations improves the sensory appeal of the fermented milks thereby increasing the sales of fermented dairy foods. Fruit addition also increases the micronutrients especially iron content of the fermented milks.[172] Recently, there has been an increased trend to fortify cultured milk products with fruit juices/pulps. A new term of fruit-based functional foods, namely "juiceceuticals" is gaining a lot of interest among the consumers.

Fresh or processed fruits in various forms (viz., fruit juices, fruit pieces, fruit purees, crushed fruit, frozen fruits, fruit preserves and dehydrated or dried fruits) are generally added to fermented dairy foods. Desai et al.[39] prepared fruit yoghurt using mango, sapota, papaya pulp, and pineapple and kokum juice added to yoghurt @ 0, 10, 15, or 20% levels. Fruit yogurts added with orange juice (15%), grape juice (15%), and carrot juice (5%), with acceptable sensory quality were prepared by Misra and Kuila.[127]Similarly, *Lycii fructus, Lycii folium, Lycii cortex* at concentration of 0.5, 1.0, 2.0, 4.0, and 6.0% were added to yoghurt for enhancing its palatability and nutritional value.[27] With the increased addition of the fruit, the titratable acidity of the yogurt was increased (0.98% in 0.5% fruit addition to 1.27% in 6% fruit addition), which is a limiting factor for the fruit additions into yogurt.

Aroyeun[10] prepared cashew apple-fortified yoghurt with acceptable sensory and physicochemical quality. Fruit-based *Shrikhand* has also been prepared using apple, papaya, mango. Fruit yogurt preparations, besides imparting value addition to the fermented milks, also enable the utilization of underutilized fruits in an effective manner. Soursops (*Annona muricata* L.) are highly aromatic fruits. Ripen soursops are highly perishable, which lead to huge postharvest losses. Soursops are good source of protein and minerals. Soursop nectar at 0, 5, 10, and 15% was supplemented into stirred yoghurts to improve its nutritional quality and also to address the problem of postharvest losses of soursops. The sensory analysis of the soursop added yogurt revealed that 10% and 15% soursop added yoghurts were more acceptable than the 0 and 5% soursop-added yoghurts. These yoghurts provided high percentage daily values of zinc, phosphorus, and calcium and a good level of protein.[117]

Coconut is widely used as an ingredient in several dishes of Asia and Pacific regions. Coconut milk is rich in protein and mineral content. Nutrition content of coconut milk is higher compared to cow milk. Coconut-supplemented yogurt was prepared by optimizing ingredients and processing parameters using response surface methodology by Yaakob et al.[210] They suggested that the addition of vanilla flavor may enhance the palatability of the coconut supplemented yogurt by preventing the typical coconut flavor, which is undesirable. Banana is considered as a good-mood food. Çakmakçi et al.[20] incorporated banana pulp into goat milk yogurt at three concentrations of 10, 15, and 20% levels. Addition of banana pulp at 20% level had obtained better sensory quality than the remaining samples. Pratap et al.[150] have tried to optimize the chemical properties of frozen yoghurt by incorporating fruit pulp. Apple, banana, grapes, and mango were incorporated in the frozen yoghurt at 5, 10, 15 and 20% levels, respectively, in the form of pulp. The authors reported that addition of mango pulp at 5% level showed promising results in the chemical properties and was considered the best sample yoghurt among other samples. A study was conducted by Jayasinghe et al.[98] to investigate the possibility of developing fruit-yoghurt using an underutilized fruits in Sri Lanka: White dragon fruit and pasteurized dragon fruit juice was incorporated at different levels (5, 7.5, 10 and 12.5% w/w) into yogurt. White dragon fruit juice addition at 10% into yogurt was the most acceptable addition level into yogurt, obtaining highest sensory scores by the sensory panelists. Several other fruit yogurt preparations have also been prepared to satisfy the increasing consumer demand for health foods, which included strawberry, orange and grape juice added yogurts[78] and strawberry- and peach-enriched yogurt.[137]

Stirred yoghurts with different fruit homogenates (papaya, kiwi, pineapple and kaki) were prepared to study the effect of fruit reparation on the physicochemical and microbial population of stirred yogurt. Sugar and fruit homogenates were added @ of 5 and 10% w/w, respectively, into the yogurt. The results indicated significant differences in proteolytic activity, acidity, protein, and fibers contents among all treatments. Addition of fruit homogenates and sugar significantly reduced the viability of yogurt starter cultures. The effect was more obvious using pineapple or kiwi than other fruits. *L. delbrueckii* ssp. *bulgaricus* was much more sensitive to the fruit addition than *S. thermophilus*. Yeast and mold count was higher in fruit containing yoghurt compared with the control and increased significantly during storage at 5±1°C.[47] Pineapple pulp fortified yogurt drink was prepared by Sawant et al.[173] Incorporation of pineapple pulp into yogurt (yogurt:pineapple pulp = 94:6) was the best combination with highest sensory scores. Pineapple addition significantly decreased the fat content in yogurt, whereas it increased the total acidity content of the product. Total viable count, coliforms, and yeast and molds were higher in the experimental samples than the control sample, after 9 days of storage study at refrigeration temperature.

Due to the presence of ENs and fiber, fruits addition may help the growth of probiotic organisms employed in the fermented milks manufacture. Fruit-yogurt-like fermented milk products with living probiotic bacteria significantly shorten the duration of antibiotics-associated diarrhea and improve gastrointestinal complaints.[37] Kailasapathy et al.[101] studied the effect of commercial fruit preparations (mango, mixed berry, passion fruit, and strawberry) on the viability of probiotic bacteria, *L. acidophilus* LAFTI L10 and *B. animalis* ssp. *lactis* LAFTIs B94 in stirred yogurts during storage (35 days) at refrigerated temperature. The authors found that addition of fruit preparations had no significant effect on the viability of the probiotic strains. Jambolan (*Eugenia jambolana* Lam) is a natural phytochemical source with pharmacological applications.[13] Addition of fresh and spray-dried Jambolan fruit pulp (*Eugenia jambolana* Lam) to caprine frozen yoghurt containing probiotic strain *B. animalis* ssp. *lactis* BI-07 resulted in outstanding probiotic survival rate (97%) throughout the 90 days of frozen storage. The workers advocated that the caprine frozen yoghurt enriched with Jambolan fruit was an efficient carrier for *B. animalis* subsp. *lactis* BI-07.[17]

Being a rich source of vitamin C, guava enhances the antioxidant potential of milk and milk products upon its supplementation. Walkunde et al.[205] prepared guava fruit yoghurt from cow milk by supplementing guava at three different levels, namely 5, 19, and 15%. Guava addition at 5% level and sugar addition at 6% level resulted in desired body, smooth texture with attractive color and appearance to yogurt. Selvamuthukumaran et al.[175] prepared antioxidant rich fruit yoghurt by incorporating seabuckthorn fruit syrup. The resultant product had higher content of fat, protein, carbohydrate, and antioxidants (vitamin C, vitamin E, carotenoids, phenols, anthocyanins) when compared to a commercial yoghurt. Singh et al.[180] evaluated the effect of addition of strawberry polyphenol (0.5 mg/mL) on physicochemical properties, total phenolic and antioxidant activity of stirred dahi. Strawberry polyphenol addition resulted in a sevenfold increase in the antioxidant activity of polyphenol-enriched stirred dahi, while pH, acidity, water-holding capacity and viscosity remained comparable with the control. There was no significant difference in the antioxidant activity and total phenolic content of the product during 14 days of storage at 7–8°C. *Euterpe oleracea* Mart. fruit (Arecaceae) juice is rich source of anthocyanin and phenolic content. The antioxidant and antiradical properties of *E. oleracea* juice are well known and as well as its potential use as food ingredient and natural pigment have been previously studied. The novel natural colorants from *E. oleracea* juice could be considered as "functional" ingredients for their antioxidant and antiradical activity. *E. oleracea* juice (10% w/w) could be used as a natural functional pigment for flavoring and coloring yogurt.[30]

Cattaneo et al.[22] reported that pre-dehydrated fruit preparations may improve the textural properties of yogurt samples when compared with the fresh fruit-added yogurts. The effect of Osmodehydrofrozen fruits on sensory, physical, chemical and microbiological properties of yoghurt and its quality during storage was evaluated by Vahedi et al.[199] Their study revealed that fruit addition after fermentation resulted in higher acceptability to yogurts by the sensory panelists. Apple addition at 10% or strawberry addition at 13% into yogurt was optimized on the basis of sensory scores. Incorporation of osmodehydrofrozen fruit preparations significantly decreased the syneresis value in yogurt. Yogurt sample contained apple did not have any mold and yeast in it and coliforms were disappeared after 7 days of storage. The coliforms disappeared after 7 days of storage in samples which contained strawberry and yeasts.

Vegetables are rich in antioxidant substances and their addition to fermented milks may enhance the antioxidant activity. Carrot was incorporated into yoghurt at 5, 10, 15, and 20% levels to enhance the health status of the product. Carrot juice was added into milk before fermentation to get carrot yoghurt. Carrot yogurt containing 15% carrot juice had obtained highest acceptability when compared with the remaining samples. Addition of carrot juice suppressed the yeast, mold, and coliforms in yogurt during its storage. However, the growth of starters was not affected by the carrot addition. Yoghurt with 5, 10, 15, and 20% carrot juice showed a significant decrease in aflatoxin M1.[171]

Vegetable preparations (viz. carrot, pumpkin, broccoli and red sweet pepper) were supplemented at 10% w/w into milk for yoghurt preparation. The vegetable-fortified yoghurts were evaluated for their antioxidant status during 14 days of storage study at refrigeration temperature. Yoghurts supplemented with broccoli and sweet pepper has obtained highest 1,10-diphenyl-2-picryl-hydrazyl (DPPH) radicals-scavenging activity. All vegetable yoghurts were characterized by higher than the natural yoghurt ferric reducing antioxidant power values that were measured directly after production.[135]

7.8   FERMENTED DAIRY FOODS SUPPLEMENTED WITH CEREALS AND DIETARY FIBER

Dietary fiber may be defined as portions of plant foods that are resistant to digestion by human digestive enzyme; and this included polysaccharides and lignin. More recently, the definition has been expanded to include oligosaccharides, such as inulin, and resistant starches.[79] Fibers have been classified as soluble (e.g., Pectin) that are fermented in the colon, and insoluble fibers (e.g., wheat bran) that are only fermented to a limited extent in the colon. Vegetables and cereals are natural sources of fiber. In most countries, the recommendation for dietary fiber intake is 25 to 35 g/day. The WHO recommended an intake of 27–40 g dietary fiber per day.[28]

Dietary fiber consumption keeps the digestive system healthy, assists in balancing blood glucose levels and weight control. Also, dietary fiber may reduce the risk of diabetes, cancer, and coronary diseases.[8] Incorporation of dietary fiber into milk and milk products may greatly enhance their nutritional and physiological effects and also make them ideal food matrices for human consumption. Addition of fiber to fermented milk improves their consistency while decreasing the syneresis.[9]

Raju and Pal[154] incorporated dietary fiber preparations (inulin, soy fiber, and oat fiber) into *Misti dahi* to enhance its health attributes. Among the three dietary fibers, inulin significantly decreased viscosity and instrumental firmness and increased syneresis and work of shear values. Due to the low solubility, oat fiber was not recommended for further studies. Fruits, vegetables and cereals are natural sources of dietary fiber. Their incorporation into food products provides fiber as well as valuable micronutrients. Fermented dairy foods developed recently include: Vegetable yoghurts added with sweet potato and yam yoghurt. In the United States and Europe, yoghurts carrying whole cereal grains (e.g. wheat and oat), soy fiber, and fruits like cranberries, blackberries, raisins, blueberries, walnuts, hazelnuts, and so on are very popular under various names such as Yoplait Breakfast Yoghurt, Yoghurt Diet Meal, Fruits of the Forest, and so on. These foods are natural sources of antioxidants, namely, carotenoids, antioxidant vitamins C and E, and phenolic compounds, which are associated with several health attributes. Soluble fiber acts as prebiotic substance by enhancing the growth and survival of probiotic organisms. Prebiotic polysaccharides are successfully used in fermented milk products to maintain probiotics viability. In Japan, several companies are using various oligo-saccharides to fortify yoghurt, for example, Yakult Honsha co., Suntory Co. and Morinaga Food Industry. In Belgium, a dietary fiber-fortified fermented milk drink called Fyos containing inulin is very popular.

Cereals have a long history of use by humans. In human diet, cereals and cereal products are important source of energy, carbohydrate, protein, fiber, and micronutrients (such as vitamins, magnesium, calcium, iron and zinc). Cereals and their products may harbor a wide variety of bioactive substances, which can provide multiple health benefits. Supplementation of milk and milk products with cereals may enhance the nutritional value of milk products by enriching their mineral content. Cereals supplementation also provides fiber to milk products, which they lack. Cereals may act as fermentable substances for the growth of probiotic microorganisms.[25] Fermentation further enhances the nutritive value, palatability, and functionality of cereals by reducing the antinutritional factors.[82] Many under-utilized cereals having potential nutritional and therapeutic properties can be better utilized by supplementing them into fermented milks.

Cereal-based fermented milk products are popular in India and African countries. *Kindumu* is popular in central African region and is prepared by sun drying the mixture of fermented milk and germinated/nongerminated sorghum flour.[193] *Raabadi*, a cereal-based fermented milk beverage, is popular in rural parts of Haryana, Punjab, and Rajasthan states of India. The cereals (germinated/non-germinated) used for this purpose are barley, wheat, maize, pearl millet, and sorghum. In the traditional method, *Raabadi* is prepared by adding the cooked and cooled cereal flour to buttermilk and allowing the mixture to ferment overnight. The resulted *Raabadi* is consumed as such. Traditional process of *Raabadi* making yields a product with limited shelf-life (1 to 2 days) with unpredictable sensory quality. The traditional method uses maize as the principle cereal for its manufacture. A new technology has been developed for the preparation of *Raabadi* using three different cereals (viz., wheat, pearl millet and sorghum) to address the quality and shelf-life issues associated with the traditional *Raabadi* manufacture. In this method, germinated cereal flour and milk mixture are fermented together after pasteurization treatment using suitable starter culture. The curd obtained is broken and mixed thoroughly with pasteurized water, which has already been added with pectin, spices, and salt. The product has a keeping quality of 7 days when kept under refrigeration (7–8°C) conditions.[82] Variants like soy-based

Raabadi[63] and moth bean-based *Raabadi*[65] have also been prepared. Soy-based *Raabadi* was characterized with typical beany flavor. Mugocha et al.[133] developed a composite-finger millet and milk-based fermented beverage. Various parameters including level of finger millet gruel in skim milk, type of starter culture, and incubation temperature were optimized to develop the finger millet-dairy beverage.

7.9 LOW-CALORIE PREPARATION OF FERMENTED DAIRY FOODS

With the changing lifestyle and dietary patterns, noncommunicable diseases such as obesity, diabetes, CVDs and cancer have become major health problems worldwide. The dietary factors such as high intake of fats, sugars, and low intake of fruits and vegetables have been ascribed as reasons for the CVD mortality.[64] Nowadays, calorie consciousness is increasing among the people to prevent noncommunicable diseases. Low-fat fermented dairy products mainly yogurts, cheese varieties, dahi, and so on have already been released into the market to cater the demand of calorie conscious sector.

Fat is the principal component that is responsible for rich flavor in dairy products. Apart from calories, fat also provides desirable body and texture to the dairy foods. Nonfat or reduced fat preparations of fermented milks need careful selection of ingredients to avoid undesirable changes in sensory. Generally, skim milk or nonfat dry milk (NDM) is used to maintain the bulk of no- or low-fat fermented milks. However, the addition of NDM increases the calorific value of yogurt and the acid production by LAB may increase due to high lactose content of NDM.[102] An undesirable powdery flavor can be perceived in yogurt fortified with too much NDM.[190] Also, nonfat yogurts suffer from whey separation[70] Thickening agents such as gelatin and carrageenan may potentially reduce synersis in yogurts.[129] Exopolysaccharide producing strains of yogurt culture can also be used to improve the texture of yogurt.[75]

Low-calorie sweeteners (e.g., xylitol, sorbitol, fructose, cyclamate and saccharin,[86] aspartame and sucralose[149]) have successfully been used to prepare low-calorie yogurt. Low-calorie sweeteners did not affect the growth of yogurt starter as well as probiotic growth in the yogurt.[149] *Misti dahi* (sweetened *dahi*) contains high amount of sugar, which is a drawback

for its marketing in the present health conscious era. Artificially sweetened *Misti dahi* was prepared by Raju et al.[153] to reduce its calorific value.
Artificial sweeteners (binary blend of aspartame and acesulfame-K) were
used to completely replace cane sugar and maintain the bulk in *Misti dahi*.
Maltodextrin (bulking agent) was found to be the most suitable bulking
agent in the preparation of artificially sweetened *Misti dahi* using aspartame and acesulfame-K. However, the authors suggested following alternate strategies for diabetic people as the maltodextrin gets metabolized in
human intestines in a similar manner to cane sugar.

Streiff et al.[187] patented a process for the manufacture of reduced
calorie, lactose free, artificially sweetened yogurt. Ultrafiltration and
enzymatic hydrolysis were used to reduce the lactose content. The calorific value of yogurt was estimated to be 60–70 calories per 6 ounces.
Lactose level was reduced to less than 0.1% by weight of the product.
Thus, the product is amenable to lactose-intolerant individuals also.
Farooq and Haque[48] studied the effect of sugar esters (0.05%) of various
hydrophilic–lipophilic balances (HLB) on the textural properties of nonfat
low-calorie yogurts (12.68% TS) during 14 days of storage period. Aspartame was used (@ 200 ppm) to sweeten a skim milk-based yogurt that was
stabilized with starch (0.5%). Sugar esters, especially the stearates (HLB
value of 5 to 9), produced yogurt with better body, texture, and mouth-
feel compared with yogurts without sugar esters. Different fat substitutes
including protein- and fiber-based preparations were also used to reduce
the calorie levels in yogurt. A protein-based fat substitute (Simplesse® 100
@ 1.5% w/w of yogurt),[15, 190] Inulin (@1%), Dairy-Lo®, inulin, and their
combinations[177] were successfully incorporated into yogurt to reduce the
fat level to as low as 0.1%.

The concept of low-fat cheese manufacture dates back to the early
1990s. Strezynski[188] and Hargrove and McDonough[68] received patents for
procedures to manufacture low-fat cheeses. However, these earlier studies
could not address several drawbacks associated with the fat reduction on
the quality of cheeses. The fat content of milk used for manufacturing
low-fat cheeses generally ranges from <0.5% to 1.8%. To address the total
solids loss or to maintain the bulk of cheese after fat reduction, addition of
NDM or use of concentrated milk,[7] use of ultra-filtered milk,[123] addition
of dried ultrafiltered or microfiltered retentate[184] were employed in the
earlier studies. Alteration in temperature of cooking, time of holding during
cooking, pH at milling, and rate of salting were suggested by Johnson and

Chen[99] to prepare good-quality low-fat cheeses. Several cheese varieties with low-fat content including low-fat Feta cheese,[103] low-fat Cheddar cheese,[128] low-fat Mozzarella,[164] low-fat Process cheese,[157] Cremoso Argentino low-fat soft cheese,[214] low-fat white-brined cheeses (~ 60% fat reduction)[161] low-fat white pickled cheeses,[105] and low-fat fresh kashar cheeses (~70% fat reduction) have successfully been developed in the past.[107]

7.10   FERMENTED DAIRY FOODS SUPPLEMENTED WITH PLANT STEROL ESTERS

Phytosterols and phytostanols are important structural components of plant membranes, which play an important role in plant cell membrane function.[152] These bioactive compounds are abundant in fruits, vegetables, seeds, nuts, and vegetable oils.[141] Humans lack the ability to synthesize phytosterols; hence, they should be taken from external sources through diet. Generally, per-day consumption of phytosterols ranges from 100 to 500 mg.[141] Generally, esterified forms of phytosterols and phytostanols are used in food fortification to increase lipid solubility. Due to their structural similarity, phytosterols intake reduces the intestinal absorption of both dietary and endogenously produced cholesterol without affecting the levels of HDL-cholesterol or triglycerides.[132] Reported health benefits of phytosterols include reduction of type II diabetes, risk of stomach cancer and inhibition of tumor growth and so on.[12] Fermented dairy beverages containing phytosterols (including stanol esters) are being promoted for their cholesterol-lowering action. It has been proven that margarine and phytosterol-enriched dairy products (yoghurt and milk) are more effective in lowering cholesterol,[159] compared to cereal-enriched yoghurt and milk.

7.11   FERMENTED DAIRY FOODS SUPPLEMENTED WITH VITAMINS, IRON, AND CALCIUM

Vitamins are essential compounds to maintain the human health. Vitamins act as cofactors, which are important for several metabolic reactions in the body. Fermented milk products are good source of vitamins as the starter culture and other beneficial bacteria used may synthesize some vitamins (vitamin B) in fermented milks. However, due to the various processing treatments employed during pre-and post-manufacture and the storage conditions, the vitamin content in the fermented milks may be significantly altered. Moreover, due to the increased trend in the development of low-fat or non-fat preparations, the fat-soluble vitamin content of the fermented milks may be significantly reduced.

Fat-soluble vitamins (namely vitamin D and A) play a major role in human nutrition. Vitamin D plays a key role skeletal development, by regulating serum calcium and phosphorus concentrations in the body. Vitamin D deficiency might result in type-I diabetes, hypertension, multiple sclerosis, and some type of cancer.[77] Photosynthesis of vitamin D by skin upon its exposure to solar UV radiation has kept this vitamin in the category of non-ENs of the body.[77] However, people living in altitudes above 40°N and altitudes below 40°S for several months of the year cannot photosynthesize vitamin D. Thus, their diet must be provided with an external source of vitamin D. Studies on the stability of vitamin D in fermented milk products have been carried out by several investigators,[67,198,204] who have claimed that vitamin D is stable during processing and storage of fermented milks.

Vitamins A and C are also being successfully incorporated into fermented milk products to improve their nutritive quality and acceptability.[35,87] These vitamins at higher levels are toxic to human body; hence attention must be paid while determining their levels of addition into the food products. Generally, the pro-vitamins of vitamin A namely carotene are nontoxic, hence they can be incorporated into fermented milks without any adverse effects.[51]

Milk and milk products are poor source of iron. Fortification of milk products with iron might prevent the nutritional deficiencies prevalent in these days. Woestyne et al.[209] reported high bioavailability of iron when fortified in yogurt. However, the effect of iron addition on the quality parameters of fermented milks must be carefully evaluated before fortification. Catalytic role of iron in oxidation of fat is a major factor that decides the use of iron in milk and milk products as it affects their sensory and physicochemical characteristics thus influencing acceptability and shelf-life. Also, type of mineral source and amount to be added to the products must be carefully evaluated before fortification as they influence the stability of iron and quality of the milk products. Oxidized flavor and metallic flavor may be the two major flavor defects observed in iron-fortified fermented milks.[91]

Food matrix composition (presence of oxidizable materials) and structure (liquid or solid) may influence the iron action on the food components. Sadler et al.[169] did not observe any increase in the oxidation of milk fat due to iron addition in the cottage cheese. Kim et al.[106] did not found any undesirable changes in yogurts added with microencapsulated iron and vitamin C during three weeks of storage. The authors advocated that microencapsulation may be a promising technique for iron and vitamin C fortification into dairy products without any changes in sensory aspects.

Calcium is an important mineral, which helps in healthy bone development. Calcium deficiency results in osteoporosis, which is related to increased bone fracture risk. Ocak and Rajendram[136] suggested the use of micronized calcium in milk products to prevent undesirable changes in sensory attributes. Gerhart and Schottenheimer[56] reported that tricalcium citrate can be used in yogurts and other dairy products at concentrations of > 1 g/L.

7.12   WHEY-BASED FERMENTED BEVERAGES

Whey is a by-product of cheese industry, which contains about 55% of the milk solids including valuable minerals and whey proteins from milk. Whey proteins are nutritionally superior proteins with biological value almost equal to egg proteins. Standardized procedures for the manufacture of several useful products (like whey protein concentrates, whey protein hydrolysates, lactose and whey drinks) from whey have been developed. Whey is used to prepare fermented beverage to utilize the nutritional benefits of whey and therapeutic advantages of fermentation in a single matrix. Several brands of fermented whey beverages (like fermented whey, probiotic-fermented whey, fruit-based fermented whey, cereal-based probiotic whey beverage, and whey cheeses) have been released in the world market. Fermented whey beverages have been reported to have immunomodulatory, anticancer, antimicrobial, mineral binding, gut health promoting, hypocholesterolemic, antidiabetic, and psychomodulatory effects.[147]

7.13   SHELF-LIFE ENHANCEMENT OF FUNCTIONAL FERMENTED DAIRY FOODS

Post acidification, contamination and growth of yeasts, and molds are major drawbacks leading to whey syneresis; and undesirable sensory changes are the major storage drawbacks reported in fermented milks.[82] Thermal treatments, namely, thermization, pasteurization, and steriliza-tion were tried to enhance the shelf-life of fermented milks, which may effectively reduce the microbial (yeast and molds and unwanted bacteria) contamination. However, thermal treatments also kill the starter cultures and probiotics employed in the fermented milks manufacture, which may not be desirable in view of health benefits endowed by the consumption of fermented milks. Also, thermal treatments may decrease the vitamin content and destroy the bioactives, when added to the fermented milks. Moreover, heat treatment of fermented milks leads to whey separation due to oozing out of water from casein micelle during heating.[82]

Nonthermal preservation treatments such as addition of chemical and bio-preservatives, effective packaging, gas flushing, high-pressure processing (HPP), pulsed electric field, and the combination of two or more of these treatments (hurdle technology) have also been employed to increase the shelf-life of fermented milks. Addition of copper (as $CuSO_4$ @ 1.25 mg/kg) was effective in preserving yogurt for about 7 days at 25°C, while unaltering the starter culture growth.[66] However, chances of fat oxidation and development of metallic-off flavors may be present when copper is added to food products. However, the susceptibility of milk to metal-induced oxidation can be lowered by homogenization. Homogeni-zation can help disperse copper ions throughout the milk system and allow proteins to be incorporated into the milk fat globule membrane, which can help bind metals and prevent them from reacting with lipids.

Addition of chemical and bio-preservatives may inhibit the probiotic count in fermented milks. However, techniques such as HPP, pulsed elec-tric field, and so on have shown a big promise in controlling post-acidifi-cation besides unaltering the beneficial flora in fermented milks. Fonterra, a New Zealand-based dairy company, patented a long shelf-life (90 days) probiotic yoghurt using the application of HPP. They took large combi-nations of probiotic strains and selected those strains, which are capable of surviving a predetermined pressure, wherein the treatment pressure reduces, delays, prevents, or eliminates growth of spoilage microflora.[21] In another study, a high-pressure treatment at 400MPa for about 30 min inhibited the acidifying activity of three out of the seven strains of *S. ther-mophilus* and the HPP treatment even improved the organoleptic quality of the fermented milk by reducing the syneresis.[158]

High barrier packaging materials are used for packaging of fermented milk products containing oxygen-sensitive bioactive substances. Probiotics are generally anaerobic in nature, which have zero or very low tolerance to the presence of oxygen. The use of high barrier and active packages with oxygen absorbers to extend the shelf-life of probiotic foods and probiotic survivability has been evaluated in many research studies. Dave and Shah[35] reported that storage of probiotic yogurt in high-barrier packaging material like glass was effective in maintaining the probiotic strain *L. acidophilus* viability up to 35 days at refrigeration temperature. Similar results have been reported by Jayamanne and Adams[97] for *Bifidobacteria* survival. High-barrier multilayer films (NUPAK®) were used to maintain the viability of probiotics for longer duration (42 days) in yogurt.[126] Cruz et al.[32] reported that plastic containers with lower oxygen permeability rates showed a lower content of dissolved oxygen and a higher count of the probiotic bacteria in yogurts during refrigerated storage.

7.14 SUMMARY

Fermented milks and milk products are regarded as important dietary sources in the history of mankind due to their long history of usage and perceived health benefits. Functionality of fermented milks is inherent or it can be induced by the addition of bioactive ingredients or alterations of the existing product. Several studies have reported that fermented milks and milk products are ideal vehicles for delivery various functional ingredients. Recently, a number of functional fermented dairy foods have been developed by incorporating a range of bioactives, viz, probiotics, prebiotics, herbal preparations, phytosterols, vitamins, minerals, fiber, cereals and so on. Low-calorie fermented dairy foods have also been manufactured by employing techniques like reduction of fat levels and addition of low-calorie sweeteners and bulking agents.

Incorporation of functional ingredients into fermented dairy foods poses several challenges. Several bioactives may undesirably change the sensory, physicochemical, and textural properties of fermented dairy foods. Most of the bioactives like probiotics and omega-3 fatty acids are sensitive to heat and light. Sometimes, milk components may interact with functional ingredients (e.g., reaction of milk fat with iron, interaction of milk proteins with herbal components) leading to the loss of their bioavailability, and so on. Also, processing operations and storage conditions may also affect the bioavailability of functional ingredients in fermented dairy foods. Microencapsulation, nano/micro-emulsions, micronization, and so on along with adoption of nonthermal treatments and effective packaging interventions have successfully been employed to maintain the bioavailability of functional ingredients in fermented dairy foods.

A large portion of fermented dairy foods available in the today's market are based on general studies in nutritional science rather than sound scientific evidence based on clinical studies. The retention of activity or bioavailability of functional ingredients present in the fermented dairy foods till their consumption is uncertain. Future research should be focused on these areas to address the issues of scientific evidence on health claims and retention or maintenance of bioavailability.

KEYWORDS

- Aloe Vera
- altered foods
- Amul
- Anthotyros
- Ashwagandha
- Aspartame
- Ayran
- *Bifidobacteria*
- calcium
- cereals
- cheese
- *chhash*
- coconut
- commercial probiotic milks
- *Cuminum cyminum*
- Cyclamate
- Dahi
- dietary fiber
- enriched foods
- essential oil
- fermented milk
- fermented whey beverage
- fiber
- fortified fermented milk
- fructose
- fruit yoghurt
- functional food
- goat milk
- HDL
- herbal cheese
- herbs

- inulin
- iron
- juiceceuticals
- *kefir*
- *koumiss*
- *kumiss*
- *kumys*
- labneh
- lactic acid bacteria
- *Lactobacillus*
- *Lassi*
- LDL
- Leuconostoc
- low-calorie fermented milk
- low-calorie sweetener
- *misti dahi*
- nonthermal preservation treatment
- phytonutrients
- phytostanols
- phytosterols
- probiotic *dahi*
- probiotic *lassi*
- probiotic yogurt
- probiotics
- pulp-fortified yogurt
- *raabadi*
- saccharin
- *shrikhand*
- sorbitol
- soy
- sucralose
- synbiotics
- vegetable yoghurts
- villi
- whey
- whey beverage
- whey cheese
- xylitol
- Yakult
- yogurt

REFERENCES

1. Ağaoğlu, S.; Dostbil, N.; Alemdar, S. The Antibacterial Efficiency of Some Herbs Used in Herby Cheese. *Yüzüncü Yıl Üniversitesi Veterinerlik Fakültesi Dergisi*, **2005,** *16*(2), 39–41.
2. Agboola, S. O.; Chan, H. H.; Zhao, J.; Rehman, A. Can the Use of Australian Cardoon (*Cynara cardunculus* L.) Coagulant Overcome the Quality Problems Associated with Cheese Made from Ultrafiltered Milk? *LWT-Food Sci. Technol.* **2009,** *42*(8), 1352–1359.
3. Agrawal, V.; Usha, M. S.; Mital, B. K. Preparation and Evaluation of Acidophilus Milk. *Asian J. Dairy Res.* **1986,** *5,* 33–38.
4. Ahmed, B.; Mital, B. K.; Garg, S. K. Effect of Magnesium and Manganese Ions on the Growth of *Lactobacillus acidophilus*. *J. Food Sci. Technol.* **1990,** *27*(4), 228–229.

5. Ahmed, Z.; Wang, Y.; Ahmad, A.; Khan, S. T.; Nisa, M.; Ahmad, H.; Afreen, A. Kefir and Health: A Contemporary Perspective. *Crit. Rev. Food Sci. Nutr.* **2013**, *53*(5), 422–434.

6. Akahoshi, R.; Takahashi, Y. Yoghurt Containing *Bifidobacterium* and Process for Producing the Same. Yakult. Japan, PCT–International Patent Application WO/1996/037113, 1996, pages 35.

7. Anderson, D. L.; Mistry, V. V.; Brandsma, R. L.; Baldwin, K. A. Reduced Fat Cheddar Cheese from Condensed Milk, I: Manufacture, Composition, and Ripening. *J. Dairy Sci.* **1993**, *76*(10), 2832–2844.

8. Anderson, J. W.; Baird, P.; Davis, R. H.; Ferreri, S.; Knudtson, M.; Koraym, A.; Waters, V.; Williams, C. L. Health Benefits of Dietary Fiber. *Nutr. Rev.* **2009**, *67*(4), 188–205.

9. Aportela-Palacios, A.; Sosa-Morales, M. E.; Velez-Ruiz, J. F. Rheological and Physicochemical Behavior of Fortified Yogurt, with Fiber and Calcium. *J. Texture Stud.* **2005**, *36*(3), 333–349.

10. Aroyeun, S. O. Optimization of the Utilization of Cashew Apple in Yogurt Production. *Nutr. Food Sci.* **2004**, *34*(1), 17–19.

11. Askerova, A.; Guseinov, I.; Azimov, A.; Dmitrieva, N.; Shamsizade, R. Manufacture of the Carbonated Fermented Milk Beverage, Airan. USSR Patent 1796122, 1993.

12. Awad, A. B.; Fink, C. S. Phytosterols as Anticancer Dietary Components: Evidence and Mechanism of Action. *J. Nutr.* **2000**, *130*(9), 2127–2130.

13. Ayyanar, M.; Subash-Babu, P.; Ignacimuthu, S. *Syzygium cumini* (L.) Skeels., a Novel Therapeutic Agent for Diabetes: Folk Medicinal and Pharmacological Evidences. *Complementary Ther. Med.* **2013**, *21*(3), 232–243.

14. Babu, V.; Mital, B. K.; Garg, S. K. Effect of Tomato Juice Addition on the Growth and Activity of *Lactobacillus acidophilus*. *Int. J. Food Microbiol.* **1992**, *17*(1), 67–70.

15. Barrantes, E.; Tamime, A. Y.; Muir, D. D.; Sword, A. M. The Effect of Substitution of Fat by Microparticulate Whey Protein on the Quality of Set-Type, Natural Yogurt. *Int. J. Dairy Technol.* **1994**, *47*(2), 61–68.

16. Bekar, O.; Yilmaz, Y.; Gulten, M. Kefir Improves the Efficacy and Tolerability of Triple Therapy in Eradicating *Helicobacter pylori*. *J. Med. Food* **2011**, *14*(4), 344–347.

17. Bezerra, M.; Araujo, A.; Santos, K.; Correia, R. Caprine Frozen Yoghurt Produced with Fresh and Spray Dried Jambolan Fruit Pulp (*Eugenia Jambolana Lam*) and *Bifidobacterium animalis* ssp. *lactis* BI-07. *LWT-Food Sci. Technol.* **2015**, *62*(2), 1099–1104.

18. Bhadoria, P. B. S.; Mahapatra, S. C. Prospects, Technological Aspects and Limitations of Probiotics—A Worldwide Review. *Eur. J. Food Res. Rev.* **2011**, *1*(2): 23–42.

19. Bigliardi, B.; Galati, F. Innovation Trends in the Food Industry: The Case of Functional Foods. *Trends Food Sci. Technol.* **2013**, *31*(2), 118–129.

20. Çakmakçi, S.; Çetin, B.; Turgut, T.; Gürses, M.; Erdoğan, A. Probiotic Properties, Sensory Qualities, and Storage Stability of Probiotic Banana Yogurts. *Turk. J. Vet. Anim. Sci.* **2012**, *36*(3), 231–237.

21. Carroll, T. J.; Patel, H. A.; Gonzalez-Martin, M. A.; Dekker, J. W.; Collett, M. A.; Lubbers, M. W. High Pressure Processing of Bioactive Compositions. USSR Patent 8062687 B2, 2011.

22. Cattaneo, T. M. P.; Maraboli, A.; Avitabile, Leva A.; Torreggiani, D., Dehydrofreezing in the Production of Strawberry Ingredients: Influence on the Quality Characteristics of Fruit Yoghurt. *IV International Strawberry Symposium*, 2000, pages 791-794.
23. Chang, B. J.; Park, S. U.; Jang, Y. S.; Ko, S. H.; Joo, N. M.; Kim, S. I.; Kim, C. H.; Chang, D. K. Effect of Functional Yogurt NY-YP901 in Improving the Trait of Metabolic Syndrome. *Eur. J. Clin. Nutr.* **2011**, *65*(11), 1250–1255.
24. Chapman, C. M. C.; Gibson, G. R.; Rowland, I. Health Benefits of Probiotics: Are Mixtures More Effective than Single Strains? *Eur. J. Nutr.* **2011**, *50*(1), 1–17.
25. Charalampopoulos, D.; Wang R.; Pandiella, S. S.; Webb, C. Application of Cereals and Cereal Components in Functional Foods: A Review. *Int. J. Food Microbiol.* **2002**, *79*(1), 131–141.
26. Chen, T.; Tan, Q.; Wang, M.; Xiong, S.; Jiang, S.; Wu, Q.; Li, S.; Luo, C.; Wei, H. Identification of Bacterial Strains in Viili by Molecular Taxonomy and Their Synergistic Effects on Milk Curd and Exopolysaccharides Production. *Afr. J. Biotechnol.* **2011**, *10*(74), 16969–16975.
27. Cho, I. S.; Bae, H. C.; Nam, M. S. Fermentation Properties of Yogurt Added by *Lycii fructus*, *Lycii folium* and *Lycii cortex*. *Korean J. Food Sci. Anim. Resour.* **2003**, *23*, 250–261.
28. Cho, S. S.; O'Sullivan, K.; Rickard, S. Worldwide Dietary Fiber Intake: Recommendations and Actual Consumption Patterns. In *Complex Carbohydrates in Food;* Cho S. S., Prosky L., Dreher M., Eds.; Marcel Dekker: New york, 1999; 71–112.
29. Cogulu, D.; Topaloglu-Ak, A.; Caglar, E.; Sandalli, N.; Karagozlu, C.; Ersin, N.; Yerlikaya, O. Potential Effects of a Multistrain Probiotic-Kefir on Salivary *Streptococcus mutans* and *Lactobacillus* spp. *J. Dent. Sci.* **2010**, *5*(3), 144–149.
30. Coïsson, J. D.; Travaglia, F.; Piana, G.; Capasso, M.; Arlorio, M. *Euterpe oleracea* Juice as a Functional Pigment for Yogurt. *Food Res. Int.* **2005**, *38*(8), 893–897.
31. Craig W. J. Health Promoting Properties of Common Herbs. *Am. J. Clin. Nutr.* **1999**, *70*(3), 491–499.
32. Cruz, A. G.; Cadena, R. S.; Castro, W. F.; Esmerino, E. A.; Rodrigues, J. B.; Gaze, L.; Faria, J. A. F.; Freitas, M. Q.; Deliza, R.; Bolini, H. M. A. Consumer Perception of Probiotic Yogurt: Performance of Check all that Apply (CATA), Projective Mapping, Sorting and Intensity Scale. *Food Res. Int.* **2013**, *54*(1), 601–610.
33. D'Souza, A. L.; Rajkumar, C.; Cooke, J.; Bulpitt, C. J. Probiotics in Prevention of Antibiotic Associated Diarrhoea: Meta-analysis. *Brit. Med. J.* **2002**, *324*(7350), 1361.
34. Daigle, A.; Roy, D.; Belanger, G.; Vuillemard, J. C. Production of Probiotic Cheese Using Enriched Cream Fermented by *Bifidobacterium infantis*. *J. Dairy Sci.* **1999**, *82*(6), 1081–1091.
35. Dave, R. I.; Shah, N. P. Effectiveness of Ascorbic Acid as an Oxygen Scavenger in Improving Viability of Probiotic Bacteria in Yoghurts Made with Commercial Starter Cultures. *Int. Dairy J.* **1997**, *7*(6), 435–443.
36. Dave, R. I.; Shah, N. P. Viability of Yoghurt and Probiotic Bacteria in Yoghurts Made from Commercial Starter Cultures. *Int. Dairy J.* **1997**, *7*(1), 31–41.
37. de Vrese, M.; Kristen, H.; Rautenberg, P.; Laue, C.; Schrezenmeir, J. Probiotic *lactobacilli* and *bifidobacteria* in a Fermented Milk Product with Added Fruit Preparation Reduce Antibiotic Associated Diarrhea and *Helicobacter pylori* activity. *J. Dairy Res.* **2011**, *78*(4), 396–403.

38. de-Vrese, M.; Laue, C.; Offick, B.; Soeth, E.; Repenning, F.; Thob, A.; Schrezenmeir, J. A Combination of Acid Lactase from *Aspergillus oryzae* and Yogurt Bacteria Improves Lactose Digestion in Lactose Maldigesters Synergistically: A Randomized, Controlled, Double-Blind Cross-Over Trial. *Clin. Nutr.* **2015,** *34*(3), 394–399.

39. Desai, S. R.; Toro, V. A.; Joshi, S. V. Utilization of Different Fruits in the Manufacture of Yoghurt. *Indian J. Dairy Sci.* **1994,** *47,* 870–870.

40. Desmond, C.; Fitzgerald, G. F.; Stanton, C.; Ross, R. P. Improved Stress Tolerance of GroESL Over-Producing *Lactococcus lactis* and Probiotic *Lactobacillus paracasei* NFBC 338. *Appl. Environ. Microbiol.* **2004,** *70*(10), 5929–5936.

41. Dinakar, P.; Mistry, V. V. Growth and Viability of *Bifidobacterium bifidum* in Cheddar Cheese. *J. Dairy Sci.* **1994,** *77*(10), 2854–2864.

42. Ejtahed, H. S.; Mohtadi-Nia, J.; Homayouni-Rad, A.; Niafar, M.; Asghari-Jafarabadi, M.; Mofid, V. Probiotic Yogurt Improves Antioxidant Status in Type 2 Diabetic Patients. *Nutrition* **2012,** *28*(5), 539–543.

43. Ejtahed, H. S.; Mohtadi-Nia, J.; Homayouni-Rad, A.; Niafar, M.; Asghari-Jafarabadi, M.; Mofid, V.; Akbarian-Moghari, A. Effect of Probiotic Yogurt Containing *Lactobacillus acidophilus* and *Bifidobacterium lactis* on Lipid Profile in Individuals with Type 2 Diabetes Mellitus. *J. Dairy Sci.* **2011,** *94*(7), 3288–3294.

44. Ekici, K.; Coskun, H.; Tarakci, Z.; Ondul, E.; Sekeroglu, R. The Contribution of Herbs to the Accumulation of Histamine in "Otlu" Cheese. *J. Food Biochem.* **2006,** *30*(3), 362–371.

45. Ekor, M. The Growing Use of Herbal Medicines: Issues Relating to Adverse Reactions and Challenges in Monitoring Safety. *Front. Pharmacol.* **2014,** *4*, 177–182.

46. Esteves, C. L. C.; Lucey, J. A.; Pires, E. M. V. Rheological Properties of Milk Gels Made with Coagulants of Plant Origin and Chymosin. *Int. Dairy J.* **2002,** *12*(5), 427–434.

47. Farahat, A. M.; El-Batawy, O. I. Proteolytic Activity and Some Properties of Stirred Fruit Yoghurt Made Using Some Fruits Containing Proteolytic Enzymes. *World J. Dairy Food Sci.* **2013,** *8*(1), 38–44.

48. Farooq, K.; Haque, Z. U. Effect of Sugar Esters on the Textural Properties of Nonfat Low Calorie Yogurt. *J. Dairy Sci.* **1992,** *75*(10), 2676–2680.

49. Foligné, B.; Parayre, S.; Cheddani, R.; Famelart, M. H.; Madec, M. N.; Plé, C.; Breton, J.; Dewulf, J.; Jan, G.; Deutsch, S.-M. Immunomodulation Properties of Multi-species Fermented Milks. *Food Microbiol.* **2016,** *53,* 60–69.

50. Friques, A. G. F.; Arpini, C. M.; Kalil, I. C.; Gava, A. L.; Leal, M. A.; Porto, M. L.; Nogueira, B. V.; Dias, A. T.; Andrade, T. U.; Pereira, T. M. C. Chronic Administration of the Probiotic Kefir Improves the Endothelial Function in Spontaneously Hypertensive Rats. *J. Transl. Med.* **2015,** *13*(1), 1–10.

51. Gahruie, H. H.; Eskandari, M. H.; Mesbahi, G.; Hanifpour, M. A. Scientific and Technical Aspects of Yogurt Fortification: A Review. *Food Sci. Hum. Wellness* **2015,** *4*(1), 1–8.

52. Ganguly, S.; Sabikhi, L. Fermentation Dynamics of Probiotic *Lactobacillus acidophilus* NCDC-13 in a Composite Dairy-cereal Substrate. *Int. J. Ferment. Foods* **2012,** *1*(1), 33–37.

53. Ganguly, S.; Sathish Kumar, M. H.; Singh, A. K.; Sabikhi, L. Effect of Fermentation by Probiotic *Lactobacillus acidophilus* NCDC 13 on Nutritional Profile of a Dairy-cereal Based Composite Substrate. *J. Food Nutr. Disord.* **2014**, S1–S2.

54. Gaon, D.; Doweck, Y.; Gómez, Z. A.; Ruiz, H. A.; Oliver G. Lactose Digestion by Milk Fermented with *Lactobacillus acidophilus* and *Lactobacillus casei* of Human Origin. *Medicina* **1994**, *55*(3), 237–242.

55. Gaon, D.; Garcia, H.; Winter, L.; Rodriguez, N.; Quintas, R.; Gonzalez, S. N.; Olivier, G. Effect of *Lactobacillus strains* and *Saccharomyces boulardii* on Persistent Diarrhea in Children. *Medicina* **2003**, *63*(4), 293–298.

56. Gerhart, M.; Schottenheimer, M. Mineral Fortification in Dairy. *Wellness Foods* **2013**, 428–429.

57. Ghoneum, M.; Felo, N. Selective Induction of Apoptosis in Human Gastric Cancer Cells by *Lactobacillus kefiri* (PFT), a Novel Kefir Product. *Oncol. Rep.* **2015**, *34*(4), 1659–1666.

58. Gill, H. S.; Rutherfurd, K. J. Probiotic Supplementation to Enhance Natural Immunity in the Elderly: Effects of a Newly Characterized Immunostimulatory Strain *Lactobacillus rhamnosus* HN001 (DR20™) on Leucocyte Phagocytosis. *Nutr. Res.* **2001**, *21*(1), 183–189.

59. Gilliland, S. E.; Kim, H. S. Effect of Viable Starter Culture Bacteria in Yogurt on Lactose Utilization in Humans. *J. Dairy Sci.* **1984**, *67*(1), 1–6.

60. Gobbetti, M.; Corsetti, A.; Smacchi, E.; Zocchetti, A.; De-Angelis, M. Production of Crescenza Cheese by Incorporation of *Bifidobacteria*. *J. Dairy Sci.* **1998**, *81*(1), 37–47.

61. Gomes, A. M. P.; Malcata, F. X.; Klaver, F. A. M.; Grande, H. J. Incorporation and Survival of *Bifidobacterium* sp. Strain Bo and *Lactobacillus acidophilus* Strain Ki in a Cheese Product. *Ned. Melk Zuiveltijdschr.* **1995**, *49*(2–3), 71–95.

62. Gourama, H.; Bullerman, L. B. Effects of Oleuropein on Growth and Aflatoxin Production by *Aspergillus parasiticus*. *Lebens. Wiss. Technol.* **1987**, *20*(5), 226–228.

63. Grewal, R. B.; Chauhan, B. M. Preparation, Organoleptic Acceptability and Nutritional Value of Soy Rabadi—An Indigenous Fermented Food of India. *Int. J. Trop. Agric.* **1993**, *11*(3), 172–177.

64. Gupta, R.; Misra, A.; Pais, P.; Rastogi, P.; Gupta, V. P. Correlation of Regional Cardiovascular Disease Mortality in India with Lifestyle and Nutritional Factors. *Int. J. Cardiol.* **2006**, *108*(3), 291–300.

65. Gupta, V.; Sharma, A.; Nagar, R. Preparation, Acceptability and Nutritive Value of Rabadi—A Fermented Mothbean Food. *J. Food Sci. Technol.* **2007**, *44*(6), 600–601.

66. Han, X.; Zhang, L.; Du, M.; Yi, H.; Li, J.; Zhang, L. Effects of Copper on the Post Acidification of Fermented Milk by *Streptococcus thermophilus*. *J. Food Sci.* **2012**, *77*(1), M25–M28.

67. Hanson, A. L.; Metzger, L. E. Evaluation of Increased Vitamin D Fortification in High-Temperature, Short-Time–Processed 2% Milk, UHT-Processed 2% Fat Chocolate Milk, and Low-Fat Strawberry Yogurt. *J. Dairy Sci.* **2010**, *93*(2), 801–807.

68. Hargrove, R. E.; McDonough F. Process of Making Low-Fat Ripened Skim Milk Cheese. US Patent 3156568, 1964.

69. Harrison, V. C.; Peat, G. Serum cholesterol and bowel flora in the newborn. *Am. J. Clin. Nutr.* **1975**, *28*(12), 1351–1355.

70. Harwalkar, V. R.; Kalab, M. Relationship Between Microstructure and Susceptibility to Syneresis in Yoghurt Made from Reconstituted Nonfat Dry Milk. *Food Struct.* **1986**, *5*(2), 13–15.

71. Hassan, F. A. M.; Helmy, W. A.; Anab, A. K.; Bayoumi, H. M.; Amer, H. Production of Healthy Yoghurt by Using Aqueous Extract of Garlic. *Arab Universities J. Agric. Sci.* **2010**, *18*(1), 171–177.

72. Hayaloglu, A. A.; Fox, P. F. Cheeses of Turkey, III: Varieties Containing Herbs or Spices. *Dairy Sci. Technol.* **2008**, *88*(2), 245–256.

73. Heller, K. J. Probiotic Bacteria in Fermented Foods: Product Characteristics and Starter Organisms. *Am. J. Clin. Nutr.* **2001**, *73*(2), S374–S379.

74. Heller, K. J.; Bockelmann, W.; Schrezenmeir, J. Cheese and its Potential as a Probiotic Food. Chapter 8. In *Handbook of Fermented Functional Foods;* Franworth, E. R., Ed.; CRC Press: USA, 2008; pp 203–226.

75. Hess, S. J.; Roberts, R. F.; Ziegler, G. R. Rheological Properties of Nonfat Yogurt Stabilized Using *Lactobacillus delbrueckii* ssp. *bulgaricus* Producing Exopolysaccharide or Using Commercial Stabilizer Systems. *J. Dairy Sci.* **1997**, *80*(2), 252–263.

76. Ho, J. N.; Choi, J. W.; Lim, W. C.; Kim, M. K.; Lee, I. Y.; Cho, H. Y. Kefir Inhibits 3T3-L1 Adipocyte Differentiation Through Down-regulation of Adipogenic Transcription Factor Expression. *J. Sci. Food Agric.* **2013**, *93*(3), 485–490.

77. Holick, M. F. Sunlight and Vvitamin D for Bone Health and Prevention of Autoimmune Diseases, Cancers, and Cardiovascular Disease. *Am. J. Clin. Nutr.* **2004**, *80*(6), 1678S–1688S.

78. Hossain, M. N.; Fakruddin, M.; Islam, M. N. Quality Comparison and Acceptability of Yoghurt with Different Fruit Juices. *J. Food Process. Technol.* **2012**, *2012*.

79. Howlett, J. F.; Betteridge, V. A.; Champ, M.; Craig, S. A. S.; Meheust, A.; Jones, J. M. The Definition of Dietary Fiber. Discussions at the Ninth Vahouny Fiber Symposium: Building Scientific Agreement. *Food Nutr. Res.* **2010**, *54*.

80. http://www.prnewswire.com/news-releases/global-functional-food-and-nutraceuticals-market-2014-2020-benefits-origin-ingredients-analysis-of-the-168-billion-industry-300037668.html. (accessed 18/5/2015.).

81. Huseini, H. F.; Rahimzadeh, G.; Fazeli, M. R.; Mehrazma, M.; Salehi, M. Evaluation of Wound Healing Activities of Kefir Products. *Burns* **2012**, *38*(5), 719–723.

82. Hussain, S. A.; Garg, F. C.; Pal, D. Effect of Different Preservative Treatments on the Shelf-life of Sorghum Malt Based Fermented Milk Beverage. *J. Food Sci. Technol.* **2014**, *51*(8), 1582–1587.

83. Hussain, S. A.; Panjagari, N. R.; Singh, R. R. B.; Patil, G. R. Potential Herbs and Herbal Nutraceuticals: Food Applications and Their Interactions with Food Components. *Crit. Rev. Food Sci. Nutr.* **2015**, *55*(1), 94–122.

84. Hussain, S. A.; Patil, G. R.; Yadav, V.; Singh, R. R. B. Effect of Storage on Sensory Quality, pH, Wheying-off and Probiotic Count of *lassi* Supplemented with *Aloe barbadensis* Miller Juice. *Indian J. Dairy Sci.* **2015**, *68* 105–110.

85. Hussain, S. A.; Patil, G. R.; Yadav, V.; Singh, R. R. B.; Singh, A. K. Ingredient Formulation Effects on Physico-chemical, Sensory, Textural Properties and Probiotic Count of Aloe vera Probiotic Dahi. *LWT-Food Sci. Technol.* **2016**, *65*, 371–380.

86. HyvÖNen, L. E. A.; Slotte, M. Alternative Sweetening of Yoghurt. *Int. J. Food Sci. Technol.* **1983,** *18*(1), 97–112.

87. Ilic, D. B.; Ashoor, S. H. Stability of Vitamins A and C in Fortified Yogurt. *J. Dairy Sci.* **1988,** *71*(6), 1492–1498.

88. Iraporda, C.; Romanin, D. E.; Rumbo, M.; Garrote, G. L.; Abraham, A. G. The Role of Lactate on the Immunomodulatory Properties of the Nonbacterial Fraction of Kefir. *Food Res. Int.* **2014,** *62,* 247–253.

89. Irvine, S. L.; Hummelen, R.; Hekmat, S. Probiotic Yogurt Consumption may Improve Gastrointestinal Symptoms, Productivity, and Nutritional Intake of People Living with Human Immunodeficiency Virus in Mwanza, Tanzania. *Nutr. Res.* **2011,** *31*(12), 875–881.

90. Ivey, K. L.; Hodgson, J. M.; Kerr, D. A.; Thompson, P. L.; Stojceski, B.; Prince, R. L. The Effect of Yoghurt and its Probiotics on Blood Pressure and Serum Lipid Profile: A Randomised Controlled Trial. *Nutr. Metab. Cardiovasc. Dis.* **2015,** *25*(1), 46–51.

91. Jackson, L. S.; Lee, K. Micro-Encapsulated Iron for Food Fortification. *J. Food Sci.* **1991,** *56*(4), 1047–1050.

92. Jain, S.; Yadav, H.; Sinha, P. R. Probiotic Dahi Containing *Lactobacillus casei* Protects Against *Salmonella enteritidis* Infection and Modulates Immune Response in Mice. *J. Med. Food* **2009,** *12*(3), 576–583.

93. Jain, S.; Yadav, H.; Sinha, P. R.; Kapila, S.; Naito, Y.; Marotta, F. Anti-allergic Effects of Probiotic Dahi Through Modulation of the Gut Immune System. *Turk. J. Gastroenterol.* **2010,** *21*(3), 244–250.

94. Jain, S.; Yadav, H.; Sinha, P. R.; Naito, Y.; Marotta, F. Dahi Containing Probiotic *Lactobacillus acidophilus* and *Lactobacillus casei* has a Protective Effect Against *Salmonella enteritidis* Infection in Mice. *Int. J. Immunopathol. Pharmacol.* **2008,** *21*(4), 1021–1029.

95. Janer, C.; Peláez, C.; Requena, T. Caseinomacropeptide and Whey Protein Concentrate Enhance *Bifidobacterium lactis* Growth in Milk. *Food Chem.* **2004,** *86*(2), 263–267.

96. Jarvis, B. Mold and Mycotoxins in Moldy Cheeses. *Microbiol. Aliments Nutr.* **1983,** *1,* 187–191.

97. Jayamanne, V. S.; Adams, M. R. Survival of Probiotic *Bifidobacteria* in Buffalo Curd and Their Effect on Sensory Properties. *Int. J. Food Sci. Technol.* **2004,** *39*(7), 719–725.

98. Jayasinghe, O.; Fernando, S.; Jayamanne, V.; Hettiarachchi, D. Production of a Novel Fruit-Yoghurt Using Dragon Fruit (*Hylocereus Undatus* L.). *Eur. Sci. J.* **2015,** *11*(3), 208–215.

99. Johnson, M. E.; Chen, C. M. Technology of Manufacturing Reduced-Fat Cheddar Cheese. Chapter 21, In *Chemistry of Structure–Function Relationships in Cheese;* Malin, E. L., Tunick, M. H., Eds.; Springer: New York, 1995; pp 331–337.

100. Jones, M. L.; Martoni, C. J.; Tamber, S.; Parent, M.; Prakash, S. Evaluation of Safety and Tolerance of Microencapsulated *Lactobacillus reuteri* NCIMB 30242 in a Yogurt Formulation: A Randomized, Placebo-Controlled, Double-Blind Study. *Food Chem. Toxicol.* **2012,** *50*(6), 2216–2223.

101. Kailasapathy, K.; Harmstorf, I.; Phillips, M. Survival of *Lactobacillus acidophilus* and *Bifidobacterium animalis* ssp. *lactis* in Stirred Fruit Yogurts. *LWT–Food Sci. Technol.* **2008,** *41*(7), 1317–1322.

102. Kalab, M.; Allan-Wojtas, P.; Phipps-Todd, B. E. Development of Microstructure in Set-style Nonfat Yogurt—A Review. *Food Struct.* **1983**, *2*(1), 7–10.
103. Katsiari, M. C.; Voutsinas, L. P. Manufacture of Low-fat Feta Cheese. *Food Chem.* **1994**, *49*(1), 53–60.
104. Kaushal, D.; Kansal, V. K. Probiotic Dahi Containing *Lactobacillus acidophilus* and *Bifidobacterium bifidum* Alleviates Age-Inflicted Oxidative Stress and Improves Expression of Biomarkers of Ageing in Mice. *Molecularbiol. Rep.* **2012**, *39*(2), 1791–1799.
105. Kavas, G.; Oysun, G.; Kinik, O.; Uysal, H. Effect of Some Fat Replacers on Chemical, Physical and Sensory Attributes of Low-Fat White Pickled Cheese. *Food Chem.* **2004**, *88*(3), 381–388.
106. Kim, S. J.; Ahn, J.; Seok, J. S.; Kwak, H. S. Microencapsulated Iron for Drink Yogurt Fortification. *Asian Australasian J. Anim. Sci.* **2003**, *16*(4), 581–587.
107. Koca, N.; Metin, M. Textural, Melting and Sensory Properties of Low-Fat Fresh Kashar Cheeses Produced by Using Fat Replacers. *Int. Dairy J.* **2004**, *14*(4), 365–373.
108. Köksoy, A.; Kılıç, M. Effects of Water and Salt Level on Rheological Properties of Ayran, a Turkish Yoghurt Drink. *Int. Dairy J.* **2003**, *13*(10), 835–839.
109. Kumar, M.; Mital, B. K.; Garg, S. K. Effect of Papaya Pulp Addition on the Growth of *Lactobacillus acidophilus*. *J. Food Saf.* **1989**, *10*(1), 63–73.
110. Landge, U. B.; Pawar, B. K.; Choudhari, D. M. Preparation of Shrikhand Using Ashwagandha Powder as Additive. *J. Dairying Foods Home Sci.* **2011**, *30*(2), 1–8.
111. Lee, J. H.; Yoon, Y. H. Characteristics of Aloe Vera Supplemented Liquid Yoghurt Inoculated with *Lactobacillus casei* YIT 9018. *Korean J. Anim. Sci.* **1997**, *39*(1), 93–100.
112. Leporanta, K. Viili and Långfil–Exotic Fermented Products from Scandinavia. Valio Foods & Functionals, 2003. *http://www.valio. fi/portal/page/portal/valiocom/Valio_ Today/Publications/valio_foods_functionals05102006130335/2003.* Accessed on August 1, 2017.
113. Leroy, F.; De Vuyst, L. Lactic Acid Bacteria as Functional Starter Cultures for the Food Fermentation Industry. *Trends Food Sci. Technol.* **2004**, *15*(2), 67–78.
114. Levy, J. The Effects of Antibiotic Use on Gastrointestinal Function. *Am. J. Gastroenterol.* **2000**, *95*(1), S8–S10.
115. Liu, Z.-m.; Xu, Z.-y.; Han, M.; Guo, B.-H. Efficacy of Pasteurised Yoghurt in Improving Chronic Constipation: A Randomised, Double-Blind, Placebo-Controlled Trial. *Int. Dairy J.* **2015**, *40*, 1–5.
116. Lourens-Hattingh, A.; Viljoen, B. C. Yogurt as Probiotic Carrier Food. *Int. Dairy J.* **2001**, *11*(1), 1–17.
117. Lutchmedial, M.; Ramlal, R.; Badrie, N.; Chang-Yen I. Nutritional and Sensory Quality of Stirred Soursop (*Annona muricata* L.) Yoghurt. *Int. J. Food Sci. Nutr.* **2004**, *55*(5), 407–414.
118. Makino, S.; Ikegami, S.; Kume, A.; Horiuchi, H.; Sasaki, H.; Orii, N. Reducing the Risk of Infection in the Elderly by Dietary Intake of Yoghurt Fermented with *Lactobacillus delbrueckii* ssp. *bulgaricus* OLL1073R-1. *Brit. J. Nutr.* **2010**, *104*(07), 998–1006.
119. Malpeli, A.; Taranto, M. P.; Cravero, R. C.; Tavella, M.; Fasano, V.; Vicentin, D.; Ferrari, G.; Magrini, G.; Hébert, E.; deValdez, G. F. Effect of Daily Consumption of

Lactobacillus reuteri CRL 1098 on Cholesterol Reduction in Hypercholesterolemic Subjects. *Food Nutr. Sci.* **2015,** *6*(17), 1583–1585.

120. Mann, G. V.; Spoerry, A. Studies of a Surfactant and Cholesteremia in the Maasai. *Am. J. Clin. Nutr.* **1974,** *27*(5), 464–469.

121. Maragkoudakis, P. A.; Zoumpopoulou, G.; Miaris, C.; Kalantzopoulos, G.; Pot, B.; Tsakalidou, E. Probiotic Potential of *Lactobacillus* Strains Isolated from Dairy Products. *Int. Dairy J.* **2006,** *16*(3), 189–199.

122. Martinez-Villaluenga, C.; Frías, J.; Gómez, R.; Vidal-Valverde, C. Influence of Addition of Raffinose Family Oligosaccharides on Probiotic Survival in Fermented Milk During Refrigerated Storage. *Int. Dairy J.* **2006,** *16*(7), 768–774.

123. McGregor, J. U.; White,C. H. Effect of Enzyme Treatment and Ultrafiltration on the Quality of Lowfat Cheddar Cheese. *J. Dairy Sci.* **1990,** *73*(3), 571–578.

124. Merenstein, D.; Murphy, M.; Fokar, A.; Hernandez, R. K.; Park, H.; Nsouli, H.; Sanders, M. E.; Davis, B. A.; Niborski, V.; Tondu, F. Use of a Fermented Dairy Probiotic Drink Containing *Lactobacillus casei* (DN-114 001) to Decrease the Rate of Illness in Kids: The Drink Study, a Patient-Oriented, Double-Blind, Cluster-Randomized, Placebo-Controlled, Clinical Trial. *Eur. J. Clin. Nutr.* **2010,** *64*(7), 669–677.

125. Metchnikoff, E. *The Prolongation of Life;* Putnam ,1908, pp 98.

126. Miller, C. W.; Nguyen, M. H.; Rooney, M.; Kailasapathy, K. The Influence of Packaging Materials on the Dissolved Oxygen Content of Probiotic Yoghurt. *Packaging Technol. Sci.* **2002,** *15*(3), 133–138.

127. Misra, A. K.; Kuila, R. K. Use of *Bifidobacterium bifidum* for the Manufacture of Bio-Yoghurt and Fruit Bloyoghurt. *Indian J. Dairy Sci.* **1994,** *47,* 192–198.

128. Mistry, V. V. Low Fat Cheese Technology. *Int. Dairy J.* **2001,** *11*(4), 413–422.

129. Modler, H. W.; Kalab, M. Microstructure of Yogurt Stabilized with Milk Proteins. *J. Dairy Sci.* **1983,** *66*(3), 430–437.

130. Mohania, D.; Kansal, V. K.; Shah, D.; Nagpal, R.; Kumar, M.; Gautam, S. K.; Singh, B.; Behare, P. V. Therapeutic Effect of Probiotic Dahi on Plasma, Aortic, and Hepatic Lipid Profile of Hypercholesterolemic Rats. *J. Cardiovasc. Pharmacol. Ther.* **2013,** *8*(5), 490–497.

131. Moises, *Lactobacillus casei* and *Lactobacillus acidophilus* in the treatment of bladder superficial tumors: follow-up during 36 months. 2012, pages 12.

132. Moreau, R. A.; Whitaker, B. D.; Hicks, K. B. Phytosterols, Phytostanols, and Their Conjugates in Foods: Structural Diversity, Quantitative Analysis, and Health-Promoting Uses. *Prog. Lipid Res.* **2002,** *41*(6), 457–500.

133. Mugocha, P. T.; Taylor, J. R. N.; Bester, B. H. Fermentation of a Composite Finger Millet-Dairy Beverage. *World J. Microbiol. Biotechnol.* **2000,** *16*(4), 341–344.

134. Nabavi, S.; Rafraf, M.; Somi, M. H.; Homayouni-Rad, A.; Asghari-Jafarabadi, M. Effects of Probiotic Yogurt Consumption on Metabolic Factors in Individuals with Nonalcoholic Fatty Liver Disease. *J. Dairy Sci.* **2014,** *97*(12), 7386–7393.

135. Najgebauer-Lejko, D.; Grega, T.; Tabaszewska, M. Yoghurts with Addition of Selected Vegetables: Acidity, Antioxidant Properties and Sensory Quality. *Acta Scientiarum Polonorum Technologia Alimentaria* **2014,** *13*(1), 35–42.

136. Ocak, E.; Rajendram, R., Fortification of Milk with Mineral Elements. Chapter 17. In *Handbook of Food Fortification and Health;* Preedy, V. R., Srirajaskanthan, R., Patel, V. B. Eds.; Springer: New York, 2013; pp 213–224.

137. Oliveira, A.; Alexandre, E. M. C.; Coelho, M.; Lopes, C.; Almeida, D. P. F.; Pintado, M. Incorporation of Strawberries Preparation in Yoghurt: Impact on Phytochemicals and Milk Proteins. *Food chem.* **2015**, *171*, 370–378.

138. Ong, L.; Henriksson, A.; Shah, N. P. Development of Probiotic Cheddar Cheese Containing *Lactobacillus acidophilus, Lb. casei, Lb. paracasei* and *Bifidobacterium* spp. and the Influence of These Bacteria on Proteolytic Patterns and Production of Organic Acid. *Int. Dairy J.* **2006**, *16*(5), 446–456.

139. Ostadrahimi, A.; Taghizadeh, A.; Mobasseri, M.; Farrin, N.; Payahoo, L.; Gheshlaghi, Z. B.; Vahedjabbari, M. Effect of Probiotic Fermented Milk (Kefir) on Glycemic Control and Lipid Profile in Type 2 Diabetic Patients: A Randomized Double-Blind Placebo-Controlled Clinical Trial. *Iran. J. Public Health* **2015**, *44*(2), 228–230.

140. Østlie, H. M.; Helland, M. H.; Narvhus, J. A. Growth and Metabolism of Selected Strains of Probiotic Bacteria in Milk. *Int. J. Food Microbiol.* **2003**, *87*(1), 17–27.

141. Ostlund, R. E. Phytosterols in Human Nutrition. *Ann. Rev. Nutr.* **2002**, *22*(1), 533–549.

142. Özer, D.; Akin, S.; Özer, B. Effect of Inulin and Lactulose on Survival of *Lactobacillus acidophilus* LA-5 and *Bifidobacterium bifidum* BB-02 in Acidophilus-Bifidus yoghurt. *Food Sci. Technol. Int.* **2005**, *11*(1), 19–24.

143. Panesar, P. S.; Shinde, C. Effect of Storage on Syneresis, pH, *Lactobacillus acidophilus* Count, *Bifidobacterium bifidum* Count of Aloe Vera Fortified Probiotic Yoghurt. *Curr. Res. Dairy Sci.* **2012**, *4*(1), 17–23.

144. Parvez, S.; Malik, K. A.; Ah Kang, S.; Kim, H. Y. Probiotics and Their Fermented Food Products are Beneficial for Health. *J. Appl. Microbiol.* **2006**, *100*(6), 1171–1185.

145. Pawan, R.; Bhatia, A. Systemic Immunomodulation and Hypocholesteraemia by Dietary Probiotics: A Clinical Study. *J. Clin. Diagnostic Res.* **2007**, *6*, 467–475.

146. Peng, D.; YaDong, Z.; Ying, M.; Yuan, G.; Dong, L. Production and Function of Dealcohol Yogurt Beverage. *Chin. Dairy Ind.* **2010**, *38*(1), 26–28.

147. Pescuma, M.; Hébert, E. M.; Mozzi, F.; De Valdez, G. F. Functional Fermented Whey-Based Beverage Using Lactic Acid Bacteria. *Int. J. Food Microbiol.* **2010**, *141*(1), 73–81.

148. Pimentel, T. C.; Garcia, S.; Prudencio, S. H. Probiotic Yoghurt with Inulin-Type Fructans of Different Degrees of Polymerization: Physicochemical and Microbiological Characteristics and Storage Stability. *Ciências Agrárias* **2012**, *33*(3), 1059–1069.

149. Pinheiro, M. V. S.; Castro, L. P.; Hoffmann, F. L.; Penna, A. L. B. Estudo comparativo de edulcorantes em iogurtes probióticos. *Revista do Instituto de Laticínios Cândido Tostes* **2002**, *57*, 142–145.

150. Pratap, Y. S. M.; Chandra, R.; Shukla, S.; Ali, M. N. Optimization of the Chemical Properties of Frozen Yoghurt Supplemented with Different Fruit Pulp. *Pharm. Innovation J.* **2015**, *4*(2), 56–58.

151. Pszczola, D. E. The ABCs of Nutraceutical Ingredients. *Food Technol.* **1998**, *52*(3), 30–37.

152. Quilez, J.; Garcia-Lorda, P.; Salas-Salvado, J. Potential Uses and Benefits of Phytosterols in Diet: Present Situation and Future Directions. *Clin. Nutr.* **2003**, *22*(4), 343–351.

153. Raju, P. N.; Pal, D. Effect of Bulking Agents on the Quality of Artificially Sweetened Misti Dahi (Caramel Colored Sweetened Yoghurt) Prepared from Reduced Fat Buffalo Milk. *LWT Food Sci. Technol.* **2011**, *44*(9), 1835–1843.

154. Raju, P. N.; Pal, D. Effect of Dietary Fibers on Physico-chemical, Sensory and Textural Properties of Misti Dahi. *J. Food Sci. Technol.* **2014,** *51*(11), 3124–3133.

155. Rather, S. A.; Pothuraju, R.; Sharma, R. K.; De, S.; Mir, N. A.; Jangra, S. Anti-obesity Effect of Feeding Probiotic Dahi Containing *Lactobacillus casei* NCDC 19 in High Fat Diet-Induced Obese Mice. *Int. J. Dairy Technol.* **2014,** *67*(4), 504–509.

156. Rattray, F. P.; O'Connell, M. J. Fermented Milks: Kefir, In: *Fuquay, J.W.; Patrick, F.F.; Paul, L.H.M. (Eds.), Encyclopedia of Dairy Sciences,* Academic Press, USA, 2011, pages 518–524.

157. Raval, D. M.; Mistry, V. V. Application of Ultrafiltered Sweet Buttermilk in the Manufacture of Reduced Fat Process Cheese. *J. Dairy Sci.* **1999,** *82*(11), 2334–2343.

158. Reps, A.; Warminska-Radyko, I.; Krzyzewska, A.; Tomasik, J. Effect of High Pressure on *Streptococcus salivarius* subsp. *thermophilus. Milchwissenschaft* **2001,** *56*(3), 131–133.

159. Richelle, M.; Enslen, M.; Hager, C.; Groux, M.; Tavazzi, I.; Godin, J.-P.; Berger, A.; Métairon, S.; Quaile, S.; Piguet-Welsch, C. Both Free and Esterified Plant Sterols Reduce Cholesterol Absorption and the Bioavailability of β-Carotene and α-Tocopherol in Normocholesterolemic Humans. *Am. J. Clin. Nutr.* **2004,** *80*(1), 171–177.

160. Rizzardini, G.; Eskesen, D.; Calder, P. C.; Capetti, A.; Jespersen, L.; Clerici, M. Evaluation of the Immune Benefits of Two Probiotic Strains *Bifidobacterium animalis* ssp. *lactis, BB*-12® and *Lactobacillus paracasei* ssp. *paracasei, L. casei* 43® in an Influenza Vaccination Model: A Randomised, Double-Blind, Placebo-Controlled Study. *Brit. J. Nutr.* **2012,** *107*(06), 876–884.

161. Romeih, E. A.; Michaelidou, A.; Biliaderis, C. G.; Zerfiridis, G. K. Low-Fat White-Brined Cheese Made from Bovine Milk and Two Commercial Fat Mimetics: Chemical, Physical and Sensory Attributes. *Int. Dairy J.* **2002,** *12*(6), 525-540.

162. Rosenthal, I.; Rosen, B.; Bernstein, S. Phenols in Milk. Evaluation of Ferulic Acid and Other Phenols as Antifungal Agents. *Milchwissenschaft* **1997,** *52*(3), 134–138.

163. Roy, D. Technological Aspects Related to the Use of *Bifidobacteria* in Dairy Products. *Le lait* **2005,** *85*(1–2), 39–56.

164. Rudan, M. A.; Barbano, D. M.; Kindstedt, P. S. Effect of Fat Replacer (Salatrim®) on Chemical Composition, Proteolysis, Functionality, Appearance, and Yield of Reduced Fat Mozzarella Cheese. *J. Dairy Sci.* **1998,** *81*(8), 2077–2088.

165. Saarela, M.; Rantala, M.; Hallamaa, K.; Nohynek, L.; Virkajärvi, I.; Mättö, J. Stationary-Phase Acid and Heat Treatments for Improvement of the Viability of Probiotic *Lactobacilli* and *Bifidobacteria. J. Appl. Microbiol.* **2004,** *96*(6), 1205–1214.

166. Saarela, M. H.; Alakomi, H. L.; Puhakka, A.; Mättö, J. Effect of the Fermentation pH on the Storage Stability of *Lactobacillus rhamnosus* Preparations and Suitability of in vitro Analyses of Cell Physiological Functions to Predict it. *J. Appl. Microbiol.* **2009,** *106*(4), 1204–1212.

167. Sabikhi, L.; Mathur, B. N. Selection of a Suitable Strain of *Bifidobacterium bifidum* for the Manufacture of Probiotic Edam Cheese. *Indian J. Dairy Sci.* **2000,** *53*(2), 112–122.

168. Sadeghi, E.; Akhondzadeh, B. A.; Misaghi, A.; Zahraii, S. T.; Bohlouli, O. S. Evaluation of Effects of *Cuminum cyminum* and Probiotic on *Staphylococcus aureus* in Feta Cheese. *J. Med. Plants* **2010,** *2*(34), 131–141.

169. Sadler, A. M.; Lacroix, D. E.; Alford, J. A. Iron Content of Baker's and Cottage Cheese Made from Fortified Skim Milks. *J. Dairy Sci.* **1973**, *56*(10), 1267–1270.

170. Sadrzadeh-Yeganeh, H.; Elmadfa, I.; Djazayery, A.; Jalali, M.; Heshmat, R.; Chamary, M. The Effects of Probiotic and Conventional Yoghurt on Lipid Profile in Women. *Brit. J. Nutr.* **2010**, *103*(12), 1778–1783.

171. Salwa, A. A.; Galal, E. A.; Neimat, A. E. Carrot Yoghurt: Sensory, Chemical, Microbiological Properties and Consumer Acceptance. *Pak. J. Nutr.* **2004**, *3*(6), 322–330.

172. Sanchez-Segarra, P. J.; García-Martínez, M.; Gordillo-Otero, M. J.; Díaz-Valverde, A.; Amaro-Lopez, M. A.; Moreno-Rojas, R. Influence of the Addition of Fruit on the Mineral Content of Yoghurts: Nutritional Assessment. *Food Chem.* **2000**, *71*(1), 85–89.

173. Sawant, P.; Kumar, D.; Patil, V.; Ingale, Y.; Sawant, D. Physico-chemical, Sensory and Microbial Quality of Yoghurt Drink Fortified with Pineapple Pulp. *Int. J. Food Ferment. Technol.* **2015**, *5*(1), 59–61.

174. Saxena, S. N.; Mital, B. K.; Garg, S. K. Effect of Casitone and Fructose on the Growth of *Lactobacillus acidophilus* and its Survival During Storage. *Int. J. Food Microbiol.* **1994**, *21*(3), 271–276.

175. Selvamuthukumaran, M.; Farhath, K. Evaluation of Shelf Stability of Antioxidant Rich Seabuckthorn Fruit Yoghurt. *Int. Food Res. J.* **2014**, *21*(2), 759–765.

176. Senok, A. C.; Ismaeel, A. Y.; Botta, G. A. Probiotics: Facts and Myths. *Clin. Microbiol. Infect.* **2005**, *11*(12), 958–966.

177. Seydim, Z. B. G.; Sarikus, G.; Okur, O. D. Effect of Inulin and Dairy-Lo® as Fat Replacers on the Quality of Set Type Yogurt. *Milchwissenschaft* **2005**, *60*(1), 51–55.

178. Shah, N. P. Probiotic Bacteria: Selective Enumeration and Survival in Dairy Foods. *J. Dairy Sci.* **2000**, *83*(4), 894–907.

179. Shin, H. S.; Lee, J. H.; Pestka, J. J.; Ustunol, Z. Viability of *Bifidobacteria* in Commercial Dairy Products During Refrigerated Storage. *J. Food Prot.* **2000**, *63*(3), 327–331.

180. Singh, R.; Kumar, R.; Venkateshappa, R.; Mann, B.; Tomar, S. K. Studies on Physicochemical and Antioxidant Properties of Strawberry Polyphenol Extract-Fortified Stirred Dahi. *Int. J. Dairy Technol.* **2013**, *66*(1), 103–108.

181. Smith-Palmer, A.; Stewart, J.; Fyfe, L. The Potential Application of Plant Essential Oils as Natural Food Preservatives in Soft Cheese. *Food Microbiol.* **2001**, *18*(4), 463–470.

182. Songisepp, E.; Kullisaar, T.; Hütt, P.; Elias, P.; Brilene, T.; Zilmer, M.; Mikelsaar, M. A New Probiotic Cheese with Antioxidative and Antimicrobial Activity. *J. Dairy Sci.* **2004**, *87*(7), 2017–2023.

183. Srinivas, D.; Mital, B. K.; Garg, S. K. Utilization of Sugars by *Lactobacillus acidophilus* Strains. *Int. J. Food Microbiol.* **1990**, *10*(1), 51–57.

184. St-Gelais, D.; Roy, D.; Audet, P. Manufacture and Composition of Low Fat Cheddar Cheese from Milk Enriched with Different Protein Concentrate Powders. *Food Res. Int.* **1998**, *31*(2), 137–145.

185. Stankiewicz, J.; Steinka, I. Influence of Cheese Curds Supplementation with Aloes on Maintenance of *Staphylococcus* Population. *Pol. J. Food Nutr. Sci.* **2007**, *57*(4), 503–506.

186. Stanton, C.; Gardiner, G.; Lynch, P. B.; Collins, J. K.; Fitzgerald, G.; Ross, R. P. Probiotic Cheese. *Int. Dairy J.* **1998**, *8*(5), 491–496.

187. Streiff, P. J.; Hoyda, D. L.; Epstein, E. Process for the Production of Low Calorie Yogurt. US Patent 4956186, 1990, pages 14.

188. Strezynski, G. J. Process of Making Low-Fat Ripened Skim Milk Cheese. US Patent 3156568, 1964, pages 14.

189. Tamime, A. Y.; Deeth, H. C. Yogurt: Technology and Biochemistry. *J. Food Prot.* **1980,** *43*(12), 939–977.

190. Tamime, A. Y.; Kalab, M.; Muir, D. D.; Barrantes, E. The Microstructure of Set-Style, Natural Yogurt Made by Substituting Microparticulate Whey Protein for Milk Fat. *Int. J. Dairy Technol.* **1995,** *48*(4), 107–111.

191. Tarakçi, Z.; Coskun, H.; Tuncturk, Y. Some Properties of Fresh and Ripened Herby Cheese, a Traditional Variety Produced in Turkey. *Food Technol. Biotechnol.* **2004,** *42*(1), 47–50.

192. Tarakci, Z.; Temiz, H.; Ugur, A. The Effect of Adding Herbs to Labneh on Physicochemical and Organoleptic Quality During Storage. *Int. J. Dairy Technol.* **2011,** *64*(1), 108–116.

193. Tatsadjieu, N. L.; Etoa, F. X.; Mbofung, C. M. F. Drying Kinetics, Physicochemical and Nutritional Characteristics of *Kindimu*, a Fermented Milk-Based-Sorghum-Flour. *J. Food Technol. Afr.* **2004,** *9*(1), 17–22.

194. Tharmaraj, N.; Shah, N. P. Selective Enumeration of *Lactobacillus delbrueckii ssp. bulgaricus, Streptococcus thermophilus, Lactobacillus acidophilus, bifidobacteria, Lactobacillus casei, Lactobacillus rhamnosus,* and *Propionibacteria. J. Dairy Sci.* **2003,** *86*(7), 2288–2296.

195. Tillisch, K.; Labus, J.; Kilpatrick, L.; Jiang, Z.; Stains, J.; Ebrat, B.; Guyonnet, D.; Legrain–Raspaud, S.; Trotin, B.; Naliboff, B. Consumption of Fermented Milk Product with Probiotic Modulates Brain Activity. *Gastroenterology* **2013,** *144*(7), 1394–1401.

196. Tsiraki, M. I.; Savvaidis, I. N. Effect of Packaging and Basil Essential Oil on the Quality Characteristics of Whey Cheese *Anthotyros. Food Bioprocess Technol.* **2013,** *6*(1), 124–132.

197. Turkish Food Codex, *Communiqué on Fermented Milk.* Communication No. 2001/21, 2008, pages 21.

198. Upreti, P.; Mistry, V. V.; Warthesen, J. J. Estimation and Fortification of Vitamin D3 in Pasteurized Process Cheese. *J. Dairy Sci.* **2002,** *85*(12), 3173–3181.

199. Vahedi, N.; Tehrani, M. M.; Shahidi, F. Optimizing of Fruit Yoghurt Formulation and Evaluating its Quality During Storage. *Am.-Eurasian J. Agric. Environ. Sci.* **2008,** *3*, 922–927.

200. Villena, J.; Racedo, S.; Agüero, G.; Alvarez, S. Yoghurt Accelerates the Recovery of Defence Mechanisms Against *Streptococcus pneumoniae* in Protein-Malnourished Mice. *Brit. J. Nutr.* **2006,** *95*(3), 591–602.

201. Vimala, Y.; Kumar, P. D. Some Aspects of Probiotics. *Indian J. Microbiol.* **2006,** *46*(1), 1–7.

202. Vinderola, C. G.; Mocchiutti, P.; Reinheimer, J. A. Interactions Among Lactic Acid Starter and Probiotic Bacteria Used for Fermented Dairy Products. *J. Dairy Sci.* **2002,** *85*(4), 721–729.

203. Vlieger, A. M.; Robroch, A.; van Buuren, S.; Kiers, J.; Rijkers, G.; Benninga, M. A.; te Biesebeke, R. Tolerance and Safety of *Lactobacillus paracasei* ssp. *paracasei*

in Combination with *Bifidobacterium animalis* ssp. *lactis* in a Prebiotic Containing Infant Formula: A Randomized Controlled Trial. *Brit. J. Nutr.* **2009,** *102*(6), 869–875.

204. Wagner, C. L.; Greer, F. R. Prevention of Rickets and Vitamin D Deficiency in Infants, Children, and Adolescents. *Pediatrics* **2008,** *122*(5), 1142–1152.

205. Walkunde, T. R.; Kamble, D. K.; Pawar, B. K. Sensory Quality of Yoghurt from Cow Milk by Utilizing Guava Fruit. *Asian J. Anim. Sci.* **2009,** *3*(2), 99–102.

206. Wang, J.; Chen, X.; Liu, W.; Yang, M.; Zhang, H. Identification of *Lactobacillus* from Koumiss by Conventional and Molecular Methods. *Eur. Food Res. Technol.* **2008,** *227*(5), 1555–1561.

207. Wang, S. Y.; Chen, H. C.; Liu, J. R.; Lin, Y. C.; Chen, M.-J. Identification of Yeasts and Evaluation of Their Distribution in Taiwanese Kefir and Viili Starters. *J. Dairy Sci.* **2008,** *91*(10), 3798–3805.

208. WHO/FAO. Evaluation of Health and Nutritional Properties of Powder Milk and Live Lactic Acid Bacteria. *Food and Agriculture Organization of the United Nations and World Health Organization Expert Consultation Report,* 2001, 1–34.

209. Woestyne, M. V.; Bruyneel, B.; Mergeay, M.; Verstraete, W. The Fe^{2+} Chelator Proferrorosamine A is Essential for the Siderophore-Mediated Uptake of Iron by *Pseudomonas roseus fluorescens. Appl. Environ. Microbiol.* **1991,** *57*(4), 949–954.

210. Yaakob, H.; Ahmed, N. R.; Daud, S. K.; Malek, R. A.; Rahman R. A. Optimization of Ingredient and Processing Levels for the Production of Coconut Yogurt Using Response Surface Methodology. *Food Sci. Biotechnol.* **2012,** *21*(4), 933–940.

211. Yadav, H.; Jain, S.; Sinha, P. R. Antidiabetic Effect of Probiotic Dahi Containing *Lactobacillus acidophilus* and *Lactobacillus casei* in High Fructose Fed Rats. *Nutrition* **2007,** *23*(1), 62–68.

212. Yildiz, F. Overview of yogurt and other fermented dairy products. Chapter 1. *In: Yildiz, F. (Ed.), Development and Manufacture of Yogurt and Other Functional Dairy Products.* CRC Press, USA, 2016, pages 1–46.

213. Yildiz, F. Microbiology and Technology. Chapter 5. In *Development and Manufacture of Yogurt and Other Functional Dairy Products; Yildiz, F., Ed.;* CRC Press: USA, 2016; pp 143–164.

214. Zalazar, C. A.; Zalazar, C. S.; Bernal, S.; Bertola, N.; Bevilacqua, A.; Zaritzky, N. Effect of Moisture Level and Fat Replacer on Physicochemical, Rheological and Sensory Properties of Low Fat Soft Cheeses. *Int. Dairy J.* **2002,** *12*(1), 45–50.

PART III

Issues, Challenges, and Specialty Topics in Food Science

CHAPTER 8

WHOLE WHEAT FLOUR: STABILITY ISSUES AND CHALLENGES

NAVNEET SINGH DEORA

Applied Science Department CP-12A, Nestle R&D Centre India Pvt. Ltd., Sector-8, IMT Manesar, Gurgaon 122050, India, Tel.: +91-7042307007, E-mail: navneet.deora@rd.nestle.com

CONTENTS

8.1 INTRODUCTION

In the current scenario, millers and food Industries face many challenges in relation to the whole wheat flour (WWF) as well as their derived products. Among the miller and food manufacturers, one of the major concerns is with respect to the limited shelf-life of flour as compared to the refined flour.[8] In the tropical countries, producing the stable WWF is more challenging due to high storage temperatures and long travel distances from the mill to food-manufacturing plants, which tend to accelerate the lipid degradation resulting into rancidity issues.[22]

The purpose of this chapter is to address the key issues in the milling, shelf-life, key biomarkers of WWF, and finally to discuss the strategies to overcome the existing challenges and issues.

8.2 WWF MILLING: ISSUES AND CHALLENGES

The AACC International has defined WWF as being prepared from wheat (other than durum) so that the proportions of the intact grain, the bran, germ, and endosperm remain unaltered. WWF contains substantially more vitamins, minerals, antioxidants, and other nutrients than regular wheat flour, since these compounds are concentrated in the outer portions of the grain. Some of these nutrients are replaced in the enrichment process of wheat flour, which is mandatory in 64 countries worldwide,[30] although many nutritional components are still lower, especially minerals and dietary fiber.[12]

WWF possesses several unique challenges to the milling and related industries. For instance, whereas milling procedures for traditional flours have been well-established, whole grain flours (WGFs) are produced by a variety of techniques and result in flours with widely different particle sizes and functionalities.[3] Furthermore, WWF contains more enzymatic activity,[54] lipids,[24,64] and antioxidants[53] than wheat flour, which can affect end-use[39] and storage properties.[6,8,66]

8.3 RESEARCH STUDIES ON THE WW DEGRADATION

In one of the past works, the effect of storage temperature (4, 20, and 30°C) on WGF storage for up to 12 months was studied.[57] Authors found significantly higher fat acidity caused by (partial) hydrolysis of triacylglycerols (TAG) yielding free fatty acids (FFAs) in samples stored at higher temperature as compared to samples stored at 4°C.

Within the context of WWF storage, the presence of nonpolar lipids is one of the important factors to consider.[22] Nonpolar lipids predominantly occur in the germ and aleurone tissues of the wheat kernel and consist of FFA, monoacylglycerols (MAG), diacylglycerols (DAGs), and, mainly, TAG.[45] Wheat milling transfers approximately 50% of the TAG of germ to flour.[42] During WGF storage, lipase (LA; EC 3.1.1.3) hydrolyzes part of the TAG lipids,[12] yielding DAG, MAG, and FFA. The latter can be responsible for what is referred to as acid rancidity. Acid rancidity can decrease WGF functional properties and affect the product such as noodles leading to onset of rancidity. In addition, FFA occurring as such are more susceptible to enzymatic oxidative conversion than when they are esterified to glycerol and, hence, present as part of TAG.[11] Released polyunsaturated FFA can be oxidized by wheat lipoxygenase (LOX; EC 1.13.11.12), which is mainly located in the germ and bran. It has been observed that the oxidation of lipid during WWF storage is relatively slower than the preceding lipid hydrolysis, because of the fact that WWF contains higher levels of antioxidants such as vitamin E, which protect the lipid oxidation.[39] With reference to the loss in the sensory acceptability, it has been stated that FFA/acid value is the major cause that leads to typical off-flavor due to lipid rancidity. Indeed, accumulation of FFA and volatile compounds related to subsequent lipid oxidation often gives it an off taste.[26,27]

Within a similar context, oat processors execute groat roasting, kilning, or steaming treatments before further processing to control the lipid oxidation. The lipid content of oat groats varies from as low as 3.1% to as high as 11.6%, with most cultivars containing 5–9% lipids.[46] As is the case for WWF, the main factor limiting the storage and handling possibilities of raw oats is lipolysis followed by oxidation, which leads to rancid off-flavors and, hence, inadequate shelf life.[25] Heat inactivates enzymes in the kernel, such as LA, LOX, and peroxidase.[9] By implementing oat heat-processing technology into wheat flour production systems, it may well be possible to control rancidity through inactivation of LA. Some of the studies related to the heat processing are discussed in this section.

In one of the works, wheat bran was roasted at 175°C for 40 min. Authors noticed that LA, LOX, and protease activities were reduced by 40, 100, and 50%, respectively. Similar work was carried out to compare the effect of dry heat, steam, and microwave treatments of wheat bran on both its LA. It was found that all treatments effectively decreased LA activity. Optimal conditions (sample size, 203 g of bran), in as much as further heating did not decrease LA activity, were 25 min of dry heating at 175°C, 60 s of microwaving (1000 W), or 60 s of steaming. Microwave and steam treatments were more effective at decreasing LA activity than dry heat (93 and 96, versus 74% LA activity decrease, respectively).

During the storage of WWF containing heat-treated bran, lower levels of FFAs were found than in WWF containing control bran, with the rates of FFA release in each of the treated samples correlating well with the measured LA activity.

In one of the studies, the rate of hydrolytic and oxidative degradation of lipids during WWF storage was evaluated.[16] The rancidity was measured in terms of the rate of oxygen uptake. It was found that there was a linear relationship between the rate of oxygen uptake with storage at 20°C. This trend was positively correlated with the increases in unesterified fatty acids. Also the rate of oxygen uptake was found to be positively correlated with the increase in the relative humidity. The increase in relative humidity was not due to microbial growth.

In terms of oxygen uptake with different grain components, the rate of deterioration of bran over 30 days at 20°C was five-fold greater than that of a relatively pure germ fraction, but was only half the rate of a 5:1(w/w) blend of bran and germ. It was concluded that increases in the rate of O_2 uptake and fatty acid content were highly correlated. Fine milling of bran, germ, and blends increased the rate of deterioration and enhanced the observed synergistic effect between bran and germ on both O_2-uptake and fatty acid increases. Lipid analyses of bran-germ blends indicted that during storage there is a relatively slow (over several weeks) release of fatty acids, catalyzed by a lipolytic enzyme in the bran component of whole meal, but that lipid oxidation occurs rapidly (within minutes) when excess water is added, facilitating LOX-catalyzed peroxidation of polyunsaturated fatty acids.

In one of the unique studies, when the flour was hydrated there was relatively rapid and extensive oxidation. Due to this oxidation, O_2 depended LOX-catalyzed degradation of polyunsaturated fatty acids was found out to occur. It was observed that even though reaction is initiated by TG acyl hydrolase (LA) activity in the bran component of whole meal, degradation due to oxidation depends primarily upon the LOX activity, mainly in the germ, and on a supply of TG from germ, endosperm, and outer layers of wheat grain.

Since the oxidation of lipids was found out to be more progressive in case of hydrated WWF, selection of key lipid oxidation products (e.g., peroxide value, thiobarbituricacid test or O_2-uptakeof dry materials) may not be suitable for the products that are hydrated before processing. Direct measurement of the TG LA activity, measurement of unesterified fatty acids (FFAs, fat acidity) or the most rapid and direct measurement of rates of O_2-uptakeby aqueous suspensions are likely to be more useful predictors of potential rancidity. However, this issue needs future investigation to establish firm relationship between the oxygen uptake and rancidity development.

8.4   LIPID DEGRADATION DURING STORAGE

Despite being a minor constituent of wheat flour, endogenous lipids contribute substantially to flour functionality. For example, it has been reported that bread made from defatted flour is inferior to bread with endogenous lipids, even when shortening is added during the mixing process.[22,45] Upon mixing flour with water, lipids cannot be extracted from dough with common solvents due to their binding with gluten proteins, which is essential for proper gluten development.[20] Lipids begin to break down in WWF by hydrolytic rancidity, which can be followed by oxidative rancidity. These changes can occur enzymatically or nonenzymatically and affect flour quality (Fig. 8.1).

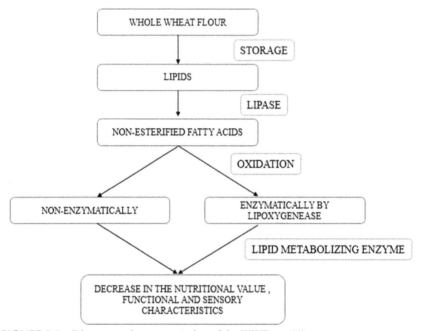

FIGURE 8.1   Diagrammatic representation of the WWF rancidity.

8.4.1 HYDROLYTIC RANCIDITY IN WWF

It has been documented in the past that the action of LA the hydrolytic rancidity leads to the WWF rancidity.[12] LA (EC 3.1.1.3) hydrolyzes TAGs to nonesterified fatty acids and diglycerides, monoglycerides, and eventually glycerol; thus, the release of nonesterified fatty acids in WWF is related to LA activity. It has been stated that the wheat LA is mainly concentrated in the bran fraction of the germ.[54] It is also termed as "wheat germ lipase" that catalyzes de-esterification of triacetin and other artificial water-soluble substrates and thus is technically an esterase. True LA activity (i.e., activity on water-insoluble substrates) of wheat germ (WG) LA is likely a result of contamination with LA from wheat bran.[56] The enzymatic activity of the LA is related to the moisture content of the WWF. For example, it has been documented that the maximum activity of LA is mainly exhibited in wheat at about 17% moisture content. However, LA activity continues at about 50% of maximum of moisture contents commonly observed in flour during storage (10–14%). This property makes LA unique among hydrolytic enzymes. Thus, it could be deduced that LA only requires a catalytic amount of water to act, whereas excessive amounts of water protect the lipid from being exposed to the catalytic site of the enzyme and reduce activity.[47]

Hydrolytic rancidity in WWF leads to a decrease in sensory quality and functional properties of WGF.[22] In one of the studies, sensory evaluation of WWF for 11 weeks demonstrated that hydrolytic rancidity was inversely related to the acceptability of bread made from these flours.[16] WWF with a high content of nonesterified fatty acids has been described as musty, bitter, and rancid. Products of hydrolytic rancidity have an effect on baking quality.[17] At low concentrations, nonesterified polyunsaturated fatty acids have a positive effect on loaf volume through co-oxidation of gluten protein sulfhydryl groups during mixing. However, at high concentrations nonesterified polyunsaturated fatty acids affect dough mixing by reducing lipid-binding capacity of gluten.[41] This reduces gas-holding capacity and elasticity of gluten. Interestingly, saturated fatty acids seem to have no effect on dough and baking properties. In addition to direct effects on bread quality, nonesterified polyunsaturated fatty acids are substrates for LOX, an enzyme that generates oxidation products that decrease the quality and acceptability of WWF.[31]

8.4.2 OXIDATIVE RANCIDITY IN WWF

The oxidation of the lipids in WWF takes place either enzymatically[12,54] or through the process of autoxidation.[61] Through the action of LOX (EC 1.13.11.12), the lipid oxidation occurs enzymatically.[15,58,62] LOX in wheat is located in the germ and bran of the grain. It consists of a group of isozymes with a molecular mass of 110 kDa and optimal activity at pH between 4.5 and 6.0. LOX attacks the methylene group between two double bonds in polyunsaturated fatty acids, preferentially nonesterified polyunsaturated fatty acids.[65] Autoxidation can occur by nonenzymatic reaction of grain lipids with atmospheric oxygen. Under both mechanisms, lipid oxidation involves addition of oxygen to polyunsaturated fatty acids thus forming hydroperoxides,[1] followed by fissure of the carbon chain into smaller volatile compounds (e.g., epoxyaldehydes, ketones, lactones, furans).[17] Lipid oxidation during storage of WWF is a slower process than lipid hydrolysis. This is because, unlike LA, LOX exhibits very little activity at moisture contents typically found during storage, and because WWF contains high levels of protective antioxidants. Despite being a slower process than lipid hydrolysis during storage, lipid oxidation can contribute substantially to the loss of product quality. While minimally active in dry flour, LOX becomes active when stored flour is mixed with water and rapidly oxidizes nonesterified fatty acids present in the flour from the action of LA.[17]

Oxidation of lipids leads to a decrease in the nutritional quality and consumer acceptability of WWF and WWF-based products. Lipid oxidation reduces nutritional quality through the loss of essential fatty acids, although, more significantly, reduced nutritional quality is affected through co-oxidation of other flour components. Free radicals that are generated can denature proteins and convert essential amino acids into unavailable derivatives. LOX activity also causes significant losses of carotenoids and vitamin E. Consumer acceptability of WWF declines as a result of lipid oxidation, which can generate undesirable odor components that affect sensory acceptability of WWF-based products.

8.5   DEVELOPMENT OF RANCIDITY IN WG: A CASE STUDY

The WG (embryonic axis and *scutellum*) represents about 2.5–3.8% of total seed weight and plays a critical role during storage of WWF.[7] The germ contains about 10–15% lipids, 26–35% proteins, 17% sugars, 1.5–4.5% fiber, and 4% minerals, as well as significant quantities of bioactive compounds such as tocopherols (300–740 mg/kg dry matter), phytosterol (24–50 mg/kg), policosanols (10 mg/kg), carotenoids (4–38 mg/kg), thiamin (15–23 mg/kg), and riboflavin (6–10 mg/kg).[58] The fraction of lipids in the germs mainly comprises oleic, linoleic, and R-linolenic acids. Depending on the type of wheat and method of milling, the key fatty acid is the unsaturated linoleic acid (C18:2), which constitutes 52–57% (w/w) of the fatty acids.[67]

Application of WG is mainly for the fodder use, however it has been estimated that the minor part of the annual production of 16 million tons is also used for human consumption. Because of its unfavorable baking properties and susceptibility to oxidation, the germ is removed from the endosperm during milling. High LA and LOX activities as well as a high content of unsaturated oil are characteristics of WG. Even slight oxidation may cause a destruction of essential fatty acids and vitamins. Oxidation may be prevented and shelf-life prolonged by inactivating the enzymes by heat shock or by removing the oil fraction from the WG by extraction or combined techniques. Extrusion cooking [21] and microwave heating have been reported to be rapid and effective methods for inactivating LA.

Thermal processing is the most common treatment for stabilizing whole germ (WG). The water content (WC) of WG stabilized with steam heating is increased to a very high level, which restricts its industrial application in the oil industry. Microwave, spouted bed, and other dry heating technologies have been tried to improve the storage stability of WG. However, nonuniform temperature distribution and time-consuming heating limit the wide application of microwave and traditional convection heating methods, respectively, in WG stabilization. On the other hand, WC is an important factor for food storage; most foodstuffs have a critical WC below which the rate of quality spoilage is negligible. Scientists have clearly identified that water activity (A_w) is correlated well with the deterioration of oil stability due to enzyme activity. The reaction rate of hydrolysis accelerates with increase in A_w, with the reaction being extremely slow at very low activities. Enzymes almost lose their activity below an A_w of about 0.3. However, it was found that if foodstuffs were overdried (WC less than 2–3%), they became very susceptible to autoxidation.

In one of the earlier works, development of rancidity in WG was analyzed by Headspace Gas Chromatography (HS-GC) and Sensory Analysis (Polymer Solutions, Christiansburg, VA).[59] Volatile compounds in stored WG were evaluated using dynamic HS-GC and sensory analysis. Preliminary comparisons were also made between freshly prepared WG and WG subjected to microwave heating at 45 and 55°C prior to storage at room temperature. The progress of oxidation was followed in untreated WG for 4 weeks and in heat-treated WG for 7 weeks by HS-GC and sensory evaluation. Significant changes in rancid odor and flavor were observed in the untreated WG after 3 weeks, whereas no corresponding differences were observed in the microwave-heated WG after 7 weeks of storage. Identification of a total of 36 volatile compounds was performed. The major volatiles were hexanal, R-pinene, 1-hexanol, and 3-carene. In addition to the analysis of a short period of storage, 30 volatile compounds were identified from the headspace of WG stored for > 1 year.

WG is quite susceptible to rancidity due to the presence of lipolytic enzymes, mainly LA and LOX. Rancidity, with a resultant loss of quality and acceptability, can occur in products derived from the milled wheat flour possibly due to hydrolysis and oxidation degradation reactions. Quality stability of WG during storage is limited mainly by hydrolytic and oxidative rancidity development. Due to the loss in functional, nutritional, and sensory quality during WG storage, several strategies have been attempted to inactivate the LA and LOX for restraining the rapid enzymatic rancidity of WG.

8.6   IMPACT OF PARTICLE SIZE DISTRIBUTION ON THE WWF STABILITY: A CASE STUDY

In the recent past, there has been considerable interest to evaluate the role of particle size and its relationship to the WWF stability as well as products derived out of it. Among the available technology, two emerging technologies are superfine- and ultrafine-grinding techniques, both of which are useful tools to decrease the particle size and improve the reactive surface of fine powder.

Superfine grinding can produce fine powder on a microscale whereas ultrafine grinding is able to reduce the particles in the submicron range (100–1000 nm). Recently, superfine and ultrafine grinding technologies have been used with food materials to reduce fiber size and to improve the quality of powders related to the nutritional and cooking properties of products. As for wheat products, the two techniques have been applied for grinding wheat bran. One of the authors found that ultrafine grinding could effectively pulverize the wheat bran fiber particles to submicron scale. Authors observed that the there was a considerable decrease in the particle size as there was redistribution of fiber components from insoluble to soluble fractions. In one of the similar studies, the influence of ultrafine grinding temperatures on wheat bran was evaluated. The results indicated that the intrinsic characteristics of bran had significant influence on its grinding behavior as compared to the grinding device used in the experiment. Furthermore, patented technologies for wheat bran and reconstituted WWF have been produced by the fine-grinding technique.

It has been shown in the past that particle size of the bran fraction in WWF has a significant influence on functional properties of the flour as derived products. In general, with reference to bread, it has been observed that larger wheat bran particles (mean particle size [MZ] of more than about 500 mm) lead to higher water absorption and loaf volume compared with finer bran particle sizes (MZ≤500 mm).

In one of the recent studies, effects of superfine grinding on the quality characteristics of WWF and its raw noodle product were evaluated.[44] It was found that four particle size distributions of whole-wheat flour (WWF) with MZ of 125, 96, 72, and 43 mm were obtained by superfine grinding. Starch damage and Farinograph water absorption were significantly affected by the reduction of particle size, while dough development time, stability, tolerance index, and time to breakdown of WWF were little changed. These changes were further observed in the microstructural analysis of the noodles. The results demonstrated that the superfine grinding technique could improve the structural characteristics of WWF by reducing the particle size.

Thus, the studies show that particle size distributions have a significant effect on the whole properties as well as on the product derived out of it. However, still there is a need for additional studies to determine the relationship between the particle size distribution and WWF stability.

8.7 TYPE OF MILLING AND IMPACT ON WHOLE WHEAT STABILITY

In selection of the WGF, one of the prime considerations should be milling process. For example, milling technique may have a greater impact on whole wheat (WW) bread quality than the quality of wheat used for producing the flour or the formulation of the bread itself.[33] Therefore, the impact of milling is not only limited to the flour but also extended to the product derived from the raw material.

Currently stone and roller mills (RMs) are the two predominant techniques for grinding WGFs. WGFs could also notionally be produced with an impact or hammer mill, however these techniques are rarely used. In the recent past, work has been carried out to understand the effect of different milling methods on the WWF. In one such case, effect of milling was studied on the chemical composition of WWF. In order to assess the effect of different types of milling methods on protein and lipid composition of WWF, two types of wheat varieties belonging to strong and weak wheat type were selected and milled in different mills such as plate, hammer (HM), stone, and RMs. The temperatures generated during grinding of wheat in stone, plate, HM, and RMs were 90, 85, 55 and 35°C, respectively. The studies on SDSPAGE indicated degradation in proteins of WWF obtained from stone and plate mills, especially in the high molecular glutenin regions. Greater loss of total amino acids was also observed in the above milled flours when compared with that of HM- and roller-milled samples. Free lipid content was lower in flours milled in stone and plate mills when compared with that of flours milled in other mills. Unsaturated fatty acid content, particularly linolenic acid, was lower in stone milled flour (1.3%) followed by plate mill (2.2%), HM mill (2.8%), and roller flour mill (3.8%). The trends in these values as influenced by different milling methods remained similar both in the weak and strong wheat types.

In one of the recent studies, chapatti-making quality of WWF (*atta*) obtained by various processing techniques was evaluated.[29] Wheat was processed in chakki (CM), hammer, disk (DM), pin (PM), and RM with an objective of quality characterization of WWF (*atta*) in relation to chapatti making quality. Results indicated that atta produced from RM was cooler and had retained more moisture. Ash content was not significantly influenced by different grinders. However, acid insoluble ash was higher at 0.063% for the atta produced from CM. Variation in damaged starch was observed in the order of 15.99, 13.76, 11.76, 10.16, and 9.1% for CM, HM, DM, PM, and RM-ground atta, respectively. Farinograph water absorption of CM ground atta was the highest at 85% and least for RM ground atta (71.5%). Overall quality of chapatti prepared from CM atta scored higher and had better texture and desirable wheaty aroma. Studies revealed that atta quality parameters and its chapatti-making quality were varied with processing techniques.

8.7.1   STONE MILLING

Stone mills are the oldest attrition mills used for making WGFs, which simultaneously use compression, shear, and abrasion to grind wheat kernels between two stones and produce a theoretical extraction rate of 100%.[33] Traditional stone (*Chakki*) mills consist of two dressed stone disks (one stationary, other rotating). Stone mills generate considerable heat due to friction and due to this it undergoes higher maintenance cost (high abrasion) and subsequently lower shelf life.

It has been found that the high abrasion in the stone mill leads to considerable damage to starch, protein, and unsaturated fatty acids in comparison with other milling techniques.[48] This leads to the low stability of the WWF using the stone mill. Breakdown of the unsaturated fatty acids in the WWF normally results into rancidity. It has been observed that there is higher breakdown of the unsaturated fatty acids in the stone mill as compared to the roller flour mills. This could be possibly due to different types of forces (compression/shear/attrition) that act during the process of grinding.

There are certain reports that describe the effect of type of milling on the mycotoxins content of the wheat flour. In one of published works it was found that stone milling reduced vomitoxin and zearalenone content in flours, compared with the use of the roller-mill system.

8.7.2 ROLLER MILLING

The process of RM involves separation of the endosperm from the bran and germ followed by gradual size reduction of endosperm. In this process, wheat is passed through a series of corrugated and smooth rollers accompanied by sifting between stages. Producing flour that fulfills the requirement for being whole grain is achieved by blending bran and germ back with the endosperm flour in the naturally occurring proportions. Feeding the bran and germ milling streams with the endosperm flour stream is most often achieved in a continuous process, rather than collecting all fractions in separate bins and recombining at the end of milling. In this case, production of WWF would not involve additional capital cost beyond what is required for regular RM. Sometimes WWF is made by physically separating flour millstreams and then recombining at the end of the milling process. This is usually done when the bran will undergo some post-milling such as ultrafine grinding or heating. In these cases, extra capital costs would be required for the post-milling, plus equipment for recombining the fractions. When producing WWF on RM, number of conditions are different from those used for wheat flour:[32]

- First: Conditioning (tempering) is less important when milling WWF. While wheat flour relies on proper conditioning to facilitate endosperm and bran separation, this is not required for WW milling. Thus, in theory no conditioning should be required, although many mills will add 12% moisture to soften the grain and improve efficiency in terms of the energy required to produce the flour. Efficiency can also be improved by tightening the roll gap and using more open scalp covers to increase the break release, as well as changing some of the smooth rolls to corrugate during reduction.[48]

- The purifier air valves should also be adjusted so that the bran and germ are not rejected but are returned to the reduction system.
- There are several noteworthy advantages of making WGF from RMs as opposed to stone mills.
- The amount of grinding and reduction at each roll can be adjusted to accommodate variations in raw materials, which makes RM both economical and flexible.
- The use of selective corrugations and differential speeds subjects the endosperm fraction to minimal shear and compressive forces during the grinding and reduction, which allows less heat to build on reduction rolls and results in less destruction to chemical components in the flour.
- Third advantage of making WGFs from RMs is that wheat bran and germ can be separated from the endosperm fraction and subjected to further processing such as heating or fine grinding to affect the storage or functional properties of the flour.

8.7.3 JET MILLING

Nowadays, alternative milling procedures and micronizing technologies are tested in order to produce flours with enhanced functional properties, which are suitable for making new edible products or for improving the properties of the current ones. Milling technologies focused on producing finer flours with improved properties are getting increased attention.[10, 51] Jet milling is one such latest technological development that aims at the production of super fine flours by accelerating the particles in a high-velocity air stream, the size reduction being the result of interparticle collisions or impacts against solid surface.[34,51]

The particles impact at high velocities produces superfine powders and reduces the size of all aggregates. It is a fluid energy impact-milling technique, commonly used to produce particle sizes lower than 40 μm, which are greatly appreciated in the chemical, pharmaceutical, and mineral industry. In food applications, smaller particle size results in faster starch digestion. Small particles have high surface-to-volume ratio increasing the access of enzymes to the interior of the particle taking advantage of the absence of intact cell walls. An increased surface area of food materials could increase the rate of water absorption of materials, improving solubility of dry products and increase site accessibility for chemical reactions (e.g., oxidation, digestion, flavor release, catalyst, and enzyme activity). Jet milling combined with air classification has been successfully used to separate starch from protein in order to produce starch-rich fine flours.

Furthermore, differential scanning calorimetry showed lower gelatinization enthalpy values of 6.76–7.09 and 9.92–10.12 mJ/mg for the doughs (flour:water::60:40) of fine flours and their coarse flour counterparts, respectively. Overall, particle size of wheat flour seems to have an impact on dough mechanical and starch gelatinization properties. Therefore, there is a consensus that particle size reduction promotes changes in the majority of physicochemical properties due to the increase of a particle's surface area, although it must be assessed if there is a critical point that leads to an increase of damaged starch.

The higher the specific surface area per weight unit, the higher is the rate of hydration and water absorption. Generally, starch granules become physically injured with milling's shearing and scrapping, that is, starch damage occurs which could also increase water holding capacity. Moreover the production of ultrafine powders from cereals flours may present benefits to human health. Jet milling may be useful for modifying or improving functionality and availability of bioactive compounds. However studies relating to the impact of jet milling on the storage stability of WWF are not sufficient. This area needs future investigation.

8.8 DISTRIBUTION OF ENZYMES IN WHEAT FLOUR

Wheat flour contains several technologically important enzymes such as amylases, proteases, LOX, polyphenol oxidase, and peroxidase.[54] In the wheat grain, *alpha*-amylase is located mainly in the pericarp with small quantities present in the aleurone layer and the seed coat. Protease is concentrated in the endosperm, germ, and aleurone layer.[14,50] The scutellum and embryo are rich in LOX. Polyphenol oxidase and per-oxidase are predominant in bran layers. Although these enzymes are inactive during storage of grain and flour, yet when water is added they become active and play a significant role in determining the functional attributes of flour.

The aim of roller flour milling is the gradual reduction of the wheat kernel through a series of break and reduction rolls. This results in the production of different types of flour streams containing endosperm, bran and germ in varying proportions. Thus, the knowledge about the enzyme distribution in the different steams of whole mill can possibly help to design appropriate treatment of different streams so as to minimize the overall enzymatic activity thereby resulting to higher flour stability.

The levels of activity of various enzymes differ in different flour mill streams and hence their functional properties are also different. Several reports are available on the protein and ash contents in different flour mill streams,[43,49,63] and a few reports are available on the suitability of various mill streams for different bakery products. Some information is available on the activity of individual enzymes in mill streams. It has been reported in the past regarding the distribution of *alpha*-amylase in various mill streams of soft winter wheat.[23] It has also been reported that proteolytic activity was found to be higher in tail end reduction streams as compared to other flour streams during the wheat milling operations.[13] In one of the report, polyphenol oxidase activity was found out to be linearly correlated with ash content and it was also reported that PPO was most active in bran- and germ-rich milling fractions. The distribution of LOX and peroxidase in flour streams has also been reported. Knowledge about the distribution of the enzymes can be useful in the preparation of blends either by selecting or omitting particular stream/streams for the preparation of specific mill fractions for use in different product category. This area need to be explored in the near future.

8.9   DIFFERENTIATION OF WHOLE GRAIN FROM REFINED WHEAT

Differentiation of a whole grain product from a refined grain product is very challenging. This is also important in the context of identification of the key markers that can differentiate between the WGF and refined flour. Since the food labels can often be misleading, the whole Grains Council has created an official packaging symbol (the Whole Grain Stamp) to help consumers. The 100% whole grain stamp assures that food contains a full-serving or more of whole grain in each labeled serving and that all the grain used is whole grain, whereas the basic Whole Grain Stamp indicates the products contain at least 8 g of whole grain per labeled serving.

In one of the recent studies, differentiation of whole grain from refined wheat (RW) (*T. aestivum*) flour was done using lipid profile of wheat bran, germ, and endosperm with ultrahigh-performance liquid chromatography–high-resolution accurate-mass (UHPLC-HRAM; UBM Life Sciences, Ellesmere Port Cheshire, UK) multistage Mass Spectrometry.[19] A comprehensive analysis of wheat lipids from milling fractions of bran, germ, and endosperm was performed using UHPLC-HRAM multistage mass spectrometry with electrospray ionization and atmospheric pressure chemical ionization (APCI) in both positive and negative modes. About 155 lipid compounds, including FFAs, oxylipins, alk(en)ylresorcinols (ARs), γ-oryzanol, sphingolipids, triglycerides (TGs), diglycerides (DGs), phospholipids, and galactolipids, were characterized from the three milling fractions. Galactolipids and phospholipids were proposed to be potential discriminatory compounds for refined flour, whereas γ-oryzanols, ARs, TGs, and DGs could distinguish WWF from refined flour based on principal component analysis (PCA). These key compounds could be also measured during the WWF storage.

In one of the similar works, differentiation of bread made with whole grain and RW (*T. aestivum*) flour was done using LC/MS-based chromatographic fingerprinting and chemometric approaches.[18]Fuzzy chromatography mass spectrometric (FCMS) fingerprinting method combined with chemometric analysis was established to differentiate between breads made from WW flour and RW flour. The chemical compositions of the bread samples were profiled using UHPLC-HRAM multistage mass spectrometry with APCI in positive ionization mode. PCA and soft independent modeling of class analogy (SIMCA) of the FCMS fingerprints revealed the components responsible for the chemical differences between WW and RW flour/bread samples. Alk(en)ylresorcinols (ARs) have been demonstrated to be the most important markers for differentiation between WW and RW flour/breads. Diglycerides (DGs), and phosphatidylethanolamine (PE) also been shown contributed significantly to the classification. In this way, there was no significant difference observed between the bread crumb and crust. It was concluded that SIMCA, using WW modeling, could be a potent and robust tool for authentication of WW breads.

8.10   BIOCHEMICAL MARKERS FOR THE ASSESSMENT OF WHEAT GRAIN TISSUE PROPORTIONS IN MILLING FRACTIONS

Identification of the key biomarkers in the whole grains would be enormously helpful in differentiating the WWF from refined flour. Also, these biomarkers could be tracked during the WW storage so as to understand the changes within WWF. This section discusses some of the recent studies in relation to the biomarkers in whole grain tissues.

Numerous epidemiological studies have demonstrated the health benefits of consuming more whole-grain foods.[4,36,60] However, all the wheat grain parts are not health-promoting, for example, the outermost parts have been shown to concentrate the majority of the grain contaminants, such as microorganisms, mycotoxins, pesticide residues and heavy metals.[28] On the other hand, the wheat aleurone layer has been shown to have great nutritional interest, and to concentrate most of the minerals and vitamins of the wheat grain. It has been reported that wheat aleurone layer contains interesting proportions of proteins, β-Glucan, phenolic compounds and other phytochemicals (lignans, sterols).

It has been found that antioxidant and phenolic acids are concentrated in the aleurone layer of wheat bran, and the higher the proportion of aleurone material in wheat fractions, the higher is the antioxidant capacity.[40] Aleurone-rich fractions exhibited better in vitro digestibility and colonic fermentability than wheat bran. The digestibility of minerals, protein and non-starch polysaccharide is much higher in bran fractions rich in aleurone than in fractions rich in pericarp and testa. These studies suggest that it could be interesting to produce aleurone-rich fractions for use as food ingredients. As a consequence, new processes should be developed to exploit all the nutritional benefits of whole grain and to produce new wheat foods and wheat-based ingredients with enhanced nutritional quality. For example, depending on the desired product, a process can aim at discarding the pericarp to obtain whole grains containing less contaminants, while other processes may be developed in order to produce bran fractions highly concentrated in aleurone material. It is often difficult to exactly monitor the distribution of the different grain tissues among fractions during processing, as no simple method exists to quantify the respective proportions of these tissues in fractions. However, the monitoring of tissue proportions in the different fractions is essential, as it allows to control the quality of the products and to consequently adapt the processes. Therefore, quantitative tools are needed. Different compounds can be measured to evaluate bran contamination in wheat flours and fractions. Ash content and measurement of flour color are widely used in the milling industry as indicators of flour purity. The amino acid composition of the various tissues was also studied but did not allow the quantification of the grain tissues in milling fractions.

Using the concentration of ferulic acid to quantify bran in flours and semolinas/alkylresorcinols have more recently been shown to be good markers of wheat bran content in foods. Such analyses may be useful to evaluate the total outer layer content but they do not allow distinguishing between the different outer tissues. Another way to evaluate the different grain tissues in wheat fractions consists of the use of specific fluorescence properties of the outer layers. Indeed, the aleurone cell walls display blue fluorescence under UV-light due to the presence of ferulic acid, whereas the pericarp shows green fluorescence under blue light.

Based on these fluorescence properties, commercial equipment has been developed to determine the amount of aleurone in flours. Multispectral fluorescence image analysis of grain sections coupled with classification techniques has also been developed to more precisely quantify the proportions of the different parts of the grain, but has not yet been applied to powdery samples. These imaging methods can be good tools for online use, but their main disadvantage would be their lack of specificity, as they do not allow quantifying other tissues than aleurone and pericarp. Moreover, all these methods allow the determination of bran proportions in flours during milling, but they may not be adaptable to other fractionation systems (such as progressive abrasion or bran fractionation). In the past work, biochemical analyses of isolated wheat grain tissues were carried out and the differences in chemical composition between these tissues to assess the histological composition of technological fractions were determined. Compounds such as phenolic acids, phytic acid, and starch as biochemical markers were used. Indeed, these compounds were shown to be either present exclusively in one part of the grain (starch in starchy endosperm), or present in greater amounts in one particular tissue (phytic acid in aleurone cell contents and some phenolic acids in the cell walls). These biochemical markers were used to determine the amounts of aleurone layer and pericarp in flours and other milling fractions, and in the samples obtained from a wheat bran fractionation process.

Different markers for aleurone cell walls and aleurone cell contents have been used to assess the histological composition of bran fractions and to evaluate the dissociation and the accessibility of aleurone cellular components. This method can provide accurate quantification of the histological composition of samples and is versatile, as it can be refined and adapted depending on the type of sample analyzed, from either bran or whole-grain fractionation. However, it did not allow the quantification of testa (this tissue was deduced by subtraction and thus was perhaps overestimated), and it neither allowed the detection of the germ. Having a marker for WG would nevertheless be very useful as it is either a part of the grain that needs to be excluded to avoid lipid oxidation and rancidity, or a nutritionally interesting by-product that could be followed during fractionation processes in order to get germ-rich fractions.

8.11   CURRENT METHODS TO ENHANCE WHEAT FLOUR STABILITY

8.11.1   ANTIOXIDANTS

During the storage of WWF, development of rancidity has been observed as early as 2–14 days subsequently after the milling operation. Also, some authors have suggested in the past the limits 15–60 days for WWF storage. High levels of vitamin E (tocopherols, tocotrienols) in whole-wheat flour provide antioxidant protection during the first 22 days of storage and efficiently contrast the peroxidation process, but a longer storage period (10 months) induced the degradation of up to 24% of the total vitamin E amount. The role of antioxidant compounds, such as polyphenols, is important to preserve nutritional properties of whole-wheat flour during storage. In the production of WGF, the selection of the milling process is a key factor as it exerts a great impact on the quality of the final whole-wheat flour. Thus the milling operation has to be optimized produce the WWF with higher antioxidant activity.

Cereals have been known to contain a high amount of hydroxycinnamic acid (HCA) derivatives that render potential health benefits.[1,39] Commercial processing of cereals may lead to products with low-value fractions such as hulls and polish waste. In general, hulls are removed prior to production operations. However, these low-value fractions may serve as potential sources of natural antioxidants at relatively high concentrations.[40] In oats, antioxidant compounds are mostly concentrated in the bran as compared to that in the endosperm as shown by in vitro assays. Oat pearling fractions containing different levels of bran layers have higher antioxidant activity and total phenolic content (TPC) compared to those of the flour extracts. Also, bran extracts of different wheat varieties exhibits significant antioxidant properties against free radical scavenging and metal ion chelation. Moreover in the FISH model system, different varieties of whole grains exhibit antioxidant activity against lipid peroxidation. For example, the "Akron" variety of wheat was found out to be highly effective in scavenging 2,2-diphenyl-1-picrylhydrazyl (DPPH) radical and chelating Fe(II). TPC, scavenging of DPPH radical and chelation of Fe(II) were significantly influenced by agronomic practices and environmental conditions. The antioxidant properties of whole grains, bran and aleurone layer of a Swiss red wheat variety was studied using free radical scavenging and metal ion chelation capacity. Thus, aleurone, bran, and grains differed significantly in their antioxidant potential, TPC and phenolic acid composition. Moreover, the aleurone layer exhibited the highest antioxidant activity, TPC and content of phenolic acids.

Ferulic acid has been reported to be the predominant phenolic acid accounting for approximately 57–77% of total phenolic acids present in wheat on a dry weight basis.[40] Ferulic acid content was positively correlated with scavenging of free radicals and TPC and hence may be used as a potential marker of wheat antioxidants. Plant phenolic compounds including phenolic acids, flavonoids and anthocyanins, among others, have also been recognized as conferring stability against autoxidation of vegetable oils. There is much interest in the use of crude phenolic extracts from fruits, herbs, vegetables, cereals and other plant materials in the food industry because they have been shown to retard oxidative degradation processes, especially those of lipids thereby improving the quality and nutritional value of food.

Study on antioxidant and free radical scavenging activities of WW and milling fractions[37] indicated that the milling of wheat afforded several fractions, namely bran, flour, shorts and feed flour. In addition, semolina was the end product of durum wheat milling. Among different milling fractions, the bran had the highest phenolic content while the endosperm possessed the lowest amount and this was also reflected in free radical and reactive oxygen species scavenging capacity, reducing power and iron(II) chelation capacity of different milling fractions in the two cultivars. This study demonstrated the importance of bran in the antioxidant activity of wheat; hence consumption of WW grain may render beneficial health effects.

8.11.2   IRRADIATION

Cereals and cereal products, like semolina (*rava*), refined (*maida*) and whole-wheat flour (*atta*), commonly in pre-packed form, are sold in the retail market. However, their shelf-life is restricted to 6 ± 8 weeks because of insect infestation. In tropical countries, adverse climatic conditions of high temperature and humidity result in insect proliferation, even in sealed pouches. The conventional method of fumigation that is used for disinfestation of grains is not suitable for sealed pouches because of nonpenetration of the fumigant through the packaging material. Extensive research at the Bhabha Atomic Research Centre, India—has shown the effectiveness of low-dose gamma radiation (up to 1.00 kGy) to achieve insect disinfestation of wheat, Basmati rice, and rava, and so on. (It may be described as the residues of milled material).

Gamma radiation (0.2 ± 1.00 kGy) destroys all the metamorphic stages of insects and sterilizes the adults of 32 known granary insects. As irradiated food is wholesome and nutritionally adequate, the FAO/WHO/IAEA Joint Expert Committee on Food Irradiation has unconditionally cleared foods irradiated up to 10 kGy as safe for human consumption. Insect disinfestation of wheat and wheat products by gamma irradiation (1.00 kGy) has been approved in eight countries. In India, draft rules amending the Prevention of Food Adulteration Act, 1954 to permit irradiation of 14 food products including *rava*, *atta* and *maida* have been modified by the Government, for the commercial application of the technology. In one of the recent studies, extension of shelf-life of whole-wheat flour by gamma radiation has been proposed.

The effect of low-dose gamma irradiation (0.25 ± 1.00 kGy) on prepacked WWF (*atta*) has been assessed in terms of physicochemical properties, nutritional quality, chapatti-making quality and sensory attributes. Semipilot scale storage studies on irradiated pre-packed whole-wheat flour revealed that there was no adverse effect of irradiation and storage up to 6 months of whole-wheat flour treated at doses up to 1.00 kGy on total proteins, fat, carbohydrates, vitamin B1 and B2 content, color index, sedimentation value, dough properties, total bacterial and mold count. Storage of wheat flour resulted in slight increase in moisture, FFAs, damaged starch, reducing sugars, and slight decrease in gelatinization viscosity. However, irradiation had no effect on any of these parameters. Irradiation at 0.25 kGy was sufficient to extend the shelf-life of *atta* up to 6 months without any significant change in the nutritional, functional attributes. Chapattis made from irradiated atta (0.25 kGy) were preferred even after 6 months storage, compared with the control.[2,38]

8.11.3   EFFECTS OF DIFFERENT MILLING PROCESSES ON WWF STABILITY

Selecting the milling process that will be used is the key consideration in producing WGF. The four predominant techniques for grinding WGFs are stone mill (SM), RM, ultra-fine mill and HM. The HM causes the product to be heated up and to lose moisture. Stone mills generate considerable heat due to friction, resulting in damage to starch, protein, and unsaturated fatty acids. The process of RM involves separation of the endosperm from the bran and germ followed by gradual size reduction of endosperm. Producing flour that fulfills the requirement for being whole grain is achieved by blending bran and germ back with the endosperm flour in the naturally-occurring proportions. In comparison with stone mills, RM is more economical and flexible, less heat production and thus less destruction to chemical components. A third advantage of making WGFs from RMs is that wheat bran and germ can be separated from the endosperm fraction and subjected to further processing or post-milling such as heating or ultrafine grinding to affect the storage or functional properties of the flour. On the other hand, there are considerate studies about the post-milling processes for WWF, which including the twin-screw extrusion, the heat treatment of the bran and germ and the ultra-fine grinding processes.

8.11.4   PRECONDITIONING WITH ADDITIVES

When WWF is stored, LA hydrolyzes lipids into nonesterified fatty acids (NEFA). NEFA are oxidized nonenzymatically during storage or enzymatically by LOX when the flour is mixed with water. These processes result in a decrease in the nutritional value and functional and sensory characteristics of WWF. More stable WWF would be desirable because it would not require such careful control of storage time and conditions and may result in higher quality bakery products. Researchers have focused on inhibiting LA to stabilize WWF. Heat treatments are employed most commonly;

however, this would require elevated costs in a large-scale operation, and the exposure to high temperature can initiate autoxidation and lead to non-enzymatic spoilage. Another strategy to stabilize WWF could involve the addition of LA inhibitors. Metal salts are known to affect the activity of LAs from other cereals and oilseeds. In semipurified extracts from wheat bran or rice bran, lipolytic activity has been reduced or activated by $CaCl_2$, FeNa–ethylenediaminetetraacetic acid (FeNa–EDTA), and NaCl.

Wheat is commonly conditioned by adding water prior to milling (approved method: 26-10.02 by the AACC International, 2012). Others have replaced water with solutions to impart certain desirable characteristics. For instance, in one of the past works, tempering water was replaced with ozonized water or acetic acid solution (1%) to reduce microbial load in flour. It was hypothesized that adding the salts in this fashion would allow the metal ions to diffuse into the kernel (along with the water) and interact and ultimately inhibit the LA enzyme more readily and more practically than spraying the solutions on the bran after milling. Therefore, if salts added during wheat conditioning at levels that are typically found in baked goods formulations could substantively inhibit LA and thus improve WWF functionality during storage. It was found that while no salt treatments completely inhibited LA, salt solutions applied during wheat conditioning significantly influenced activity of the enzyme. This dictated the salt's effectiveness in stabilizing lipids and maintaining functionality during storage. Thus, the application of the salt during conditioning could be probed for the inactivation of the enzymes.

8.11.5   HEAT/STEAMING TREATMENT

Heat treatment can enhance lipid stability in WWF by LA inactivation and antioxidant retention. In a study on dry heat, steam, and microwave treatments in decreasing LA activity whereas retaining antioxidant activity, bran was heat-treated in 230-g batches using four levels (exposure times) for each of the three treatment methods. LA activity and antioxidant activity were quantified for all treatment combinations. None of the treatments significantly decreased antioxidant activity; the levels determined to be optimal were 25 min of dry heat, 60 s of microwave (1000 W), and 60 s of steam. These treatments effectively decreased LA activity by 74, 93, and 96%, respectively. Optimum treatments were evaluated for acceptance using a consumer sensory panel during a 12-month storage period. No significant differences in acceptance were found between the control and any of the samples either at baseline or after storage. This suggests that WWF can be stabilized against lipolysis by utilizing the treatments without decreasing antioxidant activity, and manufacturers may utilize these treatments without risking decreased consumer acceptance.

Effect of wheat grain steaming and washing on LA Activity in WGF demonstrated that lipase activity in WGF can be reduced effectively by steam treating wheat grains prior to milling. Steaming grains for 180 s sufficiently inactivated lipase, peroxidase, endoxylanase, and part of the α-amylase without altering the WGF gelatinization profile. Moreover, treatment of separate WGF fractions demonstrated that the (free) lipase activity is mainly, if not only, present in the bran fraction and offers another possible solution for obtaining WGF with longer shelf life. Finally, because washing grains did not reduce the lipase activity in WGF, authors of this chapter conclude that lipase is mainly located within the bran rather than associated with the kernel outside.

8.11.6   FLAMELESS CATALYTIC INFRARED TREATMENT

Flameless catalytic infrared radiation (FCIR) is a new technology developed by catalytic drying technologies. In flameless infrared emitters, propane or natural gas chemically react at the surface of a platinum catalyst, below gas ignition temperatures, delivering peak radiant energy in the range of 2.8–7 mm. The water molecule shows peak radiation absorption bands at 3, 4.5, and 6 mm.[18] The characteristic of water-absorbing infrared radiation directly in the range of 3–7 mm has been applied in rapid drying of cereal commodities and in deactivating peroxidase in carrots. Till date, FCIR has not been employed in stabilizing WG or other cereal brans for extending their shelf life. In one of the earlier studies, the effect of flameless catalytic infrared treatment on rancidity and bioactive compounds in WG oil was conducted.[35] FCIR technology was used to inhibit lipase and LOX activities of WG to extend its shelf life. Moreover, the influence of FCIR heating on some quality characteristics of WG oil was assessed. Results reflect that FCIR treatment could effectively decrease the water activity (A_w) and exert a damping effect on lipase and LOX activities of WG within a short time. The WG sample heated 35 cm below the emitter for 6 min with FCIR obtained an excellent stabilization effect. Under this condition, the residual WC and A_w were 3.18% and 0.186, respectively, and the corresponding relative lipase and LOX activities were 7.94 and 14.33%, respectively. The FFA content and peroxide value of this WG sample at 40°C remained below 4.65% and 3.15 meq O_2 per kg WGO, respectively for 60 days. The optimal A_w for WG storage is about 0.186. No significant change in main fatty acid but a significant decrease in tocopherol content and oxidative stability was observed when compared to raw WGO. In addition, no significant darkening was observed in WGO extracted from all treated WG samples.

8.12 RECENT ADVANCES IN WHEAT BRAN AND STRATEGIES OF STABILIZATION

The use of wheat bran in the food and feed industry has increased distinctly and visibly over the last decade.[5,52,55] Annually, data from different sources suggest that wheat bran results as a by-product of the milling industry and is mainly used for animal feed. Since this approach has a lower valorization potential than applications in food, either directly or after tailored processing, the milling industry is increasingly interested in finding new strategies of wheat bran utilization.

Bran is rich in natural antioxidants, phytoestrogens and lignans, which may exert many beneficial effects to various body functions. However, native or even mechanically pretreated bran poses several disadvantages, which complicate its application in food systems. Among these, negative sensory attributes such as the bitter taste are an important issue contributing to low consumer acceptance. Moreover, its pronounced water-holding capacity negatively affects rheological properties of dough, thus leading to bread of low volumes and low elasticity. For these reasons, technologies need to be explored to overcome these issues.

Various treatments such as milling, heating, extraction, extrusion, and fermentation seem to offer some potential to improve the applicability of bran. In principle, the milling step is very effective for reduction of the particle size that disrupts the stable fiber matrix resulting in a significantly improved mouth-feel. However, a bigger surface also enables the development of hydrolytic and oxidation processes thereby leading to reduced shelf-life.

Thermal treatments including extrusion procedures may aid in the improvement of shelf-life due to the inactivation of endogenous enzymes and contaminating microorganisms. There are, however, some side-effects as the color darkens, and the hardness of the extrudate necessitates an additional milling step. Solutions are to be sought to reduce the level of undesired components in order to allow enhanced application of bran in baked goods (Fig. 8.2).

FIGURE 8.2   Different strategies for functionalization of wheat bran.

For example, the use of enzymes or fermentation cultures to modify these parameters is optional. Fermentation may induce an anti-stalling effect, retarded starch digestibility and further improves the bio-accessibility of minerals and bioactive compounds. On the other hand, increased water activity necessary for solid state fermentation may lead to fungal spoilage. Thus, a subsequent drying step is necessary to ensure sufficient shelf-life.

8.13   SUMMARY

There are many challenges and issues on the stability of WWF that are faced by food industries and millers. Lipid degradation by the action of enzymatic as was well nonenzymatic pathways seems to be one of the leading causes of WWF stability. Both hydrolytic as well as oxidative rancidity seems to play a major role in reducing the life of WWF. Among different components of the wheat grain, WGs seem to be highly susceptible to deterioration due to the presence of lipolytic enzymes. Also the distribution of enzymes in wheat grain is important and milling techniques should be adopted to remove the lipolytic enzymes, which are mainly concentrated in the bran as well as germ section of the wheat grain.

Among the available techniques of milling, producing WWF using RM seems to be advantageous as opposed to stone mill due to lower temperature as well as better control over entire operations. Jet milling also seems to be promising. However additional research has to be carried out to explore its potential. Identification of the key biomarkers is also important to differentiated products made from WWF from the regulatory perspectives. However this area needs more time to mature before it can be added as regulatory norms. Among the available methods to enhance WWF stability, retention of antioxidant during milling as well as heat treatment to inactivate enzymes seems to be promising. FCIR seems to be a promising technology, however its need future investigation. Different strategies for functionalization of wheat bran need to be explored in the future to develop WWF with addition of stabilized bran.

KEYWORDS

- alk(en)ylresorcinols (ARs)
- alpha-amylase antioxidant
- bran
- degradation
- diglycerides (DGs)
- endosperm
- enzymes
- extrusion
- free fatty acids (FAs)
- flour
- germ
- infrared
- jet milling
- lipase
- lipoxygenase
- milling
- milling technique
- nonpolar lipids
- oxidation
- oxylipins
- particle size
- phospholipids
- polyphenol oxidase
- rancidity
- refined flour
- roller mill
- sphingolipids
- stability
- stone mill
- superfine grinding
- triglycerides (TGs)
- ultrafine grinding
- whole wheat flour
- γ-oryzanol

REFERENCES

1. Adom, K. K.; Sorrells, M. E.; Liu, R. H. Phytochemicals and Antioxidant Activity of Milled Fractions of Different wheat Varieties. *J. Agric. Food Chem.* **2005,** *53*(6), 2297–2306.
2. Agundez, A. Z.; Fernández R. M.; Arce, C.; Cruz Z.; Melendrez, R.; Chernov,V.; Barboza F. Gamma Radiation Effects on Commercial Mexican Bread Making Wheat Flour. *Nucl. Instrum. Methods Phys. Res. Section B: Beam Interactions Mater. Atoms* **2006,** *245*(2), 455–458.
3. Alam, S.; Ullah S.; Saleemullah, M.; Riaz, A. Comparative Studies on Storage Stability of Ferrous Iron in Whole Wheat Flour and Flat Bread (naan). *Int. J. Food Sci. Nut.* **2007,** *58*(1), 54–62.
4. Ampatzoglou, A.; Atwal, K. Increased Whole Grain Consumption Does Not Affect Blood Biochemistry, Body Composition, or Gut Microbiology in Healthy, Low-Habitual Whole Grain Consumers. *J. Nut.* **2015,** *145*(2), 215–221.
5. Apprich, S.; Tirpanalan, T.; Hell, J.; Reisinger, M.; Böhmdorfer, S. Wheat Bran-Based Biorefinery, 2: Valorization of Products. *LWT-Food Sci. Technol.* **2014,** *56*(2), 222–231.
6. Bahrami, N.; Bayliss, D.; Chope, G.; Penson, S.; Perehinec, T. Cold plasma: A New Technology to Modify Wheat Flour Functionality. *Food Chem.* **2016,** *202*(6), 247–253.
7. Barnes, P. Lipid Composition of Wheat Germ and Wheat Germ Oil. *Fette, Seifen, Anstrichmittel* **1982,** *84*(7), 256–269.
8. Bhat, N. A.; Wani, I.; Hamdani, A.; Gani, G .; Masoodi, F. Physicochemical Properties of Whole Wheat Flour as Affected by Gamma Irradiation. *LWT-Food Sci. Technol.* **2016,** *71*(3), 175–183.
9. Burnette, D.; Weaver, S. H. Marketing, Processing, and Uses of Oat for Food. *Oat Sci. Technol.* **1992** (oatscienceandte), 247–263.
10. De la Hera, E.; Rosell, C. Particle Size Distribution Affecting the Starch Enzymatic Digestion and Hydration of Rice Flour Carbohydrates. *Carbohydr. Polym.* **2013,** *98,* 421–427.
11. Delcour, J.; Hoseney, R. *Principles of cereal science and technology;* AACC International. Inc.: St. Paul, MN, USA, 2010; pp 229–235.
12. Doblado M.; Rose, D. Key Issues and Challenges in Whole Wheat Flour Milling and Storage. *J. Cereal Sci.* **2012,** *56*(2), 119–126.
13. Dornez, E.; Gebruers, K.; Wiame, S.; Delcour, F. Insight into the Distribution of Arabinoxylans, Endoxylanases, and Endoxylanase Inhibitors in Industrial Wheat roller Mill Streams. *J. Agr. Food Chem.* **2006,** *54*(22), 8521–8529.
14. Engel, C.; Heins, J. The Distribution of the Enzymes in resting Cereals II. The Distribution of the Proteolytic Enzymes in Wheat, Rye, and Barley. *Biochim. Biophys. Acta* **1947,** *1,* 190–196.
15. Fierens, E.; Helsmoortel, J. Changes in Wheat *(Triticum aestivum L.)* Flour Pasting Characteristics as a Result of Storage and Their Underlying mechanisms. *J. Cereal Sci.* **2015,** *65,* 81–87.

16. Galliard, T. Hydrolytic and Oxidative Degradation of Lipids During Storage of Wholemeal Flour: Effects of Bran and Germ Components. *J. Cereal Sci.* **1986,** *4*(2), 179–192.

17. Galliard, T. The Effects of Wheat Bran Particle Size and Storage Period on Bran Flavour and Baking Quality of Bran/Flour Blends. *J. Cereal Sci.* **1988,** *8*(2), 147–154.

18. Geng, P.; Harnly, J. M.; Chen, P. Differentiation of Bread Made with Whole Grain and Refined Wheat *(T. aestivum)* Flour Using LC/MS-Based Chromatographic Fingerprinting and Chemometric Approaches. *J. Food Comp. Ana.* **2016,** *8*(2), 21–34.

19. Geng, P.; Harnly, M.; Chen, P. Differentiation of Whole Grain from Refined Wheat *(T. aestivum)* Flour Using Lipid Profile of Wheat Bran, Germ, and Endosperm with *UHPLC-HRAM* Mass Spectrometry. *J. Agric. Food Chem.* **2015,** *63*(27), 6189–6211.

20. Goesaert, H.; Delcour, J. Wheat Flour Constituents: How They Impact Bread Quality, and How to Impact Their Functionality. *Trends Food Sci. Technol.* **2005,** *16*(1), 12–30.

21. Gómez, M.; González, J; Oliete, B. Effect of Extruded Wheat Germ on Dough Rheology and Bread Quality. *Food Bioprocess Technol.* **2012,** *5*(6), 2409–2418.

22. Hansen, L. Sensory Acceptability is Inversely Related to Development of Fat Rancidity in Bread Made from Stored Flour. *J. Acad. Nut. Diet.* **1996,** *96*(8), 792.

23. Hatcher, D.; Kruger, J. Distribution of Polyphenol Oxidase in Flour Millstreams of Canadian Common Wheat Classes Milled to Three Extraction Rates. *Cereal Chem.* **1993,** *70*(5), 51–51.

24. Hatcher, D.; Kruge J. Simple Phenolic Acids in Flours Prepared from Canadian Wheat: Relationship to Ash Content, Color, and Polyphenol Oxidase Activity. *Cereal Chem.* **1997,** *74*(3), 337–343.

25. Heinio, R.; Lehtinen, P. Differences Between Sensory Profiles and Development of Rancidity During Long-Term Storage of Native and Processed Oat. *Cereal Chem.* **2002,** *79*(3), 367.

26. Heiniö, R.; Noort, M. Sensory Characteristics of Wholegrain and Bran-Rich Cereal Foods–a Review. *Trends Food Sci. Technol.* **2016,** *47*(3), 25–38.

27. Hemdane, S.; Leys, S. Wheat Milling By-Products and Their Impact on Bread Making. *Food Chem.* **2015,** *187*(5), 280–289.

28. Hemery, Y.; Lullie, P. Biochemical Markers: Efficient Tools for the Assessment of Wheat Grain Tissue Proportions in Milling Fractions. *J. Cereal Sci.* **2009,** *49*(1), 55–64.

29. Inamdar, A.; Prabhasankar, P. Chapati Making Quality of Whole Wheat Flour (atta) Obtained by Various Processing Techniques. *J. Food Process. Preserv.* **2015,** *39*(6), 3032–3039.

30. Initiative for *Global Progress.* http://www.ffinetwork.org/global_progress/December, 2012, p 110 (accessed Sept 9, 2016).

31. Katina, K.; Heiniö, H. Optimization of Sourdough Process for Improved Sensory Profile and Texture of Wheat Bread. *LWT-Food Sci. Technol.* **2006,** *39*(10), 1189–1202.

32. Kent, N. *Kent's Technology of Cereals: An Introduction for Students of Food Science and Agriculture;* Elsevier, New York; 1994; pp 23–56.

33. Kihlberg, I.; Johansson, L. Sensory Qualities of Whole Wheat Pan Bread—Influence of Farming System, Milling and Baking Technique. *J. Cereal Sci.* **2004,** *39*(1), 67–84.

34. Letang, C.; Samson, M.; Lasserre, F.; Chaurand, Y.; Abecassis, J. Production of Starch with Very Low Protein Content from Soft and Hard Wheat Flours by Jet Milling and Air Classification. *Cereal Chem.* **2002,** *79*(4), 535–543.

35. Li, B.; Chen, H. Effect of Flameless Catalytic Infrared Treatment on Rancidity and Bioactive Compounds in Wheat Germ Oil. *RSC Adv.* **2016,** *6*(43), 37265–37273.

36. Liu, R. H. Whole Grain Phytochemicals and Health. *J. Cereal Sci.* **2007,** *46*(3), 207–219.

37. Liyana, P.; Shahidi, F. Antioxidant and Free Radical Scavenging Activities of Whole Wheat and Milling Fractions. *Food Chem.* **2007,** *101*(3), 1151–1157.

38. Marathe, S.; Machaiah, J. Extension of Shelf-life of Whole-Wheat Flour by Gamma Radiation. *Int. J. Food Sci. Technol.* **2002,** *37*(2), 163–168.

39. Masisi, K.; Moghadasian, M. Antioxidant Properties of Diverse Cereal Grains: A Review on in Vitro and in Vivo Studies. *Food Chem.* **2016,** *196*(3), 90–97.

40. Mateo A.; Berg, V. Ferulic Acid from Aleurone Determines the Antioxidant Potency of Wheat Grain *(Triticum aestivum L.). J. Agric. Food Chem.* **2008,** *56*(14), 5589–5594.

41. Miller, B.; Kummerow, F. The Disposition of Lipase and Lipoxidase in Baking and the Effect of Their Reaction Products on Consumer Acceptability. *Cereal Chem.* **1948,** *25*(4), 391–398.

42. Morrison, W.; Hargin, D. Distribution of Soft Wheat Kernel Lipids Into Flour Milling Fractions. *J. Sci. Food Agric.* **1981,** *32*(6), 579–587.

43. Nelson, P.; McDonald, C. Properties of Wheat Flour Protein in Flour from Selected Mill Streams. *Cereal Chem.* **1977,** *54*(6), 1182–1191.

44. Niu, M.; Hou, G.; Wang, L. Effects of Superfine Grinding on the Quality Characteristics of Whole-Wheat Flour and its Raw Noodle Product. *J. Cereal Sci.* **2014,** *60*(2), 382–388.

45. Pareyt, B.; Finnie, M.; Putseys, J. Lipids in Bread Making: Sources, Interactions, and Impact on Bread Quality. *J. Cereal Sci.* **2011,** *54*(3), 266–279.

46. Peterson, D.; Hahn, H. Oat Avenanthramides Exhibit Antioxidant Activities in Vitro. *Food Chem.* **2002,** *79*(4), 473–478.

47. Poutanen, K. Enzymes: An important Tool in the Improvement of the Quality of Cereal Foods. *Trends Food Sci. Technol.* **1997,** *8*(9), 300–306.

48. Prabhasankar, P.; Rao, P. Effect of Different Milling Methods on Chemical Composition of Whole Wheat Flour. *Eur. Food Res. Technol.* **2001,** *213*(6), 465–469.

49. Prabhasankar, P.; Sudha, M.; Rao, P. Quality Characteristics of Wheat Flour Milled Streams. *Food Res. Int.* **2000,** *33*(5), 381–386.

50. Preston, K.; Kruger, K. Location and Activity of Proteolytic Enzymes in Developing Wheat Kernels. *Can. J. Plant Sci.* **1976,** *56*(2), 217–223.

51. Protonotariou, S.; Mandala, K.; Rosell, C. Jet Milling Effect on Functionality, Quality and in Vitro Digestibility of Whole Wheat Flour and Bread. *Food Bioprocess Technol.* **2015,** *8*(6), 1319–1329.

52. Prückler, M.; Siebenhandl, S.; Apprich, S.; Höltinger, S. Wheat Bran-Based Biorefinery 1: Composition of Wheat Bran and Strategies of Functionalization. *LWT-Food Sci. Technol.* **2014,** *56*(2), 211–221.

53. Ragaee, S.; Abdel, A. Antioxidant Activity and Nutrient Composition of Selected Cereals for Food Use. *Food Chem.* **2006,** *98*(1), 32–38.

54. Rani, K.; Rao,U.; Leelavathi, K.; Rao, P. Distribution of Enzymes in Wheat Flour Mill Streams. *J. Cereal Sci.* **2001,** *34*(3), 233–242.

55. Reisinger, M.; Tirpanalan, O.; Huber, F.; Kneifel, W. Investigations on a Wheat Bran Biorefinery Involving Organosolv Fractionation and Enzymatic Treatment. *Bioresour. Technol.* **2014,** *170*(2), 53–61.

56. Rose, D J.; Ogden, M. Enhanced Lipid Stability in Whole Wheat Flour by Lipase Inactivation and Antioxidant Retention. *Cereal Chem.* **2008,** *85*(2), 218–223.

57. Salman, H.; Copeland, L. Effect of Storage on Fat Acidity and Pasting Characteristics of Wheat Flour. *Cereal Chem.* **2007,** *84*(6), 600–606.

58. Shiiba, K.; Negishi, Y. Purification and Characterization of Lipoxygenase Isozymes from Wheat Germ. *Cereal Chem.* **1991,** *68*(2), 115–122.

59. Sjövall, O.; Virtalaine, T. Development of Rancidity in Wheat Germ Analyzed by Headspace Gas Chromatography and Sensory Analysis. *J. Agric. Food Chem.* **2000,** *48*(8), 3522–3527.

60. Slavin, J. Whole Grains and Human Health. *Nut. Res. Rev.* **2004,** *17*(01), 99–110.

61. Tsuzuki, W.; Suzuki, Y. Effect of Oxygen Absorber on Accumulation of Free Fatty Acids in Brown Rice and Whole Grain Wheat During Storage. *LWT-Food Sci. Technol.* **2014,** *58*(1), 222–229.

62. Wallace, M.; Wheeler, E. Lipoxygenase from Wheat: Reaction Characteristics. *J. Agric. Food Chem.* **1975,** *23*(2), 146–150.

63. Wang, L.; Flores, R. Effect of Different Wheat Classes and Their Flour Milling Streams on Textural Properties of Flour Tortillas. *Cereal Chem.* **1999,** *76*(4), 496–502.

64. Wang, L.; Luo, Y. Effect of Deoxynivalenol Detoxification by Ozone Treatment in Wheat Grains. *Food Control* **2016,** *66*, 137–144.

65. Wang, S.; Toledo, M. Inactivation of Soybean Lipoxygenase by Microwave Heating: Effect of Moisture Content and Exposure Time. *J. Food Sci.* **1987,** *52*(5), 1344–1347.

66. Zhang, H.; Zhang, Y. Retention of Deoxynivalenol and its Derivatives During Storage of Wheat Grain and Flour. *Food Control* 65(5)**2016.**

67. Zou, Y.; Yang, M.; Zhang, G. Antioxidant Activities and Phenolic Compositions of Wheat Germ as Affected by the Roasting Process. *J. Am. Oil Chem. Soc.* **2015,** *92*(9), 1303–1312.

CHAPTER 9

PHYSICOCHEMICAL PROPERTIES AND QUALITY OF FOOD LIPIDS

AMIT KUMAR MUKHERJEE

Department of Food Technology, Haldia Institute of Technology, P. O. HIT (Hatiberia), Haldia 721657, West Bengal, India, Tel.: +91-9477290235, E-mail: mukherjee2001@gmail.com

CONTENTS

9.1   INTRODUCTION

Edible oils and fats are produced from plant seeds, carcasses of land, and marine animals and marine fish. Depending on the source of lipids, their composition and physical behavior markedly vary in food. Lipid substances are distinctly hydrophobic in nature with lower specific gravity than water. The lipids impart characteristic flavor and taste in our food. However, some modified fats like shortenings used in bakery and salad oils possess a bland flavor. The desirable aroma and flavor of fried and cooked foods are much due to the development of flavor by the reaction of fat with the food ingredients than due to the flavor of fat itself. While dealing with edible fats and oils and the food products derived from them, we frequently use two terms, namely, "saturated fats" and "unsaturated fats." A fat becomes saturated or unsaturated due to its fatty acid composition in the triacylglycerol (TAG). Since no natural fat or oil contains TAGs made up of only saturated or only unsaturated fatty acids, we cannot grossly classify them in that manner without deciding some basis for such differentiation.

Soybean oil contains some TAGs entirely made up of unsaturated fatty acids. However, this oil is also a mixture of different TAG molecules with variable fatty acid composition including a number of saturated fatty acids in them. A saturated fat is a semisolid mass due to its high level of saturated fatty acids, whereas unsaturated fats are liquids at room temperature. By observing the texture and the physical state of existence, an apparent differentiation is done between the two types. The sources of saturated fats are butter, margarine, and animal fats such as lard, tallow. Unsaturated fats are soybean oil, cottonseed oil, sunflower oil, canola oil, fish liver oil, and so on. A TAG containing higher percentages of saturated fatty acids esterified with the glycerol is called a saturated fat and that containing higher percentages of unsaturated fatty acids is unsaturated fat. The degree of unsaturation of the fatty acids in triglycerides or TAG molecules entirely governs the physical and chemical behaviors of the fat sample. Since the glycerol fraction in these molecules has no role other than holding three similar or dissimilar fatty acid molecules and only taking part in reactions in its free state of occurrence, the nature of the fatty acids almost entirely dominates the lipid chemistry. The physical properties of the natural fat samples are also not an exception to that.

The fats differ from oils only in their physical states of occurrence. The oils are liquids at the temperature of storage or below 25°C, which sometimes referred to as "room temperature." The fats, on the other hand, are solids at 25°C. A solid fat has yield stress higher than 100 Pa at 25°C. TAG (also called triglyceride) molecules are further classified according to the types of fatty acids present in them. A "simple triglyceride" contains all three fatty acids identical to the TAG, whereas in a "mixed triglyceride" all three fatty acids are different. A refined edible sample of lipids contains about 98–99% TAGs with some minor quantities of other constituents.

In this chapter, the author will frequently call edible lipids as fats or oils and that would not necessarily isolate these two states unless they completely differ in their molecular compositions. The oil is a fat at lower temperature and a fat can produce an oil on melting down at a higher temperature. The crude fats, extracted from plant or animal cells, contain various lipid substances like monoacylglycerols and diacylglycerols (MAGs and DAGs), phospholipids, nonphosphorylated lipids, hydrocarbons, waxes, free fatty acids (FFAs), pigments, sterols, and so on. None of them is a polymer as no repetitive monomer exists in their structures. Therefore, lipids are groups of hydrophobic nonpolymeric macromolecular organic compounds unlike polysaccharides and protein macromolecules. Polysaccharides and proteins have much larger molecular masses than lipids.

The average molecular mass of sunflower, rapeseed, and olive oils is 876, 992, and 857, respectively.[1] The molecular mass of a TAG is dictated by the chain length of the fatty acids. The fatty acids are divided into four classes with respect to the number of carbon atoms in their hydrocarbon chains: short-chain (C_4 to C_6), medium-chain (C_8 to C_{14}), long-chain (C_{16} to C_{18}), and very long-chain (C_{20} or more) fatty acids. As stated earlier, crude fats and oils are mixtures of different cellular lipids. These are TAGs, DAGs, MAGs, free fatty acids, phospholipids, waxes, hydrocarbons, sterols, different pigments, metal ions, fat soluble vitamins, vitamin precursors and other finely dispersed cellular nonlipid components, which had escaped into the fats or the oils during extraction. Composition and percentage of such components in the extracted fats and oils vary depending upon the source.

The MAGs and DAGs are not desirable components in edible fats, as they have a tendency to arrest water (like common emulsifiers) and to reduce the shelf-life of the fat. The free fatty acids are another type of undesirable components in edible lipids, which impart acidity to the oil. On prolonged storage, they also encourage rancidity of fats and develop an unpleasant odor to the oil. Prooxidant character of these two types of components has been investigated.[9,13] A rancid fat develops toxicity due to the formation of peroxides and hydroperoxides and off-flavor due to the generation of decomposition products of peroxides. Metal ions such as copper, iron, calcium, and magnesium contaminate fats and oils during their extraction from the biological cells. These bivalent metal cations are cellular components and are cofactors for various enzymes. They remain attached to the phospholipids making the later nonhydratable. Copper and iron act as catalysts for autooxidation of oils and fats and their presence is grossly undesirable if the fat or oil sample is intended for long storage and for the manufacture of bakery shortening and cream which are subjected to aeration.

Pigments extracted with fats and oils have a mixed impact on human health. Such pigments are mainly carotenoids and chlorophyll. Carotenoids with hydroxyl-groups in their structure are known as xanthophylls. These pigments impart yellow (soybean oil), red (palm oil), or greenish (olive oil) color to the crude oil. The color of the carotenoid pigments is due to the conjugated double bonds (chromophores) present in them in an isoprenoid structure. The xanthophylls produce much deeper colors due to the presence of auxochromes (–OH group) in addition to the chromophores. The color of oils is undesirable when a secondary processing is applied to them after refining for the manufacture of margarine, shortening, cream and frying fats. It also leads to consumers' dissatisfaction when they purchase intensely colored cooking oil for the kitchen. However, a compromise is always done as there is little scope for 100% removal of pigments from the vegetable oil without damaging its triglyceride portion. Some carotenoids (e.g., α- and β-carotenes) are precursors of vitamin A. Carotenoids, due to having isoprenoid structures consisting of a series of conjugated double bonds, are highly susceptible to oxidation. Chlorophyll pigment is present in olive and other oils. Chlorophyll color is subsided by the carotenoids in many oils to give them yellow to reddish appearance. Chlorophyll has porphyrin structure with a central magnesium atom. The color of chlorophyll is intensified by heat and acid due to the formation of pheophytin. Chlorophyllides and pheophorbides impart phototoxicity and photosensitive dermatitis. Chlorophyll is sensitizer of light, which helps in photooxidation. This pigment is completely decomposed into its derivatives during processing of vegetable oils. Gossypol is a polyphenolic terpenoid aldehyde and is responsible for the intense red color of cottonseed oil.

Gossypol is toxic and imparts clinical disorders such as weakness, apathy, and reproductive malfunctioning in human.[4] This polyphenolic aldehyde destroys amino acids of foods. Gossypol color and chlorophyll breakdown products are difficult to remove by the conventional bleaching operation done on oils.

Fat is energy producer for our body. Among the three proximate constituents of food, namely, carbohydrate, protein and fat, the latter produces the highest amount of energy (9.2 kcal/g) on burning. Fat is accumulated in adipose tissues, kidney, liver, and omentum of animals. The high-energy value of fat is due to the fact that a single long-chain fatty acid in the triglyceride molecule can generate many acetyl-coA molecules on decomposition—a process known as β-oxidation. This, in due course of anther subsequent transformation into the TCA (tricarboxylic acid) cycle, produces ATP (adenosine triphosphate) molecules. The fats are stored in our body in almost pure form and are devoid of water unlike carbohydrates and proteins. This happens because of their hydrophobic nature. Therefore, 1 g of fat represents a higher amount of pure substance on a dry basis as compared to 1 g of carbohydrate or 1 g of protein.

The triglycerides are also known as "neutral lipids" for their nonpolar behavior, whereas phospholipids are "polar lipids." Despite being the most energy-efficient component, fat is not the first choice for the metabolic energy in our body. Fats are metabolized, next to the carbohydrates, for supplying our body needs of energy only under carbohydrate-depleting conditions and during starvation. A wide range of fatty acids of 4 to 22 carbon atoms present in the triglycerides are consumed by us in the form of various foods.

This chapter covers discussions on the physicochemical behavior of edible fats and oils and their importance in maintaining the quality of those lipid substances suitable for human consumption.

9.2 CLASSIFICATION OF LIPIDS

From the structural perspective, lipids are broadly classified into four groups, namely, (I) Fats, (II) Waxes, (III) Phospholipids, (IV) Nonphosphorylated lipids and (V) Free fatty acids. The lipids of these broader groups are further divided into various categories according to the differences in the substituent groups present in these compounds (Fig. 9.1).

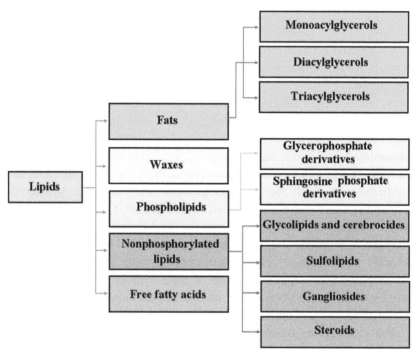

FIGURE 9.1 Classification of lipids: Lipids are divided into five groups. Fats are divided into three categories, phospholipids into two, and nonphophorylated lipids into four categories.

9.2.1 TRIACYLGLYCEROLS

The TAGs are glycerol-esters of short- (C_4 to C_6), medium- (C_8 to C_{14}), long- (C_{16} to C_{18}), and very long-chain (C_{20} and above) fatty acids. They are major desirable components of any edible fat or oil. They are "simple lipids." The TAGs from different sources (plants or animals) are highly viscous, hydrophobic substances with varying melting points. They are also classed as "neutral lipids." A visible fat sample contains mainly triglyceride molecules. The lipids of plant origin are highly colored substances, as they dissolve pigments of plants when extracted from the oil-bearing cells. A natural sample of oil or fat does not contain a cluster of a single type of molecule of the same chemical structure and molecular mass, but is a mixture of different TAGs comprising of different fatty acids esterified with the glycerol. As the name suggests, they are the tri-esters of three fatty acids and one glycerol molecule. When all three fatty acids are same in a TAG, the latter is known as "simple TAG." A "mixed TAG" molecule contains at least two different fatty acids. The TAGs differ from the phospholipids in having no phosphorous atom in the molecule and in being completely non-polar under all conditions of temperature, pH, and hydration. They are lighter (lower specific gravity) than water and float on an aqueous surface creating an extended interfacial area of contact with aqueous phase. The saturated and high molecular weight, long-chain fatty acids increase the tendency of a TAG for being solid at room temperature (at 25°C).

9.2.2 DIACYLGLYCEROLS

The DAGs are di-esters of glycerol and are placed into the group of neutral lipids. However, due to the presence of one free –OH group of the glycerol, they have tendency to form H-bonds with water. The diacylglycerols can produce three structural isomers, namely, 1,2-DAG; 2,3-DAG; and 1,3-DAG. The 1,3-structural isomers are more stable (with higher melting point) due to the formation of intramolecular H-bonding.[8] The properties of the carbon atoms in the glycerol backbone tend to regain their characteristic, which they originally had in the free glycerol molecule when abstraction of one or more fatty acids occurs from the TAG molecule. This impact is more pronounced in the crystalline structures of MAGs and DAGs. The diacylglycerol molecule is amphiphilic in nature having extensive surface-active properties. They are used as emulsifiers in margarine, breads, rolls, peanut butter, drugs, and cosmetics. Controlled chemical and enzymatic hydrolysis reactions of TAGs produce DAGs. Both MAGs and DAGs can be produced by the solvent-free glycerolysis of TAGs. Monoacylglycerols and Diacylglycerols are not removed by degumming, alkali refining or by bleaching and pose enormous problem in the storage of oils. The water molecules arrested by them encourage microbial growth and hydrolysis in the oils. MAGs and DAGs are partially removed during deodorization. The DAGs show optical activity due to the development of chiral center in the glycerol backbone.

9.2.3 MONOACYLGLYCEROLS

The MAGs are non-ionic, amphiphilic molecules with two free hydroxyl groups (Fig. 9.2) in the glycerol molecule. The third –OH group is esterified with a fatty acid. The term "partial acylglycerols" is used for indicating DAGs and MAGs.

The central C-atom of the glycerol is chiral when MAG is either *sn*-1-MAG or *sn*-3-MAG. When the fatty acid is attached to any one of the two primary alcoholic groups, the MAG is called α -isomer. When it is in the secondary carbon atom, the isomer is β.

9.2.4 WAXES

The wax is an ester of a fatty acid (C_{10} or above) and a monohydric alcohol of higher molecular mass. A natural oil sample becomes cloudy (forms haze or turbidity) on cooling due to the presence of waxes. The waxes are of animal, vegetable, and synthetic origin. Another type of wax known as mineral wax and is made of paraffin hydrocarbons. Waxes have lesser affinity for an oil phase and separate readily on cooling. Examples are carnauba, ouricouri, montan, bee's wax, candelilla wax, spermaceti (cetin), and so on. Waxes are used in cosmetics, as lubricants, in polishes, as coating materials for food preservation, as thickeners and in ink manufacture. Rice bran wax is used in producing organogel of unique physical properties. Waxes are kneadable by the application of shear force. They have low iodine value and have melting points within 40–120°C.

The waxes in oils crystallize at high temperatures than the TAGs. Rice bran, canola, corn, sunflower, and safflower oils are subjected to dewaxing before their specific applications at low temperatures. Waxes are neutral lipids and saponifiable. They contribute to the saponification numbers of the crude fats. The wax content of edible oils may be as high as 2000 ppm (parts per million). This amount is reduced to below 10 ppm for obtaining stable and clear oil at low temperatures. Rice brand oil has comparatively higher wax content (3–4%) and needs a pre-dewaxing stage before refining. The hulls of sunflower seeds contribute to most of the waxes. Dewaxing is usually done after pre-bleaching. However, for very high wax content in the oil, a prior dewaxing is necessary. The waxes are alternative energy source for some animals, plants, and microorganisms.

9.2.5 PHOSPHOLIPIDS

Phospholipids are polar functional lipids extracted from the biological cells into the oil. The extent of phospholipids in the crude oil depends upon the type of extraction. The hexane extraction process collects higher quantity of phospholipids than the pressing operation. However, oilseeds containing lower percentage of oil cannot be pressed for an efficient extraction of oil. The pressing operation is also unable to extract most of the oils from oil-bearing materials, and a solvent extraction is needed after pressing to leave the deoiled cake with 1% or lesser amount of residual oil in it.

The phospholipids are membrane lipids and can be classified into two categories, namely, glycerophosphate derivatives and sphingosine phosphate derivatives. The glycerophosphates are di-esters of fatty acids with glycero-phosphoric acid. The sphingosine derivatives contain a nitrogenous base esterified with the phosphoric acid and an esterified fatty acid (which is the R–CO– group of sphingomyelin in Fig. 9.2).

Sphingosines do not contain glycerol; instead, they contain amino-alcohols. These lipids participate in cellular growth, differentiation and apoptosis. They are found in brain and nervous tissues of higher animals. Plasmalogens are another form of phospholipids, but they contain one long-chain (C_{12} to C_{18}) α,β-unsaturated alcohol ether-linked with the first—CH_2OH group of glycerol (at sn-1 carbon). The remaining part of the molecule resembles the structure of phosphatidylcholine or phosphatidyl-ethanolamine—the glycerophospholipids. Phospholipids are "compound lipids." Due to the presence of phosphate groups in their structures, glycerophospholipids are amphiphilic and have emulsifying properties. They are commonly known as "gums" of oil and are removed by degumming. Among the two positions of the glycerol carbons (sn-1 and sn-2) allotted for the fatty acids, the first is mostly occupied by a saturated fatty acid. The second carbon atom is preferred by the unsaturated fatty acids. Crude soybean oil contains up to 4% glycerophospholipids. As little is known about the sphingosine derivatives, which could have acted in a different way either physically or chemically during processing of edible oils and

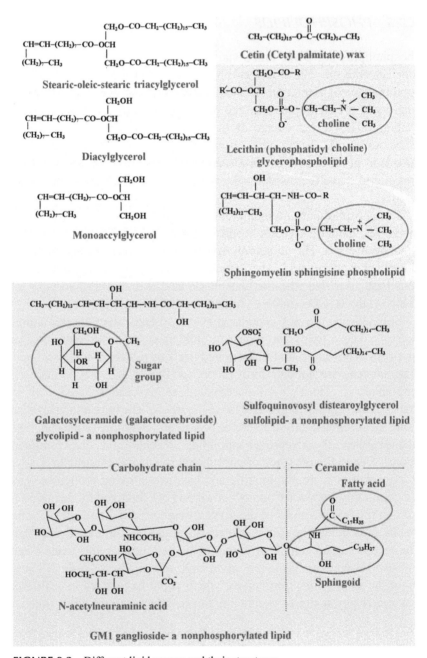

FIGURE 9.2   Different lipid groups and their structures.

fats, the term "phospholipids" is universally used for glycerophospho-lipids. The glycerophospholipids present in oils and fats in two forms. The hydratable phospholipids are easily removed by hydration of the gums following a simple water degumming operation. But the nonhydratable form present in many vegetables oils is difficult to remove unless the gum is acidulated with phosphoric or citric acid. The nonhydratable phospho-lipids are mainly associated with Ca^{++}, Mg^{++} and Fe^{++} ions and are difficult to hydrate. These bivalent metal ions inactivate the polar phosphoric acid group by chelation.

The percentage of nonhydratable phospholipids varies in different sources of oils, but usually remains within 20–45% of the total phos-pholipids. Phosphatidylinositol and phosphatidic acid are considered as nonhydratable phospholipids. On the other hand, phosphatidylcholin (leci-thin) and phosphatidylethanolamine (cephalin) are hydratable phospho-lipids. By the addition of acid, the bivalent metal cations are separated as salts and the gum becomes free. It is evident that the crude oils also contain monovalent metal cations (Na^+, K^+, etc.) and their presence does not complicate the degumming process in that manner. The three major glycerophospholipids present in soybean oil are phosphatidylcholine (about 55–56%), phosphatidylethanolamine (about 25–26%), and phos-phatidylinositol (18–19%). Hence, most of the gums in soybean oil are hydratable and they are good sources of lecithin production.

The phospholipids have a polar head and a nonpolar hydrocarbon tail in their structures. The hydrophobic tails orient themselves away from the water, while the polar head groups remain attached to the aqueous phase. Phospholipids form monolayers, bilayers, and micelles. Liposomes are similar aggregation of amphiphilic molecules with intermittent aqueous phase in the lipid bilayer produced by sonication. Liposomes have appli-cations in medical sciences. Due to the formation of clusters (reverse micelle) of several phospholipid molecules in the aqueous phase, the specific gravity of the gum increases and it is separated from the oil phase by centrifugation.

9.2.6 NONPHOSPHORYLATED LIPIDS

Nonphosphorylated lipids (another form of compound lipids) are glycolipids and cerebrosides, sulfolipids, gangliosides, and steroids.

Cerebrosides are simpler glycolipids. Metabolic disorders of cerebrosides lead to sphingolipidoses (lipid-storage disease). Glycolipids are found in the cell membrane and in nervous tissues. They contain a ceramide and one or more sugars. Ceramides are sphingosine derivatives of fatty acids. Gangliosides are complex nonphosphorylated glycophospholipids present in ganglions. The gangliosides are cerebroside derivatives with one or more N-acetylneuraminic acids (NANAs). The NANAs are sialic acids consisting of N-acetylmannose and pyruvic acid. GM1 is a brain ganglioside primarily synthesized in the endoplasmic reticulum and is modified in the golgi apparatus with the inclusion of carbohydrates.[14] Sulfolipids contain sulfate groups attached to the ceramide oligosaccharides.

The steroid compounds contain a cyclopentanoperhydrophenanthrene (CPPP) nucleus of three six-member (rings A, B, and C) rings and one five-member (ring D) ring fused together consisting of a total of 17 carbon atoms (Fig. 9.3).

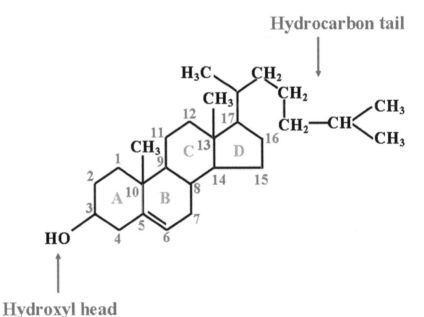

FIGURE 9.3   Structure of cholesterol.

They are derived lipids. Both plant (phytosterols) and animal (zoosterols) cells contain sterols. Brassicasterol, β-sitosterol, campesterol, avenasterol and stanols are plant sterols. Crude vegetable oils contain about 1–5 g of sterols per kg of sample. Corn, wheat germ, and rapeseed oils contain higher amount of sterols. Corn contains about 65 different sterols. The steroid compounds are removed from oil during deodorization and physical refining. Cholesterol is the major animal sterol first isolated from bile. The sterols contain a –OH group head on one side and another aliphatic hydrocarbon tail on the other side of the fused rings. This makes them amphiphilic. The sterols are nonsaponifiable lipids. Plant sterols are not absorbed in the human body and they replace cholesterol of blood by a mechanism of competitive micelle formation with the bile acids. However, oxidation products of plant sterols are not good for health. Hence, their removal is essential at the final stage of refining.

9.2.7 FREE FATTY ACIDS

Free fatty acids (FFAs) are derived (derived lipids) from the TAGs by hydrolysis and are always present in crude oils. They remain in the oil phase and are hydrophobic in nature. The quality of refined edible oils is judged by the efficient removal of free fatty acids. Refined edible oil should contain 0.05% or lesser FFA. The alkali refining process removes most of the FFAs from oil. The last traces are removed during the deodorization operation. The physical refining step (performed as an alternative to the chemical refining) is meant for the removal of FFAs. The quantity of the FFAs in extracted oils depends upon the source of oil and on the storage and handling conditions. The fatty acids are either saturated or unsaturated aliphatic monocarboxylic acids. Some oils contain hydroxyl fatty acids. For example, castor oil contains ricinoleic acid. The oil triglycerides with fatty acid substituents like –OH, C=O, cyclic hydrocarbons, and so on. are grossly inedible. Conjugated fatty acid is present in tung oil. This oil is derived from a tree *Aleuriles fordii* also known as montana. Conjugated linoleic acid (18:2, $\Delta^{9,11}$) is produced during partial hydrogenation of oils.

Natural fatty acids are even-carbon fatty acids, although odd-carbon fatty acids are also found in milk and in other animal fats. The fatty acids are produced in the cell either from acetyl-coA or from palmityl-coA. Acetyl-coA (2-carbon compound) produces fatty acids with even-carbon atoms, while palmityl-coA (3-carbon compound) produces both even and odd carbon fatty acids. In the group of saturated monocarboxylic acids, the first three members (formic or methanoic acid, acetic or ethanoic acid and propionic or propanoic acid) are colorless, pungent-smelling liquids. Those with C_4 to C_9 carbon atoms are oils with a characteristic smell of goat's butter. The fatty acids with C_{11} and higher carbons are odorless solids. Decanoic acid ($C_{10}H_{20}O_2$, capric acid) has melting point 31.6°C and possesses an unpleasant odor. The lower members of fatty acids are soluble in water. This is due to comparatively higher inductive effect ($+$ I effect) of alkyl groups in the lower members, which facilitates partial ionization of the carboxylic (–COOH) functional group and H-bond formation with water molecules. However, solubility decreases with increasing molecular mass of fatty acids. The volatility of the lower members is less and is a contrast to their low molecular masses. The lower fatty acids form cyclic dimers in water and in the vapor phase (Fig. 9.4). In the liquid phase, they exist as polymers. This behavior of low molecular mass fatty acids prevents them from being volatile. The melting points of even carbon fatty acids are higher than the odd carbon acids immediately preceding and following them in the series. The reason for this nonuniformity is uneven packing of odd fatty acids in the crystal lattice leading to greater strain in the crystal.

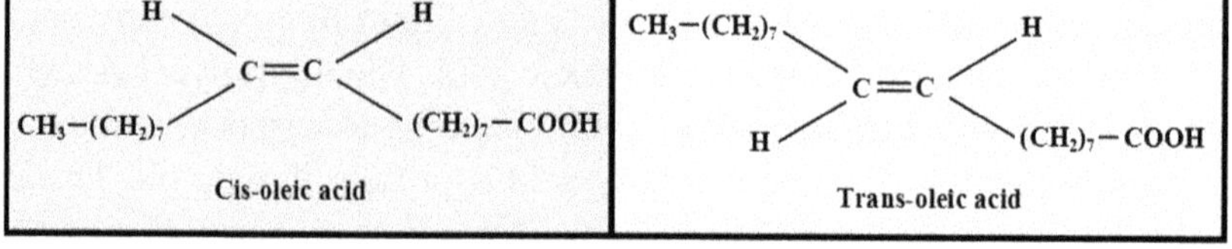

FIGURE 9.4   Acetic acid dimer and geometrical isomerism of fatty acid.

A marked degree of dissimilarities in physical properties also exists between the saturated fatty acids and their unsaturated counterparts. A C–C double bond contains a pair of π-electrons. Their presence makes the molecule more strained. The melting points of fatty acids increase with increasing molecular mass and decrease with increased unsaturation in the chain. The solubility (in organic solvents) increases with increased unsaturation and decreased chain length of fatty acids. A C–C double bond leads to geometrical isomerism (cis-trans isomerism) in the compound (Fig. 9.4). Natural fatty acids are *cis*-fatty acids. However, *trans*-fatty acids are detected in milk and animal fats. The *trans*-fatty acid content of TAGs is increased by extensive thermal treatment during refining and hydrogenation. The *trans*-fatty acids have higher melting points than the *cis*-isomers. The densities of the unsaturated acids are higher than the saturated fatty acids with same carbon atoms. The density decreases with increase in chain length. According to the position of the first double bond from the terminal methyl-group of the hydrocarbon chain, the unsaturated (and the polyunsaturated) fatty acids are named as ω (omega) acids. The positional number of the double bond is written in the suffix. Thus, linoleic acid is ω_6 acid and linolenic acid is ω_3 acid.

The above mode of classification of lipids does not include a few other lipid substances. These are lipoproteins (or proteolipids), fat-soluble pigments (chlorophylls, xanthophylls, and carotenes), fat-soluble vitamins (D, E, and K) and hydrocarbons (straight or branched chain aliphatic or alicyclic hydrocarbons). These are also important groups of hydrophobic entities remain associated with extracted fats and oils.

Lipoproteins are transport vehicles for cholesterol, TAGs, and so on to the blood plasma. Lipoproteins are of five types, namely, chylomicrons, very low-density lipoproteins (VLDL), low-density lipoproteins (LDL), high-density lipoproteins (HDL) and fatty acid-albumin complexes.[10] The albumin can form complexes with as many as 20–30 fatty acids. These are complexes of lipids and conjugated proteins. The animal fats extracted from fat-bearing tissues contain very low levels of phospholipids, carotenoids and tocopherols (<0.05%), but higher levels of cholesterol.[6] The cholesterol content of animal fat is 850–1100 mg/kg. Wackenroder extracted a pigment from carrot and named it carotene. It was later discovered that carotene is available in three different isomeric forms as α-carotene, β-carotene and γ-carotene. Carotenes contain an isoprenoid chain with conjugated double bonds. Isoprene is a diene. The β-carotene is widely distributed in plants. They also contain two rings at the terminals of the isoprenoid chain. Carotenoids with closed ring structures and without –OH groups are precursors of vitamin A (retinols). In lycopene (available in tomato, watermelon, rosehips, apricot, etc.), the rings are open and this carotenoid is not a precursor of vitamin A despite absence of any –OH group. Xanthophylls, on the other hand, contain –OH groups in their structures. All these pigments could easily be placed in the group of hydrocarbons, another minor component of crude oils, but their ability to color the oil and propagating oxidative changes made them isolated from that group.

Chlorophyll is another green pigment available in plant body parts which takes part in photosynthesis. The plant chlorophylls are chlorophyll-a and chlorophyll-b. This photosynthetic pigment contains magnesium, a metal having its impact in changing phospholipid behavior. Chlorophyll is a photosensitizer to produce singlet oxygen, which accelerates photo-oxidation of oils. Oxidation products of chlorophyll are harmful for the body. Crude oils contain about 1–100 ppm of chlorophyll. The green color of olive oil is due to its chlorophyll content. The color of the oil is removed by bleaching with activated clay or carbon, which adsorbs the pigments on its surface and is separated by filtration. Lighter color of refined oil explains about its quality to consumers. The hydrocarbons of edible fats and oils include various long-chain alkanes, alkenes, and polycyclic aromatic hydrocarbons. Squalene ($C_{30}H_{50}$), gadusene ($C_{18}H_{32}$), pristane ($C_{18}H_{28}$), zamene ($C_{19}H_{38}$), and so on are present in vegetable and marine oils. Squalene is a highly unsaturated triterpene and is abundant in fish liver oils. Tocopherols are vitamin E and they are powerful natural antioxidants.

Lipids are also classified in terms of their ability to respond to the saponification reaction. Under this scheme of classification, lipids are either saponifiable or nonsaponifiable. The saponifiable lipids are TAGs, DAGs, MAGs, waxes, free fatty acids, and phospholipids, whereas sterols, hydrocarbons and pigments are nonsaponifiable lipids.

Saponification (Fig. 9.5) is a process by which soap is prepared from long-chain fatty acids by reaction with metal ions (Na, K, Ca, Mg, Ni, Hg, Pb, etc.).

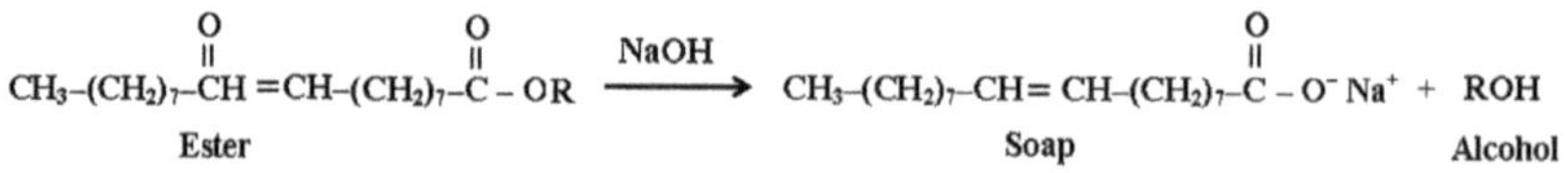

$$CH_3-(CH_2)_7-\overset{\overset{O}{\|}}{C}H=CH-(CH_2)_7-\overset{\overset{O}{\|}}{C}-OR \xrightarrow{\text{NaOH}} CH_3-(CH_2)_7-CH=CH-(CH_2)_7-\overset{\overset{O}{\|}}{C}-O^-Na^+ + ROH$$

Ester   Soap   Alcohol

FIGURE 9.5   Saponification reaction.

Phospholipids are saponified to produce metal salts of fatty acids, glycerol, inorganic phosphate salt of the metal (of alkali used) and amine or sugar. By the introduction of polar alkali metal, alkaline earth or heavy metal group into the fatty acid molecule, its hydrophobic character is lost. A polar tail and a nonpolar head are developed within the salt of fatty acid. This makes the resulting molecule an emulsifier. Soaps as emulsifiers trap the dirt in an aqueous solution within their micelles and remove them in the form of visible foams. Hard water contains metals (Ca, Mg, and Fe), which replace the metal originally attached to the soap by precipitating them as inorganic salts in solution. This reaction disturbs the emulsification process and the soap does not work well in hard water. Fatty acids are insoluble in water and are soluble in organic solvents like ethanol, ether and benzene. The neutralization reaction of fatty acids with alkali is carried out in an ethanolic medium. When triglycerides are subjected to reaction with aqueous ethanolic NaOH or KOH, they are first hydrolyzed into one molecule of glycerol and three molecules of fatty acids. These fatty acids then form the sodium or potassium salts.

Saponification is an index for determining the nature of fatty acids present in the TAGs. Fats, which undergo these reactions, should contain at least one esterified fatty acid molecule in their structures with an alcohol. Triglycerides, diglycerides, monoglycerides, phospholipids, and waxes all respond to saponification reaction. They are known as saponifiable fats, whereas sterols, hydrocarbons, and pigments do not get saponified in this manner and are nonsaponifiable fats. Many authors classify fats and oils in terms of their saponification behavior.

9.3 SOME QUALITY PARAMETERS OF EDIBLE FATS AND OILS

9.3.1 SAPONIFICATION NUMBER

Saponification number is the number of milligrams of KOH required to saponify 1 g of fat or oil. For the estimation of saponification number, calculated excess of standard ethanolic KOH (L_1 mL) is added to a weighed amount of fat (in solution) or oil and is refluxed on hot water bath for about 1 h. The amount of KOH in excess over to that required for saponification (L_2 mL) is estimated by titration with standard hydrochloric acid using phenolphthalein indicator. The difference (L_1-L_2) gives the milliliters of standard KOH actually required to saponify the weighed amount of fat (normally 1–2 g). A blank without the fat sample is also titrated against the standard acid. Saponification number of a sample of a refined fat can directly tell about the average molecular mass of the TAGs.

$$\text{Saponification number} = [56.1 \times (B - S) \times N] / W \qquad (9.1)$$

where B = volume of N-normal HCl consumed for blank, S = volume of N-normal HCl consumed for W g of fat sample, and W = weight of fat sample in g.

9.3.2 ACETYL NUMBER

It is defined as the number of milligrams of KOH required to neutralize the acetic acid liberated by saponification of 1 g of fat. This is actually a measure of hydroxy-fatty acids in a sample, which is first acetylated by acetic anhydride and then decomposed to liberate acetic acid in equivalent quantities to the hydroxylated fatty acids present in the fat.

9.3.3 ACID NUMBER

It is the number of milligrams of KOH required to neutralize free fatty acids present in 1 gram of fat.

9.3.4 CRISMER VALUE

This is a test for miscibility of a fat with standard solvent mixture. Tertiary-amyl alcohol, ethyl alcohol, and water are mixed in the proportion of 5:5:0.27 to prepare the standard solvent. The sample-solvent mixture is heated at a higher temperature to get a clear liquid and then cooled down slowly. The temperature in degree Celsius at which first turbidity appears is the Crismer value of the fat. This is an important parameter in trade in Europe.

9.3.5 IODINE NUMBER

It is the number of grams of iodine absorbed by 100 g of fat or oil. Iodine number is a measure of degree of unsaturation in a fat.

9.3.6 PARA-ANISIDINE VALUE

The test for *p*-anisidine value is done for the measurement of aldehydes present in a rancid fat produced as secondary oxidation products. These aldehydes form color with *p*-anisidine, which absorbs light at 350 nm wavelength. The *p*-anisidine value is a measure of 100 times of the optical density measured of a solution of 1 g of fat and the reagent at that wavelength. For this, 1 g of fat is dissolved in 100 mL of isooctane and is added with 0.25% *p*-anisidine solution in glacial acetic acid.

9.3.7 PERCENTAGE FREE FATTY ACIDS

It is the number of grams of FFA (as oleic acid) present in 100 g of fat.

9.3.8 PEROXIDE VALUE

Milliequivalent of peroxides present in 1 kg of fat or oil is termed as peroxide value.

9.3.9 POLENSKY NUMBER

It is the number of milliliters of 0.1 N alkali needed to neutralize volatile, water-insoluble fatty acids present in 5 g of fat.

9.3.10 REICHERT MEISSEL NUMBER

The R. M. number is the number of milliliters 0.1 N alkali required to neutralize volatile, water-soluble fatty acids in 5 g of fat.

9.3.11 TITER (OR TITER)

The titer is a measure of temperature in degree Celsius at which the oil sample begins to solidify.

9.3.12 THIOBARBITURIC ACID (TBA) VALUE

2-Thiobarbituric acid reacts with malonaldehyde (a secondary oxidation product of rancid fat) to produce red color whose absorbance is measured at 450 or at 532 nm. The later absorbance is more specific in this determination.[2]

9.3.13 TOTOX VALUE

Totox value is the algebraic sum of twice the peroxide value and the para-anisidine value of a sample.

9.3.14 SMOKE POINT

It is the temperature at which the lipid sample produces continuous, visible smoke in the presence of air.

9.3.15 FLASH POINT

The flash point is the temperature at which the oil produces sufficient vapor to produce a flash as a flame at every 30 s interval.

9.3.16 FIRE POINT

The fire point is the temperature at which the pressure of the vapor generated by the oil is sufficient to maintain a flame due to ignition even after the source flame is removed. The smoke point, flash point, and fire point tell about the presence of free fatty acids and partial glycerides in any oil.

9.3.17 REFRACTIVE INDEX

Refractive indices of oils and fats give valuable information on the nature of fatty acids in the triglycerides. For soft oils, refractive index is measured at 25°C. For animal fats, it is usually measured at 40°C. The fatty acids in TAGs with higher chain-length and high degree of unsaturation give higher refractive index values. The measurement of refractive index is done with monochromatic wavelength of light using D-line of sodium at 25°C. The refractive indices of oils are related to the iodine number [12] and are expressed as follows:

$$\eta_D \text{ at } 25°C = 1.45765 + (0.0001164 \times \text{iodine number}) \tag{9.2}$$

9.3.18 SPECIFIC GRAVITY

Specific gravity increases with increase in chain length of fatty acids and the degree of unsaturation. With increase in temperature, specific gravity of oils decreases (by 0.00035/°C in most of cases). The specific gravity of oils maintains a linear relationship with the saponification number and iodine number.

$$\begin{aligned} \text{Sp. gr. (at } 15°C) = \ & 0.8475 \\ & + (0.00030 \times \text{saponification number}) \\ & + (0.00014 \times \text{iodine number}) \end{aligned} \tag{9.3}$$

9.3.19 HEAT OF COMBUSTION

The heat of combustion of oils increases with increased degree of unsaturation and increased molecular masses of fatty acids. The heat of combustion of vegetable oils is related to the iodine number and saponification number as follows:

$$\begin{aligned} -\Delta H_c \left(\text{cal/g} \right) = \ & 11.380 - (\text{iodine number}) \\ & - (9.158 \times \text{saponification number}) \end{aligned} \tag{9.4}$$

9.3.20 SOLID FAT INDEX

The solid fat index (SFI) measurement indicates the percentage solid fat content in a sample at a certain temperature. This index is important for plastic fats used in different food processing applications. The SFI is measured by nuclear magnetic resonance (NMR) and by dilatometer.

9.3.21   COLOR MEASUREMENT

A vegetable oil has light yellow or golden yellow color. Olive oil is light green color. Colors are measured on various scales. These are Lovibond scale, Gardner color scale, FAC (Fat analysis Committee of American Oil Chemists' Society, AOCS) scale and photometric color scale. Lovibond (AOCS Official Method, Cc13e-92) method is a simpler one and is adopted in most of the industries. The Wesson Method (AOCS Method, Cc8d-55) also gives similar results. It measures red (R) and yellow (Y) shades and the color is expressed as the sum of these two shades. The Lovibond Tintometer generally consists of two 25-cm long tubes converging to an eyepiece and a standard cell (1" or 5¼"), which holds the sample for fixing the light path. A combination of standard yellow and red glasses' colors is matched with the sample color. The Gardner scale consists of a single-numbered glass color standard for the oil. The scale is calibrated from 1(lighter) to 18 (darker) for assigning a number to the oil. The lighter colors with the lower numbers tend to be yellowish. The higher numbers signify reddish colors. In the FAC scale (AOCS Method, Cc13a-43), odd numbers from 1 to 45 are given to oils in the ascending order of darkness. The spectrophotometric color (AOCS Method, Cc13c-50) scale was developed to replace the manual Lovibond system. The absorbencies of the oil are measured at 460, 550, 620, and 670 nm wavelengths and are put into an equation to obtain the color value.

9.4   PROPERTIES OF GLYCEROL

Glycerol or propane-1, 2, 3-triol is a trihydric alcohol (boiling point 290°C) with two primary (at C-1 and C-3) and one secondary (at C-2) alcoholic group. There is no chiral (asymmetric) carbon in the glycerol molecule and it has a plane of symmetry. However, introduction of one fatty acid either at the C-1 or at C-3 position makes its C-2 carbon chiral and gives rise to optical activity ($2^1 = 2$ optical isomers). Glycerol is an important part of any TAG molecule from its structural point of view. Glycerol is highly hygroscopic, colorless liquid with sweet taste. It is used as anti-freeze (melting point 17.8°C). The three hydroxyl functional groups in the glycerol molecule can form H-bonds with water. Glycerol is highly miscible with water. This polyhydric aliphatic alcohol is also able to form intra- and intermolecular H-bonds. The metabolism of glycerol occurs asymmetrically.

The "new stereochemical numbering (*sn*)" system is a special system applied to glycerol and other few molecules whose nomenclature cannot be explained properly by the conventional system of stereochemical nomenclature. The *sn*-nomenclature suggests that the top glycerol carbon (C-1) is *sn*-1. Thus C-1, C-2, and C-3 carbons of glycerol are named as *sn*-1, *sn*-2, and *sn*-3 carbons, respectively. They are also called α-, β- and α'-carbons, respectively, according to the trivial system of nomenclature. Glycerol kinase attacks the *sn*-3 carbon to produce only glycerol-3-phosphate and no glycerol-1-phosphate. In the naturally occurring L-enantiomer of glycerol, the secondary alcoholic –OH group at the *sn*-2 position is oriented towards the left and is opposite to both the terminal primary –OH groups (at *sn*-1 and *sn*-2 positions) of glycerol.

Pancreatic lipase enzyme hydrolyzes TAGs more readily by cleaving the acyl-ester linkages at *sn*-1 and *sn*-3 positions. For this region-specific nature of lipases, they are classified as either 1-, 3- or 2-specific lipases. The ester bonds in a TAG experience a considerable degree of steric hindrance due to the hydrocarbon chain length of the fatty acids. The TAGs with short-chain fatty acids are more readily hydrolyzed by the lipases than those with medium- or long-chain fatty acids.

The flexibility of the substituent groups and the number and positions of double bonds in the TAG determine the geometry of the latter.[11] Hence, similar factors also determine the geometry of the products of lipase hydrolysis. Despite both being primary alcoholic groups, the 1- and the 3-carbons of glycerol are not identical. Enzymes can easily distinguish them during reactions. Earlier research on glycerol structure had revealed that glycerol forms three staggered conformational structures, namely, α-, β- and γ-glycerol (Fig. 9.6). To distinguish between the two terminal carbon atoms in the molecule, the name L-enantiomer is designated to as a molecule with its secondary –OH group to the left side of glycerol backbone. This removes all ambiguities and firmly establishes the formation of L-glycerol-3-phosphate within the biological cells.

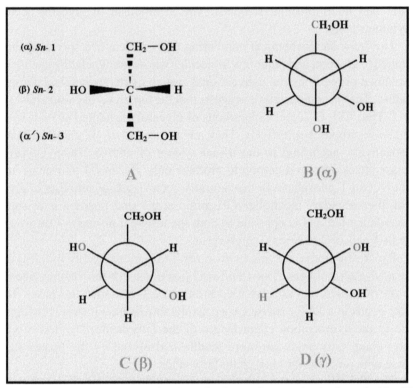

FIGURE 9.6   (A): Fischer projection formula of naturally occurring L-glycerol with –OH group placed at the left side. (B to D): Newman projection formulae of glycerol backbone-each of the two –CH$_2$OH groups rotates around the C–C bond to give three staggered conformations.

If the orientation of the oxygen of a terminal –CH$_2$OH group with respect to the –CH(OH)C– group is considered, then in the α-form that oxygen is trans to the distal (back) end carbon, β-form is trans with respect to the oxygen of the proximal (front) –CH$_2$OH group and that of the middle – CHOH– group, whereas the γ-form is trans with respect to the oxygen of the proximal –CH$_2$OH group and the hydrogen of the –CHOH– group.[3] Hence, by considering similar types of conformations from the distal carbon end, a total of 6 different conformations are possible, namely, αα, αγ, αβ, ββ, βγ, and γγ. The αα conformation was found to be the most stable with relatively higher energy due to intermolecular H-bonding. The crystalline form is made up of this conformer.

The liquid form is a combination of αα and αγ conformations.[4] When a TAG is formed, it crosses this huge amount of energy by breaking intermolecular bonding and by altering the spatial arrangement of glycerol. The overall reactivity of the two primary alcoholic groups is higher than the secondary –CHOH– group in this molecule. When heated in the presence of $KHSO_4$, glycerol is dehydrated to a pungent smelling toxic compound called acrolein. Acrolein is detected in used frying fats, which have undergone prolonged heating during frying of foods (Fig. 9.7).

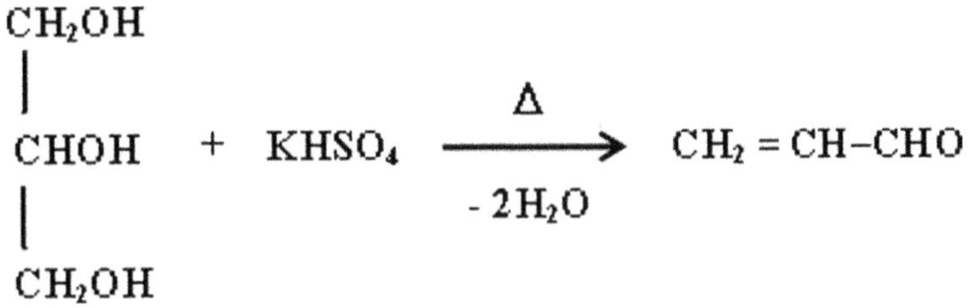

$$CH_2OH$$
$$|$$
$$CHOH \quad + \quad KHSO_4 \quad \xrightarrow[-2H_2O]{\Delta} \quad CH_2 = CH-CHO$$
$$|$$
$$CH_2OH$$

FIGURE 9.7 Production of acrolein.

9.5 CRYSTAL STRUCTURE OF TAGs

Melting down of solid substances is the property of their crystals and not of an isolated single molecule. Melting is the phenomenon where energy in the form of heat supplied from outside exceeds the lattice energy of the crystal by disturbing the solid's well-organized lattice structure. Therefore, melting breaks down the most ordered arrangement of molecules within a crystal and imparts chaotic disorder converting the solid into liquid accompanied by an increase in entropy. Fats, however, retain some degree of ordered arrangements of the solid crystalline form even after melting. This is termed as "crystalline memory effect" of the fat. This memory is entirely lost at a considerably higher temperature by extensive heating above the melting point of the crystals.

The search for a more stable form as aggregates of molecules never ends within the sample of fat. Hence, a liquefied fat, when recooled and recrystallized, tries to achieve the earlier stability. This is the reason why a sample of hydrogenated fat acquires plasticity in a votator during preparation of functional products like margarine and cocoa butter substitute. This is also related to the polymorphism characteristic (showing multiple melting points) of a crystalline fat. The TAG crystals or solid refined fats show polymorphism, which is due to the presence of more than one crystalline forms in the sample. Since the intermolecular packing arrangements within different crystals are different, their unequal stabilities toward an external heat source give rise to multiple melting points. Solid fat is a polycrystalline substance which is an aggregate of many tiny crystallites (grains) of different shapes and sizes. Fats crystallize in three different polymorphic forms, namely hexagonal, orthorhombic and triclinic. They are known as α, β', and β polymorphic forms, respectively. The hydrocarbon ends of the three fatty acids in TAG molecule orient themselves in a zigzag fashion (with each subcell or $-CH_2-$ group forming a tiny bent) in the liquid state and tend to reorient further when the oil is crystallized. Thus, they produce either a double or a triple chain-length structure (Fig. 9.8).

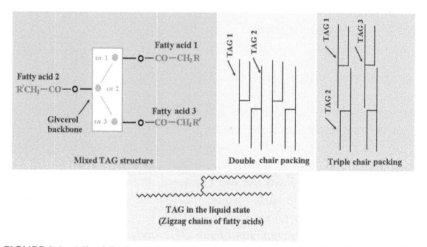

FIGURE 9.8   Mixed TAG structure and packing within a crystal lattice. R, R' and R″are hydrocarbon parts of three different fatty acids forming acyl-ester bonds at *sn*-1, *sn*-2 and *sn*-3 carbons of glycerol, respectively.

The double chain is formed with the three fatty acids in a TAG with identical chemical properties. When at least one fatty acid differs in chemical behavior from the other, the triple chain structure is developed. The structure of individual TAG molecule in the crystal resembles the form of a chair or of a tuning fork. The fatty acids orient themselves perpendicular

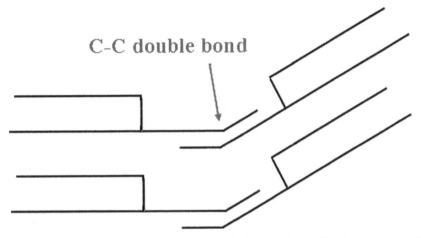

FIGURE 9.9 Triple-chair packing with kink in the hydrocarbon chain of one unsaturated fatty acid of SOS type crystal.

to the glycerol axis. When an unsaturated fatty acid is present in a mixed TAG (mostly at *sn*-2 position) and the remaining two positions (*sn*-1 and *sn*-3) are occupied by the saturated fatty acids, the triple-chair packing is produced with some bending at the terminal fatty acid residues (Fig. 9.9). This phenomenon is quite normal with stearic–oleic–stearic (SOS) type of TAGs. This is a common feature of most of the vegetable oils where *sn*-2 position is occupied by oleic acid. In general, the triple-chair packing is common with the mixed TAGs, whereas the simple TAGs arrange themselves in double-chair packing arrangement.

A double bond in fatty acid structure gives rise to steric hindrance leading to further bending of the triple-chair packing of mixed TAGs. A double chain-length structure is more compact and occurs by maximum overlap of two TAG molecules. Each TAG molecule is considered as a single chain. In the triple chain structure, no overlap occurs. During the nucleation stage of crystal growth, the property of a TAG in terms of orientation of its fatty acid moieties influence a large number of similar triglyceride molecules. The other crystals are overwhelmed by this progress and become a member of this crystal class. The nature of polymorphic crystal is influenced by two factors, namely, (1) the chain length of fatty acids in TAG and (2) the chain-length structure of the associated TAGs within the crystals. On the other hand, the melting point of solid crystalline fat is dependent on the polymorphic variety and the chain-length structure of its TAGs. Above all, it is observed that, when a single type of fatty acid is present in the TAG, the double chain-length structure is the most favored one.

When the TAGs contain at least one fatty acid different from the other two, the triple chain-length structure is produced. The α-polymorphic form is more prone to the formation of compound crystals (mixed crystals of two or more polymorphic varieties), while the β form is produced with greater degree of purity. Table 9.1 shows a comparison between the three polymorphic varieties of TAGs. Useful data are available from "Avrami Analysis" and "Fisher-Turnbull Analysis" for the detection of crystal morphology and fatty acids packing in the solid fats.

TABLE 9.1 Properties of Crystalline Polymorphs of Fat.

Property	(α) Hexagonal	(β′) Orthorhombic	(β) Triclinic
Orientation of fatty-hydrocarbon chains[a]	90°	90°–67°	67°
Crystal size (μm)	5	1–5	<50
Stability	Lowest	Intermediate	Highest
Melting point	Lowest	Intermediate	Highest
Packing density	Lowest	Intermediate	Highest
Shape of the crystal	Small platelet	Short needle	Long needle
Void space between crystals	Highest	Intermediate	Lowest
Gibbs free energy (G) in the solid phase	Highest	Intermediate	Lowest
Crystal growth rate	Highest	Intermediate	Lowest
Interplanner (d) spacing (nm)	0.415	0.42 and 0.38	0.46, 0.39, and 0.37

[a]Orientation with respect to the glycerol plane.

The glycerol is the precursor of TAGs and the membrane glycero-phospholipid syntheses. The major sites for TAG synthesis are liver and adipose tissues. Phospholipids are synthesized in the smooth endoplasmic reticulum. The fatty acids and the glycerol are activated prior to TAG synthesis. The activation of fatty acids occurs via acyl-coA formation by enzyme thiokinase. The L-glycerol-3-phosphate functions as an intermediate of TAG formation.

The chemical composition of edible fats and oils vary after extraction, refining, blending and different secondary processing (hydrogenation, inter-esterification, fractionation, etc.). The extracted lipids from plant or animal sources are considered to be natural samples of fats and oils whose composition is modified during subsequent refining stages. On the other hand, secondary processing like hydrogenation and inter-esterification, brings about intramolecular changes in the triglycerides and into their crystalline arrangements in the solid phase as well. A fractionated fat crystallizes differently than its previous form.

The knowledge of the chemistry of fats and oils is necessary to understand the changes, which take place during the course of storage and during multistage refining operations using several chemical reagents (e.g., water and acids in degumming, NaOH in alkali refining) and under variable physical conditions (e.g., high temperature in physical refining, bleaching, deodorization, etc.). The chemistry of refined fats and oils more precisely is the chemistry of fatty acids combined with the glycerol molecule through acyl-ester bonds. Since the refined edible fats are TAGs of fatty acids, the later also plays an important role in determining the physicochemical nature of the fat. In the crude fat, however, various other components (moisture, phospholipids, waxes, hydrocarbons, pigments, metals, peroxides and hydroperoxides, secondary oxidation products of lipids, sterols and low and high molecular weight free fatty acids) react in their own fashion under favorable conditions.

The edible fat consists of a mixture of different TAG molecules (95–98%) with minor quantities of other components. Restoring (rather improving) this composition is the target of all refiners. The higher the quantity of this fraction (the TAGs) extracted, the more economical is the process of refining. Even a 0.5–1% loss of TAGs induces enormous yield loss in the refining industry. The vegetable oilseeds like soybean, cottonseed, flaxseed, peanut and mustard contain oil glands in their cells. In eukaryotes, the lipid content of the cell rapidly increases from embryonic tissues (1–2%) and up to the completion of growth in the matured cells (90–95%). Storage cells of plant body contain higher amount of lipids for their future needs.

9.6   LIPID OXIDATION

Oxidation is the most important development in the fats and oils. The quality of the edible lipids largely depends on their resistance to this chemical change. An oxidized fat not only acquires off flavor but also develops toxicity due to the formation of organic peroxides and hydroperoxides.

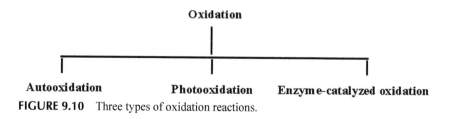

FIGURE 9.10   Three types of oxidation reactions.

A highly oxidized fat is not suitable for consumption. Broadly three types of oxidation reactions are possible: autooxidation, photooxidation, and enzyme-catalyzed lipid oxidation (Fig. 9.10). The photooxidation reaction is similar to autooxidation except that in the former reaction occurs in the presence of ultraviolet (UV) light which helps the oxidation process of unsaturated fatty acids. Lights having wavelengths in the visible region are weaker in transforming the oxygen atoms (in the diatomic molecular oxygen) from its triplet ground state to a singlet state and require sensitizers (Fig. 9.11).

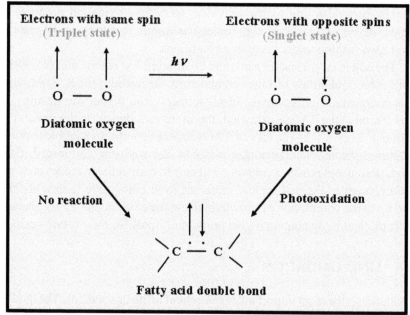

FIGURE 9.11   Formation of singlet oxygen by light and response to photooxidation.

The natural pigments present in the oil also act as sensitizers. The double bonds of unsaturated fatty acids are in their singlet spin state (electrons with opposite spins), for a reaction with the oxygen atom, the later should also be in its singlet state. The mechanism of photooxidation of lipids is different from autooxidation. Singlet oxygen directly attacks the double bonds without following a series of free radical chain reactions (as in the case of autooxidation). This type of oxidation by singlet oxygen is termed as type II oxidation (Fig. 9.12). The type II reaction (ene-reaction) proceeds via addition of two atoms of oxygen to one (Y) of the two carbons holding the olefinic bond of the fatty acid and by shifting the double bond next to other carbon (X) atom, which was initially holding the double bond.

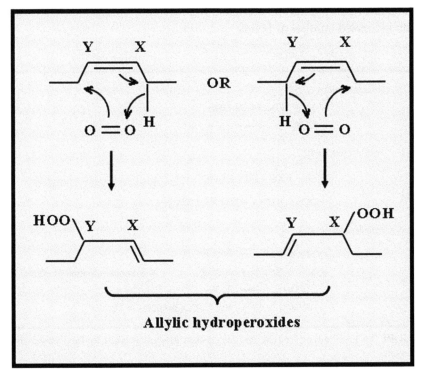

FIGURE 9.12 The type II oxidation (photooxidation) of fatty acids by singlet oxygen producing trans-allylic hydroperoxides.

The resultant products are trans-allylic hydroperoxides. Therefore, photooxidation leads to the production of both positional and geometrical isomers of the fatty acids. Chlorophyll and heme-compounds are natural sensitizers of visible light.

Autooxidation, on the other hand, is a free radical chain reaction consisting of initiation, propagation, and termination steps which occurs via a type I mechanism (Fig. 9.13). The production of free radical occurs at the initiation step by thermal dissociation, by hydroperoxide decomposition, by the metal atoms (Cu and Fe) and even by exposure of lipid to light with or without the aid of any photosensitizer. The ease of autooxidation reaction is dependent upon the availability of allylic hydrogen atom in the substrate (unsaturated fatty acids) to be donated to the peroxide free radical (ROO•). Positional (formation of conjugated diene, etc.) and geometrical isomerism are inevitable during the autooxidation reaction. The rate of autooxidation reaction is higher in the unsaturated fatty acids with increased number of bonds.

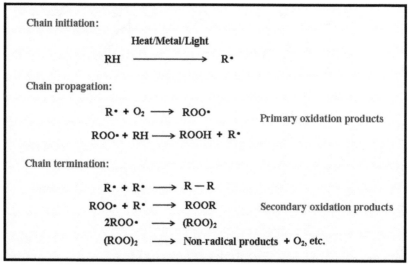

FIGURE 9.13   The type I oxidation (autooxidation) of fatty acids by free radical chain reaction.

However, photooxidation reaction is much faster than autooxidation. Photooxidation produces more number of isomers than autooxidation. Decomposition of peroxides and hydroperoxides produces various alcohols, aldehydes, semialdehydes, hydrocarbons, ketones, etc. as the secondary oxidation products. These substances, though not toxic, are highly volatile and impart objectionable odors to the edible fats and oils. Thus, oxidized fat develops both toxicity (due to peroxides and hydroperoxides) and off flavor once the oxidation reaction is started.

The termination routes of the reaction include coupling of diperoxide to produce non-radical products and oxygen, as well as radical disproportionation (not shown in the figure). By disproportionation reaction, the diperoxides generate new alkoxyl radicals (RO•, HO•, etc.) instead of producing stable products. These alkoxyl radicals further react to produce newer compounds in the chain termination stage. Ferrous (Fe^{+2}) ion catalyzes the transformation of peroxides into alkoxyl radicals, whereas, Fe^{+3} (ferric) ion produces peroxyl radicals. Both alkoxyl and peroxyl radicals may take part in the propagation of the chain reaction or may be directed otherwise to terminate the chain by producing stable products.

Enzymatic oxidation is not so pronounced in edible fats and oils unless they are contaminated by the microorganisms. Lipoxygenase enzymes are responsible for this type of oxidation. The three cellular enzymes, namely, lipoxygenase, cyclooxygenase and cytochrome P-450 enzymes function as mediators for in vivo lipid oxidation. The superoxide anion ($•O_2^-$), H_2O_2, hydroxyl radical ($•OH$) and the singlet oxygen are reactive oxygen species (ROS) in cellular lipid peroxidation.

In microorganisms, ROS is obtained from the molecular oxygen either as a product of aerobic metabolism or produced in the environment by ionizing (γ-ray) or nonionizing (UV-ray) irradiation. In higher animals, the peroxidation process is related to aging, cancer and eventually leading to even death. There are counter mechanisms in the cells to negate the impact of the toxic substances produced by these reactions. However, with growing age, such ability of cells is reduced. In vivo peroxidation of cellular lipids in eukaryotes also follows a similar free radical chain mechanism like that of autooxidation consisting of initiation, propagation and termination steps. The superoxide anion is produced initially by sharing electrons by the oxygen from the electron transport chain in mitochondria (Fig. 9.14).

In biological system, superoxide formation is the most vulnerable cause for free radical generation. Free radicals are also produced by tissue alteration, by cell lysis and release of iron and by the action of transition metals. Reduced transition metals can generate superoxide anion (Fig. 9.15). Lipoxygenase enzymes catalyze the step of conversion of polyunsaturated fatty acids (PUFAs) into their hydroperoxides using the ROS. These are related to the biosynthetic precursors of prostaglandin from enzymatic oxidation of arachidonic acid.

$$O_2 \xrightarrow[+H^+]{e^-} HO_2 \cdot \xrightarrow[-H^+]{} \cdot O_2^-$$

$$HO_2 \cdot \xrightarrow[+H^+]{e^-} H_2O_2$$

$$H_2O_2 \xrightarrow[+H^+]{e^-} \cdot OH + H_2O$$

FIGURE 9.14 Formation of ROS in the biological cells.

$$Fe^{++} + O_2 \longrightarrow Fe^{+++} + \cdot O_2^-$$

$$Cu^+ + O_2 \longrightarrow Cu^{++} + \cdot O_2^-$$

$$Fe^{++} + H_2O_2 \longrightarrow Fe^{+++} + \cdot OH + OH^-$$

$$Fe^{+++} + H_2O_2 \longrightarrow Fe^{+++} + \cdot OOH + H^+$$

FIGURE 9.15 Formation of superoxide anion from reduced transition metals.

The lipooxygenases oxidize both the free and the esterified PUFAs in the cells and produce peroxyl radicals. The PUFAs in foods are sources of essential fatty acids (linoleic and linolenic) for the body. The cellular reactions are more complex in nature and various intermediates and even the products of the reactions are utilized by the cells during their metabolism.

Any type of lipid molecule containing unsaturated fatty acids is attacked by this process.

Toxic substances are produced in the cell due to oxidation of lipids in the plasma membrane, mitochondria, microsomes and peroxisomes. The detoxification process occurs in the liver. Cytochrome P-450s (thromboxane synthetase, prostacyclin synthetase, etc.) monooxygenase system catalyze oxidations and epoxidations of PUFAs. They act as catalysts in the oxidative transformation of aldehydes into carboxylic acids and have important roles in the lipid signaling pathway. Cytochrome P-450s help other pathways which bring about similar transformations when there is a higher amount of toxic substances accumulated in the cell. Peroxidation of membrane lipid decreases membrane fluidity.

The superoxide anion (O_2^-), hydroperoxides and the derived aldehyde derivatives inhibit protein syntheses. The reactive oxygen species superoxide anion has both oxidizing and reducing properties. It is a weak oxidizing agent in aqueous solutions and oxidizes ascorbic acid and thiols. On the other hand, it readily reduces iron complexes like cytochrome c. In the aqueous solution, superoxide produces H_2O_2 and O_2. Lipid peroxides are stable at physiological temperatures and are decomposed by the transition metals and their complexes. Such reactions also occur within the cells. Metals like iron and copper can thus initiate the cellular oxidation and carry forward the reaction by disintegrating peroxides.

Although several factors affect the rate of lipid oxidation, these reactions are self-propagating and self-accelerating once they have started within a sample of fat. Several TAG and free fatty acid molecules are attacked at a time and many peroxides are produced. Multiple volatile and nonvolatile intermediates are produced within the sample, which can be detected by IR and NMR spectroscopy, GC-MS, HPLC, electron spin resonance spectroscopy, supercritical fluid chromatography and chemiluminescence spectroscopy. The peroxides are very unstable (even at 0°C and in N_2 atmosphere) toxic compounds with practically no odor or flavor. Their estimation can only assess their concentrations at a certain stage of their development.

9.6.1 FACTORS AFFECTING LIPID OXIDATION

- Nature of the lipid
- Surface and mass transfer effects

- The presence of initiator or catalyst
- Oxygen partial pressure
- Temperature
- UV light
- The presence of prooxidants and antioxidants
- Environmental and solvent effects.

The nature of the lipid, and hence, that of the fatty acids control the rate of oxidation. *Cis*-fatty acids are more susceptible to oxidation than their trans-counterparts. Conjugated fatty acids are more readily oxidized than nonconjugated acids. Since, autooxidation occurs by the abstraction of H-atom from the allylic-group, the presence of methylene ($-CH_2-$) group attached to the double bond enhances the rate of oxidation. The higher the surface area created by the substrate, the greater is the ease of oxidation.

Lipid monolayers provide extended surface and lower mass transfer resistance which enhance the rate of oxidation. Initiators such as Fe, Cu, chlorophyll, and heme-compounds increase the rate of oxidation. The carotenoids are fairly unpredictable in this regard and may either increase the rate of oxidation by creating a free radical-rich environment within the lipid sample or may scavenge peroxyl radicals (by themselves getting oxidized) and decrease the rate of oxidation reaction. The carotenoids are highly susceptible to oxidation due to their isoprenoid structures. Spontaneous free radical chain reactions for lipid oxidation occur above a certain oxygen partial pressure. An increase in water activity (A_w) of fats and oils within 0 to 0.5 was found to have inhibitory effects at the initial stage of oxidation and even during the peroxide decomposition. However, increased water activity levels attract microorganisms to grow on the sample and encourage hydrolytic changes within the TAGs producing additional quantities of free fatty acids. Temperatures 30–40°C above room temperature increase the rate of autooxidation.

UV radiation has a powerful accelerating effect on lipid oxidation. Prooxidants and antioxidants present in the lipid sample increase and decrease the rate of reaction respectively. Tocopherols have inhibitory effect on autooxidation and are natural antioxidants. Environmental factors like pH, sunlight, atmospheric oxygen, relative humidity, and the type of packaging and container surface have enormous impact on lipid peroxidation. The life-time of high energy singlet oxygen varies in different solvents. The singlet oxygen has higher energy (by 93.6 kJ) than the triplet state (ground state). Once the energy is dissipated to the solvent, the reactive oxygen species lose its ability to proceed with the chain reaction. The life-time of singlet oxygen was found to be about 2, 17 and 700 μs (micro seconds) in water, hexane, and CCl_4 respectively. Lipid emulsions have extended surface areas which facilitate diffusion of gaseous oxygen. The contact of water with oil at the interface supplies adequate quantity of protons and catalysts. Hence, the rate of oxidation of emulsions of lipids is also higher. The rate of lipid oxidation is expressed by the following equation:

$$-\frac{dO_2}{dt} = K_m[ROOH]^{0.5} \tag{9.5}$$

where K_m is the rate constant in moles of lipid oxidized per mole of sample per hour. The negative sign indicates the depletion of oxygen concentration with time.

The rate of oxygen uptake is directly proportional to the square root of peroxide concentration. Measurement of peroxide produced in a fat sample at different instances can tell about the rate of oxidation. The second step of the chain propagation in the autooxidation reaction is a slower one and is the rate-determining step.

9.7 LIPID HYDROLYSIS

Hydrolysis of TAGs and phospholipids produce fatty acids. The final products of a TAG hydrolysis are glycerol and three molecules of fatty acids. Enzymatic (lipase) hydrolysis of TAG sequentially produces diacylglycerol, monoacylglycerol and finally glycerol with the release of one fatty acid at a time. The phospholipids produce lysophospholipids and fatty acids. The phospholipase enzymes hydrolyzing phospholipids are highly stereospecific and a single type (PLA_1, PLA_2, PLC, PLD, etc.) of enzyme never cleaves both the acyl-ester bonds of a phospholipid. Lipases are less selective than phospholipases, though they are classified as 1, 3-specific and 2-specific lipases depending upon the number of glycerol carbon (sn-1, 2 or 3) holding the ester linkage they cleave.

The nonenzymatic hydrolysis reactions are favored by high temperature and high pressure. Both the enzymatic and nonenzymatic (Chemical hydrolysis) hydrolysis reactions are reversible (Fig. 9.16). The reverse step is known as esterification. The chemical hydrolysis reaction is catalyzed by acids, bases or emulsifiers. Noncatalytic hydrolysis requires moderately higher temperature and higher pressure. Most of the chemical reactions are performed at 50–60°C or higher.

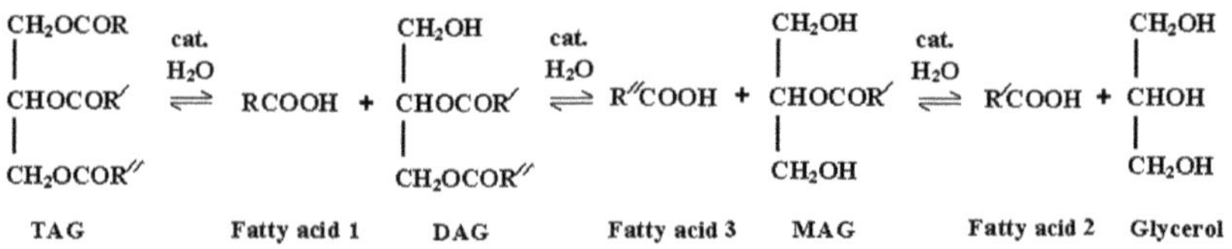

CH$_2$OCOR	cat. H$_2$O		CH$_2$OH	cat. H$_2$O		CH$_2$OH	cat. H$_2$O		CH$_2$OH
CHOCOR$'$	$\rightleftharpoons$	RCOOH +	CHOCOR$'$	$\rightleftharpoons$	R$''$COOH +	CHOCOR$'$	$\rightleftharpoons$	R$'$COOH +	CHOH
CH$_2$OCOR$''$			CH$_2$OCOR$''$			CH$_2$OH			CH$_2$OH
TAG		Fatty acid 1	DAG		Fatty acid 3	MAG		Fatty acid 2	Glycerol

FIGURE 9.16   Chemical hydrolysis reaction catalyzed by acids, bases, or emulsifiers.

Enzymatic reactions require milder temperatures. Complete chemical hydrolysis of TAGs occurs at 20–60 bars of pressure and at 250°C. This is an important chemical transformation for the production of soaps, biodiesel, and various oleochemicals. Butter, dairy products, palm oil (palmitic acid oil), coconut oil (lauric acid oil), and other edible fats containing low molecular mass fatty acids in the lipid structures produce those free fatty acids on hydrolysis. Such development in the fats and oils and in the food products produced from them imparts undesirable odor characteristics of those volatile fatty acids. In contrast to the off-flavor development in the oxidized lipids by decomposition of peroxides, this type of flavor development can even occur by the saturated fatty acids of lower molecular masses.

It has been also found that the production of free unsaturated fatty acids in larger quantities encourages autooxidation process. The hydrolysis of fats and oils induces yield loss in the refining industry while generating larger amount of soapstocks during caustic treatment. After their extraction, the moisture content of crude fats and oils and the temperature of storage should be carefully controlled for minimizing hydrolytic changes. The reversible hydrolysis and esterification reactions are sometimes discussed together under the same heads of inter-esterification and intra-esterification reactions.

9.8   FLAVOR REVERSION

Flavor reversion is an age-old and the most prevalent problem extensively related to the quality of edible fats and oils. It has more profound effect on marine and vegetable oils. All these oils contain higher percentages of PUFAs. Marine oils with a fishy flavor regains (or reverts to) the same fishy flavor after processing. Many fats and oils finally develop a flavor which was not there initially. This typical sequential change of flavor is different from the rancid flavor which is developed due to the decomposition of peroxides and hydroperoxides as a result of oxidation.

Soybean oil shows extensive reversion by changing its flavor from a beany or buttery to a grassy (or hay like) flavor, then to a painty flavor and finally to a fishy flavor. The term "reversion" was used, however, to express the phenomenon of reversion in flavor when it was first detected in the marine oils. It had been a matter of debate for many years whether the reversion and the rancidity by oxidation are interrelated or they just commence in parallel. In most of the cases, reversion takes place before the onset of rancidity. Thus, reversion is an indicator of the quality of a fat which, at a later course of time, may develop rancidity. The linolenic (18:3, $\Delta^{9,12,15}$), isolenoleic (18:2, $\Delta^{6,9}$), and linoleic (18:2, $\Delta^{9,12}$) acids were found to be responsible for reversion. Isolenoleic acid is produced by partial catalytic hydrogenation of linolenic acid or its esters. Hence, the oils containing these three fatty acids in abundance undergo reversion.

of linoleic acid and undergo reversion. Refined and hydrogenated vegetable oils, however, show lesser tendency to flavor reversion as compared to the crude nonrefined oils. The most accepted theory behind reversion of fats and oils is that a minor oxidation of unsaturated fatty acids prior to the actual oxidation causes reversion. When the oxidation is more pronounced, rancidity is detected. Compounds such as 2-pentyl furan, 2-pentenyl furan, 2-heptenal, di-n-propyl ketone, maleic aldehyde, $\Delta^{2,4}$-decadienal and acetaldehyde are isolated from reverted fats and oils. Heat, light, and traces of oxygen accelerate the reversion process. Bivalent metal (Ca^{++}, Mg^{++}, Fe^{++}, etc.) ions also favor this process.

Metal scavengers and chelators (EDTA, citric acid, etc.) can prevent the reversion of edible fats and oils. Initially, Stephen S. Chang and others (1966) had postulated (which was later established) that 2-pentyl furan produced from linoleate in reverted soybean oil contributes to its reversion flavor (Fig. 9.17). Linolenic acid also follows a similar mechanism to produce 2-pentenyl furan.[7] The reversion always tends to completion when the fat is stored under favorable conditions. A final flavor is developed in the reverted fat sample which, though not characteristic of that fat, never alters after that stage.

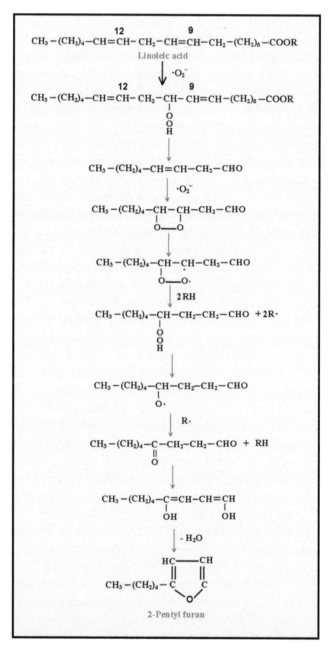

FIGURE 9.17   Conversion of linoleic acid into 2-pentyl furan during the reversion of fat. **Sunflower, safflower,** corn and palm oils contain high levels

9.9 RANCIDITY

The rancidity of fats and oils is defined as the development of unpleasant odor due to oxidation of unsaturated fatty acids. The TAGs being major components in a refined fat sample, their fatty acids are mostly attacked during the oxidation reaction. Even the fatty acids in their free form take part in this reaction. Although, several factors affect the rate of oxidation, the reaction is autocatalytic and progresses rapidly once it is started within the fat sample. A pungent odor is developed within the sample and toxic peroxides are accumulated. The peroxides have very transient existence and are decomposed into various low molecular weight volatile odoriferous substances imparting the so-called off-flavor in the fat. This oxidation process is never complete as long as there are oxidizable fatty acids present in the fat. Biscuits and other bakery products containing fats in the form of shortening agents sometimes develop rancid odor. When the water activity level is low ($A_w < 0.2$) in these products, there is lesser chance of developing such off-flavor. Rancidity is encouraged by the presence of light, oxygen, high temperature, moisture, metals, and pro-oxidants: all common accelerators of autooxidation and hydrolysis reactions. The extent of rancidity can be measured by measurement of peroxide value. Vegetable fats exceeding a value of 10 milliequivalents peroxide per kg of sample is unsuitable for consumption. For animal fats, the limit is 20 milliequivalents peroxide per kg of sample. The susceptibility of fats and oils to oxidation can also be measured by active oxygen method (AOM) test or swift stability test.

9.10 PYROLYSIS

Pyrolysis is a catalytic or noncatalytic chemical transformation by splitting of a high molecular weight compound into smaller molecules by the application of heat. The pyrolysis reaction occurs without the presence of air or oxygen. The process is analogous to that happens in a mass spectrograph. The splitting of high molecular weight compounds occurs by breakdown of chemical bonds and by the production of free radicals under favorable conditions of temperature, heating rate, and time of reaction. The pyrolysis of acyl-glycerols occurs at elevated temperatures (above 300°C) by first cleaving the acyl-ester bonds and releasing the fatty acids. The fatty acids are then decarboxylated and decomposed into various short-chain compounds. The saturated fatty acids produce ketones, whereas the unsaturated acids produce olefins by decarboxylation. The glycerol is dehydrated to acrolein and other unsaturated aldehydes. As the temperature remains far below that range, these changes normally do not occur during processing of edible fats and oils. However, frying fats may deteriorate by such transformations.

9.11 SUMMARY

The edible lipids (commonly called "fats and oils" in their purest forms) are TAG esters composed of one glycerol and three fatty acid molecules. Various other lipid components are either unintentionally passed into the refined stream of fats and oils or are separately used as food additives and in food preservation. This chapter covers discussions on the physicochemical behaviors of edible fats and oils and their importance in maintaining the quality of those lipid substances suitable for human consumption.

Since edible lipids are organic compounds extracted from plant and animal cells, these lipids have definite physical properties and are susceptible to chemical changes during extraction, refining, storage, handling, and transportation. While some changes are desirable, others may develop off-flavor, reduction in nutritional quality and even toxicity in the sample and are later transferred into the processed food materials. The physical nature of a solid fat is unique in its mode of application as a component of food in terms of its organoleptic quality, palatability, crystalline structure, softness, melting point, color, and flavor. The oils are liquid at room temperature and are means of preparation of frying fats, shortenings, margarine, salad oils, and so on by secondary processing. Hence, a proper understanding over their physicochemical properties is needed for maintenance of the quality of edible fats and oils in a processing industry. The importance of lipid substances is not only limited to their applications in food processing but also as one of the major proximate constituents in foods providing the largest amount of calorific need for our body and are sources of precursors for various hormones and fat-soluble vitamins. Besides, the lipid substances play important physiological roles in the body.

The discussion of this chapter extends to discussions and contextual analyses of different physical properties, chemical reactions and reaction parameters, mechanisms and recent developments in the field of research. Additional explanation is provided, wherever necessary for better understanding of the theories and principles related to the topics.

KEYWORDS

- Acetaldehyde
- acetyl-coA
- acetyl number
- acid value
- acrolein
- active oxygen method
- acyl-ester
- alkali refining
- amphiphilic
- animal fats
- antioxidant
- aroma
- autooxidation
- avenasterol
- Avrami analysis
- β-Sitosterol
- biodiesel
- bleaching
- brassicasterol
- butter
- campesterol
- canola
- capric acid
- carotene
- carotenoids

- cephalin
- ceramide
- cerebrocide
- chlorophyll
- cholesterol
- chylomicron
- cis-trans isomerism
- classification of lipids
- color
- compound lipid
- cottonseed
- Crismer value
- crystal growth rate
- crystal structure
- cyclooxygenase
- cytochrome P-450s
- degumming
- $\Delta^{2,4}$-decadienal
- density
- deodorization
- diacylglycerol
- dimer
- di-n-proply ketone
- edible fats and oils
- EDTA

- Emulsifier
- Emulsion
- enzyme-catalyzed oxidation
- epoxidation
- essential fatty acids
- esterification
- Fat Analysis Committee
- fats and oils
- fire point
- Fischer projection
- Fisher–Turnbull Analysis
- Flash point
- Flavor
- food lipids
- free fatty acid
- free radical
- frying oil
- Gadusene
- Ganglioside
- Gardner color scale
- Gibbs free energy
- Glycerolysis
- Glycerophosphate
- Glycolipid
- HDL
- Hexagonal
- heat of combustion
- hydrophobic
- hydrocarbon
- hydrolysis
- hydroperoxides
- interesterification
- interplanner spacing
- Iodine number
- iso-linoleic acid
- isoprenoid
- lard
- lauric acid
- LDL
- lecithin
- linoleic acid
- linolenic acid
- lipase
- lipids
- liposome
- lipoxygenasse
- Lovibond scale
- macromolecule
- malonaldehyde
- margarine
- mass transfer effect
- melting point
- metal
- metal chelator
- metal scavenger
- mixed triglyceride
- molecular mass
- monoacylglycerol
- N-acetylneuraminic acids
- neutral lipid
- Newman projection
- nonhydratable gum

- nonphosphorylated lipid
- oleochemicals
- ω acids
- organoleptic
- orthorhombic
- oxidation
- packing density
- palmitic acid
- palmityl-coA
- para-anisidine value
- percentage free fatty acid
- peroxidation
- peroxide number
- peroxides
- peroxyl radical
- phospholipase
- phospholipid
- photooxidation
- photosensitizer
- physical refining
- physicochemical properties of lipids
- phytosterol
- pigments
- plastic fat
- polar lipid
- Polensky number
- polymorphism
- polyunsaturated fatty acid
- precursor
- pristine
- prooxidant
- prostacyclin synthetase
- proteolipid
- pyrolysis
- rancidity
- rapeseed
- rate of oxidation
- refractive index
- Reichet Meissel number
- reversion
- safflower
- salad oil
- saponification number
- saturated fats
- secondary oxidation products
- secondary processing
- shortening
- sialic acid
- simple lipid
- simple TAG
- singlet oxygen
- smoke point
- soap
- solid fat index
- soybean
- specific gravity
- spectrophotometric color
- sphingisine
- squalene
- stanol

- stearic acid
- stereochemical numbering
- stereospecific
- steroid
- sterol
- sulfolipids
- sunflower
- superoxide
- surface effect
- tallow
- thiobarbituric acid value
- thromboxane synthetase
- titer
- triacylglycerol
- triclinic
- Tocopherol
- Totox value
- 2-heptenal
- 2-pentenyl furan
- 2-pentyl furan
- unsaturated fats
- UV light
- vegetable oils
- vitamin
- VLDL
- volatility
- wax
- xanthophyll
- yield stress
- zamene
- zoosterol

REFERENCES

1. Anastopoulos, G.; Zannikou, Y.; Stournas, S.; Kalligeros, S. Transesterification of Vegetable Oils with Ethanol and Characterization of the Key Fuel Properties of Ethyl Esters. *Energies* **2009,** *2,* 356–376.
2. Asakawa, T.; Matsushita, S. Thiobarbituric Acid Test for Detecting Lipid Peroxides. *Lipids* **1979,** *14*(4), 401–406.
3. Bastiansen, O. Intra-molecular Hydrogen Bonds in Ethylene Glycol, Glycerol, and Ethylene Chlorohydrin. *Acta Chem. Scand.* **1949,** *3,* 415–421.
4. Chelli, R.; Gervasio, L.; Gellini, C.; Procacci, P.; Cardini, G.; Schettino, V. Density Functional Calculation of Structural and Vibrational Properties of Glycerol. *J. Phys. Chem.* **2000,** *104*(22), 5351–5357.
5. Gadelha, I. C. N.; Fonseca, N. B. S.; Oloris, S. C. S.; Melo, M. M.; Blanco, B. S. Gossypol Toxicity from Cottonseed Products. *Sci. World J.* **2014,** 1–11.
6. Haas, M. J. Animal Fats. Chapter 5, In *Bailey's Industrial Oil and Fat Products, Volume 1; Shahidi, F. Eds.;* Wiley-Interscience: New Jersey, 2005; pp 161–212.
7. Ho, C. T.; Smagula, M. S.; Chang, S. S. The Synthesis of 2-(1-pentenyl) Furan and its Relationship to the Reversion Flavor of Soybean Oil. *J. Am. Oil Chem. Soc.* **1978,** *55*(2), 233–237.

8. Lo, S. K.; Tan, C. P.; Long, K.; Yusoff, M. S. A.; Lai, O. M. Diacylglycerol Oil Properties, Processes and Products: A Review. *Food Bioprocess Technol.* **2008,** *1*, 223–233.

9. Mistri, B. S.; Min, D. B. Prooxidant Effects of Monoglycerides and Diglycerides in Soybean Oil. *J. Food Sci.* **1988,** *53*(6), 1896–1897.

10. Satyanarayan, U.; Chakrapani, U. *Biochemistry.* Books And Allied (P) Ltd., 2009, pages 792.

11. Valenzuela, A.; Morgado, N. Trans Fatty Acid Isomers in Human Health and in the Food Industry. *Biol. Res.* **1999,** *32*(4), 273–287.

12. Wang, T. Soybean Oil. Chapter 2, In *Vegetable Oils in Food Technology: Composition, Properties and Uses;* Gunstone, F. D., Eds.; Blackwell Publishing: UK, 2002; pp 18–58.

13. Waraho, T.; Cardenia, V.; Rodriguez, E. M.; McClements, D. J.; Decker, E. A. Prooxidant Mechanisms of Free Fatty Acids in Stripped Soybean Oil-in-Water Emulsions. *J. Agric. Food Chem.* **2009,** *57*(15), 7112–7117.

14. Yu, K. R.; Tsai, Y. T.; Ariga, T.; Yanagisawa, M. Structures, Biosynthesis, and Functions of Gangliosides—An Overview. *J. Oleo Sci.* **2011,** *60*(10), 537–544.

CHAPTER 10

METHODS FOR FOOD ANALYSIS AND QUALITY CONTROL

HRADESH RAJPUT[1,*], JAGBIR REHAL[2], DEEPIKA GOSWAMI[3], AND HARSHAD M. MANDGE[4]

[1]*Sam Higginbottom Institute of Agriculture, Technology & Sciences (Formerly Allahabad Agricultural Institute) Deemed-To-Be University, Allahabad 211007, U.P., India, Tel.:+91-9454183802*

[2]*Food Science and Technology Division, Punjab Agricultural University, Ludhiana 141004, Punjab, India, Tel.: +91-9417751567, E-mail: jagbir@pau.edu.in*

[3]*ICAR—Central Institute of Post-Harvest Engineering and Technology (ICAR-CIPHET), Ludhiana 141004, Punjab, India E-mail: deepikagoswami@rediffmail.com*

[4]*College of Horticulture, Banda University of Agriculture and Technology, Banda 210001, Uttar Pradesh, India E-mail:mandgeharshad@gmail.com*

Corresponding author. E-mail: hrdesh802@gmail.com

CONTENTS

10.1 INTRODUCTION

Food is a more basic need of human than that of shelter and clothing. It provides adequately for the body's growth, maintenance, repair, and reproduction. Plant and animal origin are the sources of foods that contain the essential nutrients such as carbohydrates, fats, proteins, vitamins, and minerals. Usually after consumption, food undergoes different metabolic processes that eventually lead to production of energy, maintenance of life, and/or stimulation of growth.[3]

Food analysis is the process for both fresh and processed products by the standardized form and are those most commonly used in the laboratory. These analytical procedures are used to provide information about a wide variety of different characteristics of foods, including their composition, structure, physicochemical properties, phytochemical properties, and sensory attributes. This information is critical to our rational understanding of the factors that determine the properties of foods, as well as to our ability to economically produce foods that are consistently safe, nutritious, and desirable and for consumers to make informed choices about their diet.

Quality control is the standards which maintain the quality of the food products according to the customer's acceptability. Physical, chemical, microbiological, nutritional, and sensory parameters are used for the maintenance of nutritious food. These quality factors depend on specific attributes such as sensory properties, based on flavor, color, aroma, taste, texture and quantitative properties, namely, percentage of sugar, protein, fiber and so on as well as hidden attributes such as peroxides, free fatty acids, enzyme.[1,2,13,18] Although quality attributes are many, yet not all need to be considered at every point in time for every particular product. It is important to always determine how far relatively a factor is related to the total quality of the product. The quality attribute of a particular product is based on the composition of the product, expected deteriorative reactions, packaging used, shelf-life required, and the type of consumers.

The most important element and ultimate goal in food quality control is protecting the consumer. To ensure standardization of these procedures, food laws and regulations cover the related acts affecting the marketing, production, labeling, food additive used, dietary supplements, enforcement of good manufacturing practice (GMP), hazard analysis and critical control point (HACCP), federal laws and regulations, factory inspections, and import/export inspections.[2,8,19]

This chapter discusses attributes and parameters that are essential in food analysis and quality control.

10.2 FOOD QUALITY

10.2.1 QUALITY PARAMETERS

In order to ensure the right quality of various food products, several parameters are evaluated by different methods:

- Physicochemical and rheological parameters (Table 10.1).
- Phytochemical parameters (Table 10.2).
- Packaging materials (Table 10.3).

TABLE 10.1   Physicochemical and Rheological Parameters for Quality of Selected Food Products.

Parameter name	Instruments and chemicals used	Products
Admixture	Visual observation	Cereals, pulses
Bellier turbidity temperature	Visual	Oils
Bulk density	Calibrated graduated cylinder	Cereal, fruits, and vegetables and other products
Color on Lovibond scale	Lovibond Tintometer	Oil, fat
Crude fiber	Chemical	Most of the fruits and vegetables and cereal products
Fat or oil	Chemical and soxhlet method	Most of the food products, animal feeds
Insect infestation	Visual observation	Cereals, pulses

TABLE 10.1 *(Continued)*

Parameter name	Instruments and chemicals used	Products
Moisture	Hot-air oven	Most of the food products, animal feeds
	Vacuum oven	
	Karl Fischer titer	
	Dean and Stark	
Oil-holding	Centrifuge	Cereal products, other powder products
Optical rotation	Polarimeter	Sugar, syrup, oil and fat
Protein	Chemical and Kjeldahl method	Most of the food products, animal feeds
Refraction	Sieve test	Flour
Starch	Chemical	Starch containing products
Swelling	Centrifuge	Cereal products, other powder products
True density	Gas Pycnometer–volumetric analyzer	Cereal products, other powder products
Water-retention	Centrifuge	Cereal products, other powder products
Water-solubility index	Centrifuge	Cereal products, other powder products

TABLE 10.2   Phytochemical Parameters for Quality of Selected Food Products.

Parameter name	Instruments and chemicals used	Products
Anthocyanins	Chemical and spectrophotometer	Red color-rich fruits and vegetables and other food products
Antioxidant activity	Chemical and spectrophotometer	Most of the fruits and vegetable products
Ascorbic acid	Chemical and titration method	Most of the fruits and vegetable products
Lycopene	Chemical and spectrophotometer	Colored fruits and vegetables and other products
Total carotenoids and β-carotene	Chemical and spectrophotometer	Yellow color rich food products
Total phenols	Chemical and spectrophotometer	Most of the fruits and vegetable products

TABLE 10.3 Packaging Materials for Quality of Selected Food Products.

Parameter name	Instruments and chemicals used	Products
Heavy metals: Pb, As, Cd, Se, Ba	AAS	Colored plastics
Laquer	Physical Chemical	Tin cans
Migration tests	Chemical	Food grade plastics
Sulfide stain	Chemical	Food cans
Tin, chromium	AAS	Tin plate
Water vapor permeability	Humidity chamber	Plastics

10.2.2 BENEFITS OF FOOD ANALYSIS

Food analyses are used for the removal of toxic substances from the food products. It increases the shelf-life of the foods during storage and deteriorates the quality of the food products. Foods are safe to spoilage and microorganisms by the analyses. It improves the quality of life for people with allergies, diabetics, and other people, who cannot consume some common food elements. It adds extra nutrients such as vitamins.

10.2.3 INSTRUMENTAL TECHNIQUES FOR FOOD QUALITY

These techniques are used for the evaluation of organoleptic quality of food products. Different types of instruments are used for the analysis of food products, that is, gas chromatography, spectrophotometer, electronic nose, and so on.

10.2.3.1 GAS CHROMATOGRAPHY-OLFACTOMETRY (GC-O)

The gas chromatography method is based on sensory evaluation of the eluate (Figs. 10.1 and 10.2). Identification of aroma active compounds is possible on the basis of simultaneous use of the second detector. Mostly, second detector function performs mass spectrometer (MS) or flame-ionization detector (FID).

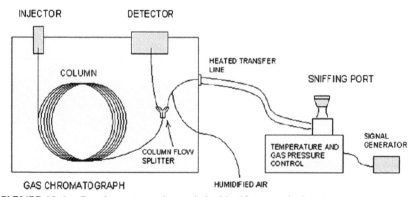

FIGURE 10.1   Gas chromatograph coupled with olfactometric detector.

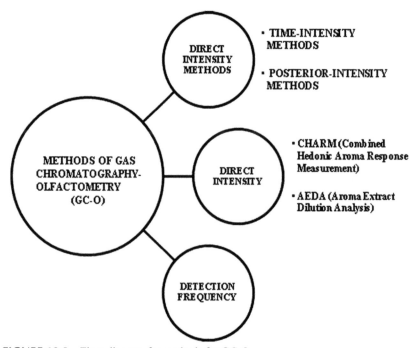

FIGURE 10.2   Flow diagram for methods for GC-O.

10.2.3.1.1 Spectrophotometer

Spectrophotometric techniques are used to measure the concentration of solutes in solution by measuring the amount of light that is absorbed by the solution in a cuvette placed in the spectrophotometer. Spectrophotometry takes advantage of the dual nature of light. Namely, light has:

- A particle nature which gives rise to the photoelectric effect.
- A wave nature which gives rise to the visible spectrum of light.

10.2.3.1.2 Electronic Nose

Electronic nose consists of electrochemical sensors (Fig. 10.3). It is selective for certain volatile compounds. The system is capable for the identification of simple and complex odors. It is used mainly in the food industry for the detection of different aromatic compounds (Fig 10.4).

10.2.4 DRAWBACKS OF FOOD QUALITY

- To affect its nutritional density, the amount of nutrients lost depending on the food and method of processing.
- By the heat treatment, vitamin C is destroyed. Fresh fruit juices have high content of vitamin C than processed fruit juice.
- Large mixing, grinding, chopping and emulsifying equipment, inherently introduce a number of contamination risks.
- Large food processors will utilize many metal within the processing stream, metals may be dangerous for our health.

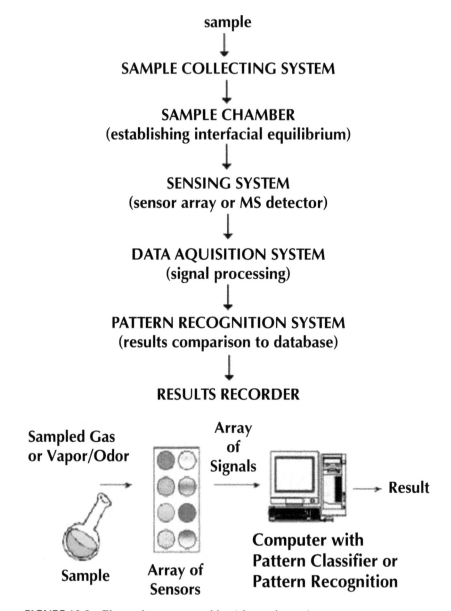

FIGURE 10.3   Electronic sensory machine (electronic nose).

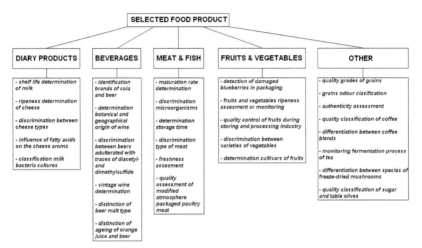

FIGURE 10.4   Food products for use in electronic sensory machine.

10.2.5   REASONS FOR ANALYZING FOODS

Foods are analyzed by scientists working in all of the major sectors of the food industry including food manufacturers, ingredient suppliers, analytical service laboratories, government laboratories, and university research laboratories.

10.2.6   GOVERNMENT REGULATIONS AND RECOMMENDATIONS

The government has designed and recommended regulations to maintain the quality of food products, ensure food companies to supply safe and wholesome food to the consumers, inform consumers about the nutritional composition of foods, enable fair competition among food companies, and to eliminate economic fraud. There are various government departments responsible for regulating the composition and quality of foods, such as the US Food and Drug Administration (FDA), the United States Department of Agriculture (USDA), the National Marine Fisheries Service (NMFS), and the Environmental Protection Agency (EPA). Each of these agencies regulates a particular sector of food industry and also publishes the documents that contain detailed information about the regulations and recommendations. These documents can be purchased from the government or can be obtained online website.

10.2.7   FOOD SAFETY

It is harmful for the consumers to consume unsafe foods. A food may be considered unsafe when it contains harmful microorganisms (e.g., Listeria, Salmonella), toxic chemicals (e.g., pesticides, herbicides) or extraneous matter (e.g., glass, wood, metal, insect matter). Therefore, food manufacturers should take all the preventive measures to eliminate these harmful substances from food. Government laboratories use analytical techniques to analyze food and detect toxic substances and also to ensure that the food is safe for consumers.

10.3   QUALITY CONTROL

To make high profit and market share, food manufacturers are continuously trying to have products of higher quality, and less expensive than the other competitive products. Analytical techniques to analyze food are required to meet these standards during and after manufacturing of food products. In a food factory, one starts with a number of different raw materials, processes them in a certain manner (e.g., heat, cool, mix, dry), packages them for consumption and then stores them. The food is then transported to a warehouse or retailer where it is sold for consumption. The important concern of the food manufacturers is to produce a final product that has all acceptable properties such as appearance, texture, flavor and shelf-life. When we purchase a product, we expect it to have same properties whenever we buy it. However, the raw materials for manufacturing of product vary from time to time, which can cause the properties of the final product to vary, often in an unpredictable way. First, they have to understand that different food ingredients and processing operations are important in determining the final properties of food so that they can be controlled during the manufacturing process of final product. This type of information can be established through research and development work. Second, they can monitor the properties of foods during production to ensure that they are meeting the specified requirements, and if a problem is detected during the production process, appropriate actions can be taken to maintain final product quality.

10.3.1 METHODS OF FOOD QUALITY CONTROL

In addition to ensuring safe and healthy food for the consumer, product manufacturers and service industries have realized that competition in a global market requires a continual and committed effort towards the improvement of product and service quality. Therefore, they follow the process improvement cycle comprising:

- PLAN (plan improvement)
- DO (implement plan for improvement)
- CHECK (analyze the collected data)
- ACT (take action)

Quality control process consists of raw materials, in-process, product and service. The major factors in process that cause variability in quality of finished product are people, equipment, and methods or technologies employed in the process. The use of proper statistical process control methods is also vital for assurance of the product quality. Usually, the value of quality characteristics is used to provide feedback on how processes may be improved. Statistical quality control consists of the following procedures:

- Finished product is measured.
- Sampling occurs for days or weeks.
- Lot is either accepted or rejected based on the information from sample.

Contrary to statistical quality control, statistical process control methods focus on identifying factors in process that cause variability in finished product, eliminating the effect of these factors before worse product is manufactured, and control charts give online feedback of information about process.[11,18] Food quality control measures have continuously been improved since the 20th century, owing largely to the implementation of

good practices, quality systems, and increased traceability in food production. Ever since microorganisms were discovered in our environment and
linked to typhoid fever and other diseases that have plagued humanity,
public health, authorities have been concerned with the accumulation of
filth and foul odors in urban areas.[18] The first early inspection systems
based on sensory evaluations were legally enforced at the beginning of
the 20th century. Initial bacteriological techniques to detect pathogenic
bacteria in foods, such as shellfish, appeared soon after.[18] From that point
on, the food and beverage industry has applied stricter product inspection
procedures and more and more effective production methods to conserve
the freshness of natural raw materials.

Today, the establishment of GMP and good hygienic practices (GHP)
in many countries has significantly reduced the risk of spoilage and pathogenic microorganisms in modern food products. In addition to comply
with national and international food regulations, food manufacturers are
required to follow international quality standards, such as ISO as well as
the HACCP system. In recent years, there has been an increasing focus
on traceability in food production.[18] This has followed public concerns
arising from cases of food contaminations and the development of foods
containing ingredients derived from genetically modified (GM) crops. In
light of increasing need for food more rapid food testing, it became clear
that the traditional microbiological detection and identification methods
for foodborne pathogens were no longer effective. Because, it was time-
consuming and laborious to perform, and was increasingly unable to meet
the demands for rapid quality control. A rapid method is generally characterized as a test giving quicker results than the standard accepted method
of isolation and biochemical and/or serological identification.[18]

10.3.1.1   ION MOBILITY SPECTROMETRY OR DIFFERENTIAL MOBILITY SPECTROMETRY

Ion mobility spectrometry (IMS) and differential mobility spectrometry
(DMS) are the methods used in the identification and quantification of
analysts with high sensitivity. The selectivity can even be increased as
necessary for the analyses of complex mixtures-using pre-separation techniques such as gas chromatography or multi-capillary columns (MCC).
The method is suitable for application in the field of food quality and

safety including storage, process and quality control as well as the characterization of foodstuffs.[23]

10.3.1.2   IMMUNOCHEMICAL METHODS

This method is based on antigen–antibody interaction. The antibodies are highly specific for the antigen, and secondly, the antigen, the antibody, or an anti-globulin may be conjugated to an enzyme that produces an intensely colored or fluorescent product in the presence of the enzyme substrate to enhance the detectability of analyze in an amplification step.

10.3.1.3   ENZYME IMMUNO ASSAY (EIA)

In recent years, the EIA using monoclonal antibodies have made available rapid and consistent microbiological detection systems. The most widely used systems employ a sandwich technique using antibody attached to a polystyrene matrix to which the sample is added. Post incubation, a second antibody, which is specific for the organism and has been tagged with an enzyme, is added. The addition of enzyme substrate to the mixture completes the EIA. The presence of the specific organism results in a colorimetric change in the enzyme substrate, which may be observed visually or with a spectrophotometer. Most EIA are very specific but lack sensitivity. Normal sensitivity has been reported to be approximately 106 org/ml.[9]

10.3.1.4   BIOSENSORS

Biosensor is usually a device or instrument comprising a biological sensing element coupled to a transducer for signal processing.[22] Biological sensing elements include enzymes, organelles, antibodies, whole cells, DNA, and tissues. There are different types, conductance bioluminescence enzyme sensors utilizing potentiometric, amperometric, electrochemical, optoelectric, calorimetric, or piezoelectric principles. Basically, all enzyme sensors work by immobilization of the enzyme system onto a transducer.[22] This technique provides sensitive and miniaturized systems that can be used to detect unwanted microbial activity or the presence of a biologically active compound, such as glucose or a pesticide in food. Immunodiagnostics and enzyme biosensors are two of the leading technologies that have had the greatest impact on the food industry.[9]

10.3.1.5   FLOW CYTOMETRY (FCM)

Specific detection of pathogenic strains can be accomplished by flow cytometry using immunofluorescence techniques, which allow microorganism detection at the single-cell level. Although this technology can be used for food samples, it requires prior isolation of the target organism to generate antibodies.[6] FCM finds wide application in milk and brewing quality control. The advantage of FCM is that it can also differentiate viable nonculturable (VBNC) form of bacteria from healthy cultivable cells.[6] This technology has the ability to detect microorganisms at relatively low concentrations in a short time, whereas multiple labeling allows the detection of different organisms or different stages in the same sample.[6]

10.3.1.6   STEPS IN POLYMERASE CHAIN REACTION

The steps involved in polymerase chain reaction (PCR) are as follows:

- Isolation of DNA from the food
- Amplification of the target sequences
- Separation of the amplification products by agarose gel electrophoresis
- Estimation of their fragment size by comparison with a DNA molecular mass marker after staining with ethidium bromide
- Finally, a verification of the PCR results by specific cleavage of the amplification products and by restriction endonuclease or southern blot. Alternatively amplification products may be verified by direct sequencing or a second PCR.[9]

10.3.1.7   PULSED-FIELD GEL ELECTROPHORESIS (PFGE)

PFGE is a restriction-based typing method that is considered by many to be the "gold standard" molecular typing method for bacteria.[9] In this electrophoretic approach, DNA fragments are separated under conditions where there is incremental switch of the polarity of the electric field in the running apparatus. This technique allows for the resolution of DNA fragments up to 800 kb in size. When DNA is restricted with a restriction enzyme, PFGE provides a DNA "fingerprint" that reflects the DNA sequence of the entire bacterial genome. PFGE is a widely accepted method for comparing the genetic identity of bacteria.[9] PFGE typing has demonstrated a high level of reproducibility for foodborne pathogens. A major advantage of this method is its universal nature making it useful in bacteria subtyping; however, its limitation is that it is time-consuming.

10.3.1.8   MAGNETIC SEPARATION

Using this technique, investigators[15,20] separated salmonella from food and fecal matter using myeloma protein and hybrid antibody (for O antigen), conjugated to a polycarbonate-coated metal bead. It has also been reported that food sample like milk, yogurt, meat, and vegetables can be tested.[10] The challenge is in detecting *Escherichia coli* is in the isolation of pathogenic strain from nonpathogenic strains. Immunomagnetic detection of listeria monosytogens has also been investigated.[21]

10.3.1.9   NEAR-INFRARED SPECTROSCOPY

Near-infrared spectroscopy (NIR) has proven to be an effective analytical tool in the area of food quality control. The key advantages of NIR spectroscopy are: (1) its relatively high speed of analysis, (2) the lack of a need to carry out complex sample preparation or processing, (3) low cost, and (4) suitability for online process monitoring and quality control. The disadvantage of this method includes the requirement for large sample sets for subsequent multivariate analysis. Recently researchers at Zhejiang University in Hangzhou, China, used Vis/NIR spectroscopy together with multivariate analyses to classify nontransgenic and transgenic tomato leaves.[7,16,24]

10.3.1.10   X-RAY

This is a relatively newer technology in food quality control. X-rays started making inroads into the food industry in the early 1990s. The driving force behind this was the increasing number of foreign bodies which could not be identified by metal detectors. Other than contaminants such as glass, bone, rubber, stone or plastic, some specific applications are also more challenging for metal detectors, such as fresh meat and poultry, or foil-wrapped products.[5] X-ray inspection has considerable advantages in many foods and beverage-processing environments in that, it is easy to install, safe and simple to use, even without previous experience. It quickly and consistently identifies substandard products, reducing product recall, customer returns and complaints, therefore protecting manufacturers' brands and most importantly, preventing ill health.

10.3.1.11   COMPUTER VISION

Computer vision system consists of four basic components: the illumination source, an image acquisition device, the processing hardware, and suitable software modules.[3] The study was focused on analyzing the relevance of computer vision techniques for the food industry, mainly in Latin America. The authors described how the use of these techniques in the food industry eliminates the subjectivity of human visual inspection, adding accuracy, and consistency to the investigation. They also reported that the technique can provide fast identification and measurement of selected objects, classification into categories, and color analysis of food surfaces with high flexibility. They mentioned that since the method was noncontact and nondestructive, temporal changes in properties such as color and image texture can also be monitored and quantified.

10.3.2   FOOD QUALITY CONTROL: CASE STUDIES

10.3.2.1   LIVESTOCK

The meat and milk characteristics are more related with human health and with some factors affecting the quality.[4] The molecular biology techniques was of great interest to the researchers as these give insight to new product certification, namely, species, breed, animal category (age, sex, and so on.). Automatic milking systems (AMS) increases milk yield and milking frequency from twice to three times or more per day requiring a minimum extra amount of labor. However, contradictory results are reported about the effects of AMS on milk quality. Several authors found that after the introduction of AMS milk quality decreased, particularly fat, proteins percentage whereas total bacterial plate count, SCC, freezing point and the amount of free fatty acids were increased significantly.[4] The system consists of a highly general hardware setting, able to support different applications, and highly modular software, easily adapted to the measurement needs of diverse food products. The main result of this application was to classify rice grains and lentils.[3] Grain quality attributes are very important for all users and especially the milling and baking industries. Other study showed the usefulness of machine vision to identify different varieties of wheat and to discriminate wheat from nonwheat components.[25] Visible light photoluminescence (PL) peaking at around $\lambda=460$ nm is characteristic of cereals, such as rice, wheat, barley, millet, flour, corn starch, peanut, under illumination of ultraviolet light at $\lambda=365$ nm.[12] Authors further reported that peak intensity of PL and distribution of PL intensity varies with variety and source of the specimens, which was found to be fitted with a Gaussian curve. Visible light PL is suggested to be potentially useful technique for the nondestructive and quick evaluation of the cereals and other starchy products. The use of amperometric nanobiosensor for the determination of glyphosate and glufosinate residues in corn and soybean samples has been mentioned.[22] The biosensor has the features of high sensitivity, fast response time (10 to 20 s) and long-term stability at 40°C (> 1 month). Detection limits were in the order of 10–10 to 10–11 M for standard solutions of herbicides and the spiked samples. The author found that herbicide analyses can be spiked on real samples of corn and soybean, corroborating that the biosensor is sensitive enough to detect herbicides in these matrices.

10.3.2.2 FRUITS, VEGETABLES, AND NUTS

Narendra et al.[17] observed that computer vision has been widely used for the quality inspection and grading of fruits and vegetables. It offers the potential to automate manual grading practices and thus to standardize techniques and eliminate tedious inspection tasks. The capabilities of digital image analysis technology to generate precise descriptive data on pictorial information have contributed to its more widespread and increased use. Method of PCR-SSCP has been used for the genetic differentiation of canned abalone and commercial gastropods in the Mexican retail market. The study was aimed at creating molecular tools that can differentiate abalone (*Haliotis* spp.), from other commercial fresh, frozen and canned gastropods based on 18S-rDNA and also identify specific abalone product at the species level using the lysine gene. It was found that the methods were reliable and useful for rapid identification of Mexican abalone products and could distinguish abalone at the species level. The methods could genetically identify raw, frozen and canned products and the approach could be used to certify authenticity of Mexican commercial products or identify commercial fraud.

Qualitative analysis of pesticide residues in fruits and vegetables has been conducted using fast, low-pressure gas chromatography—time of flight (ToF) mass spectrometry (LP-GC/MS).[14] It was demonstrated that, to increase the speed of analysis for GC-amenable residues in various foods and provide more advantages over the 40 traditional GC-MS approach, LPGC/MS on a ToF instrument should be applied as it provides high sample throughput with <10 min analysis time. The method had already been validated to be acceptable quantitatively for nearly 150 pesticides, and in this study of qualitative performance, 90 samples in total of strawberry, tomato, potato, orange, and lettuce extracts were analyzed. The extracts were randomly spiked with different pesticides at different levels, both unknown to the analyst, in the different matrices. Researchers compared automated software evaluation with human assessments in terms of false positive and negative results only to find that the result was not significantly different. Other investigated methods are robust ten-plex quantitative and sensitive ligation-dependent probe amplification method, the allergen-multiplex, quantitative ligation-dependent probe amplification (MLPA) method, for specific detection of eight allergens: sesame, soy, hazelnut, peanut, lupine, gluten, mustard, and celery. Ligated probes were amplified by PCR and amplicons were detected using capillary electrophoresis. Quantitative results were obtained by comparing signals with an internal positive control. The limit of detection varied from approximately 5–400 gene copies depending on the allergen. The method was tested using different foods spiked with mustard, celery, soy or lupine flour in the 1 to 0.001% range. Depending on the allergen, sensitivities were similar or better than those obtained with PCR.

10.4 CONCLUSIONS

Foods are the substances which provide nutritional support for the body. It may be of plant or animal origin, containing the known five essential nutrients, namely, carbohydrates, fats, proteins, vitamins, and minerals. These analytical procedures are used to provide information about a wide variety of different characteristics of foods, including their composition, structure, physicochemical properties, phytochemical properties, and sensory attributes.

Quality control is the maintenance of quality at levels and tolerance limits acceptable to the buyer while minimizing the cost for the vendor. Scientifically, quality control of food refers to the utilization of technological, physical, chemical, microbiological, nutritional and sensory parameters to achieve the wholesome food. These quality factors depend on specific attributes such as sensory properties, based on flavor, color, aroma, taste, texture and quantitative properties, namely, percentage of sugar, protein, fiber and so on as well as hidden attributes likes peroxides, free fatty acids.

The growth of diagnostic industry should result in increased rapid tests in the nearest future and this should result in improved performance. It is expected that, there will be significant economic benefits and the ability to practice proactive and risk prevention food safety programs, by 2015, the companies should be able to utilize automation technology to screen incoming raw materials and in-process parameters with near real-time information; physical, chemical or biological, while utilizing these newer methods.

It is therefore safe to conclude that food analysis and quality control is an indispensable tool in the food industry. As enumerated above, the development of adequate, effective, rapid, and sensitive food quality control systems however, faces serious challenges driven by its capital intensive nature and sophisticated adulteration. While it may seem easier for the developed nations to match quality control with adulteration techniques, to make any meaningful progress in resource-limited nations of the world, there is the need for collaborations between laboratories around the globe, just as it is necessary for regulatory agencies around the world to also collaborate both in sharing information and in technologies as well as capacity development.

Although some of these automated newer technologies are extremely rapid, there have been questions about their sensitivity. Some investigators agree that the instruments can produce results in seconds, but they opine strongly that they are not sensitive enough. In the nearest future, the mass spectrometry may be needed in food quality control. DNA-based assays, instruments, and software, all are designed to work together including the emerging area of bioinformatics; sequencing, assay design, and chemical analysis are all capable of developing new ways for food quality control in the future.

10.5 SUMMARY

Food analysis and quality control have many attributes and most of them have been mentioned in this chapter. Different food analysis methods are discussed, that is, physicochemical, phytochemical and packaging methods. Instrumental methods are also discussed in this chapter. Different types of quality control methods are also discussed to maintain the quality of food products. Food analysis and quality control processes are compulsory for all types of food industry.

KEYWORDS

- Antioxidant
- ascorbic acid
- biosensors
- fiber
- food analysis
- food quality
- food technology
- gas chromatography
- good manufacturing practices
- nondestructive
- phenol
- physicochemical
- phytochemical
- protein

REFERENCES

1. Adamson, M. W. *Food in Medieval Times*; Greenwood Publishing Group: Westport, CT, 2004; pp 64–67.

2. Adu-Amankwa, P. *Quality and Process Control in the Food Industry*; Food Research Institute: The Ghana Engineer, 1999.
3. Aguilera, J. M.; Stanley, D.W. *Microstructural Principles of Food Processing and Engineering*, 2nd ed.; Springer: New York, 1999; p 379.
4. Alessandro, N. *Evolution of Livestock Production and Quality of Animal Products.* 39th Annual Meeting of the Brazilian Society of Animal Science, Brazil, 2002; pp 486–513.
5. Ansell, T. *X-Ray a New Force in Food Quality Control*; Al Hilal Publishing & Marketing Group, 2008; p 421.
6. Comas-Riu, J.; Núria, R. Flow Cytometry Applications in the Food Industry. *J. Ind. Microbiol. Biotechnol.* **2009**, *36*(8), 999–1011.
7. Cozzolino, D.; Fassio, A.; Restaino, E.; Fernandez, E.; La. Manna, A. Verification of Silage Type Using Near-Infrared Spectroscopy Combined with Multivariate Analysis. *J. Agric. Food Chem.* **2008**, *56*(1), 79–83.
8. Gravani, R. B. *How to Prepare A Quality Assurance Plan, Food Ware Housing.* Department of Health and Human Services, Public Health Service, U.S. Food and Drug Administration, Food Science Facts for the Sanitarian, Dairy and Food Sanitation, 1986.
9. Greiner, R. U. *Modern Molecular Methods (PCR) in Food Control: GMO, Pathogens, Species Identification, Allergens*. An oral presentation given at the 7th Simpósio Latino Americano de Ciência de Alimentos (SLACA) hosted by sbCTA at the State University of Campinas, Brazil, (accessed Nov 4–7, 2007).
10. Haik, Y.; Sawfta, R.; Ciubotaru, I.; Qablan, A.; Tan, E. L. Magnetic Techniques for Rapid Detection of Pathogens. In *Principles of Bacterial Detection: Biosensors, Recognition Receptors and Microsystems*; Springer Science &Business Media, 2008; p 415.
11. http://faculty.uca.edu/~march/bio1/scimethod/. (accessed Aug 31, 2017).
12. Katsumata, T.; Suzuki, T.; Aizawa, H.; Matashige, E. Photoluminescence Evaluation of Cereals for a Quality Control Application. *J. Food Eng.* **2007**, *78*(2), 588–590.
13. Lasztity, R.; Petro-Turza, M.; Foldesi, T. History of Food Quality Standards. In *Food Quality and Standards*; Davidek, J. Ed.; Oxford Publisher: UK, 2004, vol 2; pp 10–21.
14. Lehotay, S. J.; Koesukwiwat, U.; Van der Kamp, H.; Mol, H. G.; Leepipatpiboo, N. Qualitative Aspects in the Analysis of Pesticide Residues in Fruits and Vegetables Using Fast, Low-Pressure Gas Chromatography–Time-of-Flight Mass Spectrometry. *J. Agric. Food Chem.* **2011**, *59*(14), 7544–7556.
15. Mattingly, J. A. An Enzyme Immunoassay for the Detection of all Salmonella Using a Combination of a Myeloma Protein and a Hybridoma Antibody. *J. Immunologic. Methods* **1984**, *73*(1), 147–156.
16. Michelini, E.; Simoni, P.; Cevenini, L.; Mezzanotte, L.; Roda, A. New Trends in Bioanalytical Tools for the Detection of Genetically Modified Organisms: An Update. *Anal. Bioanal. Chem.* **2008**, *392*(3), 355–367.
17. Narendra, V. G.; Hareesh, K. S. Quality Inspection and Grading of Agricultural and Food Products by Computer Vision—A Review. *Int. J. Comp. App.* **2010**, *2*(1), 43–65.
18. Raju, K. V. R.; Yoshihisa, O. *Report of the APO*. Seminar on *Quality Control for Processed Food* held in the Republic of China, 2002; 02-AG-GE-SEM-02.
19. Roe, R. S. *The Food and Drugs Act- Past, Present and Future;* Food & Drug Administration: USA, New York, 1956; pp 15–17.

20. Safarik, M.; Safarikova; Forsethe, M. J. Application of Magnetic Separation in Applied Microbiology. *J. Appl. Microbiol.* **1995,** *78,* 575–585.

21. Skjerve, E.; Rørvik, L. M.; Olsvik, O. Detection of Listeria Monocytogenes in Foods by Immunomagnetic Separation. *Appl. Environ. Microbiol.* **1990,** *56*(11), 3478–3481.

22. Songa, A. E.; Somerset, S. V; Waryo, T.; Baker, G. L. P.; Iwuoha, I. E. Amperometricnanobiosensor for Determination of Glyphosate and Glufosinate Residues in Corn and Soya Bean Samples. *Pure Appl. Chem.* **2009,** *81*(1), 123–139.

23. Vautz, W.; Zimmermann, D.; Hartmann, M.; Baumbach, J. I.; Nolte, J.; Jung, J. Ion Mobility Spectrometry for Food Quality and Safety. *Food Additive Contaminants* **2006,** *23*(11), 1064–1073.

24. Xie, L.; Ying, Y.; Ying, T. Quantification of Chlorophyll Content and Classification of Nontransgenic and Transgenic Tomato Leaves Using Visible/Near-Infrared Diffuse Reflectance Spectroscopy. *J. Agric. Food Chem.* **2007,** *55*(12), 4645–4650.

25. Zayas, I. Y.; Martin, C. R.; Steele, J. L.;Katsevich, A. Wheat Classification Using Image Analysis and Crush Force Parameters. *Am. Soc. Agric. Eng.* **1996,** *39*(6), 2199–2204.

CHAPTER 11

INTERVENTIONS OF OHMIC HEATING TECHNOLOGY IN FOODS

K. A. ATHMASELVI[1,*], C. KUMAR[2], AND P. POOJITHA[3]

[1]Department of Food Process Engineering, School of Bioengineering, SRM University, Kattankulathur, Chennai 603203, Tamil Nadu, India, Tel.: +91-9600007823

[2]Department of Physics, Government Arts and Science College for Men, Nandanam, Chennai 600035, Tamil Nadu, India, Tel.: +91-9600039748, E-mail: vckumar60@yahoo.co.in

[3]Department of Food Process Engineering, School of Bioengineering, SRM University, Kattankulathur, Chennai 603203, Tamil Nadu, India, Tel.: +91-7418956148, E-mail: poojithapushparaj@gmail.com

*Corresponding author. E-mail: athmaphd@gmail.com

CONTENTS

11.1   INTRODUCTION

Heat processing is the most common method of food preservation. Heat processing is used in most of the food industries for processing and preservation of food. Very high temperature generated during heat processing leads to organoleptic changes and the loss of nutrients. During heat processing, heat transfer to the food from the hot source takes place either due to conduction, convection, or radiation. Conventional heat processing results in heterogeneous heating, as heat has to conducted from outside to inside, by the time the thermal center reaches the desired temperature, surface may be overheated. Therefore, holding the organoleptic properties and nutritional profile of food, especially fruits and vegetables by conventional heating method, is a great dispute. These disadvantages could be overcome by the adoption of new technologies such as high-pressure technology, ohmic heating, ultrasound processing, pulsed electric field technology, irradiation, hurdle technology, and oscillating magnetic fields.

When more novel methods of heating such as microwave heating, inductive or ohmic heating are used, heat treatment for complex food fluids is substantially ameliorated. In these methods, heat is generated inside the food and depends less on thermal conduction and convection thereby causing fewer temperature gradients.[36]

Ohmic heating (which is also called Joule heating, electrical resistance heating, direct electrical resistance heating, electro-heating, or electroconductive heating) is defined as a thermal process in which alternating electric currents (AC) are passed through the foods to heat them. Heat is internally generated due to electrical resistance. The electrical energy dissolution in the product–mass gives heat.[10,51] Most food preparations contain a controlled amount of free water with dissolved ionic salts such as salts and acids; therefore, electric current can be made to pass through the food and generate heat inside the food.[30] The liquid which does not have ionic species like distilled water cannot be heated by ohmic heating system. ohmic heating was used in the early 20th century. Processing of whole fruits by ohmic heating is currently being done in Japan and the United Kingdom,[51] for the production of syrup-based fruit-salad and fruit juices.[31]

Ohmic heating has gained interest recently because the products are of a superior quality than those processed by conventional technologies.[8,33]

Microbial inactivation during ohmic heating is due to the heat generated during ohmic heating, electroporation mechanisms also help in microbial inactivation.[8] One of the most promising applications of ohmic heating is the processing of fluids containing particulates and high viscous fluids in a sterile way.[41,60] The applicability of ohmic heating depends on the electrical conductivity of the product. Temperature, applied voltage, sample concentration, food particle size, and type of pretreatment are some of the factors that can affect the electrical conductivity of food.[5,36] During ohmic heating, the voltage applied can be easily and rapidly changed, and the product temperature can also be accurately controlled.

This chapter deals with principles and applications of ohmic heating in food industry, the role of electrical conductivity during ohmic heating, and the influence of ohmic heating on ascorbic acid degradation and microbial inactivation.

11.2   ELECTRICITY

All substances are comprised of atoms that contain protons, electrons, and neutrons. A substance is said to be a conductor when it allows the electrons to pass through freely. An insulator is a substance where the electrons are tightly bound and do not move freely. This flow of electrons throughout a substance is known as electricity.

Electricity comprises electrical current, voltage, and resistance. Electrical current is measured in amperes, where 1 A is the flow of ~6×10^{18} electrons per second through a substance. Voltage is the electron pressure or a measure of the ability to move an electrical charge through a resistance (opposition to flow of electricity). Voltage can be calculated by multiplying the current and the resistance. This principle is known as Ohm's law, first proposed by the physicist George Ohm in 1827.[20]

$$V = I \cdot R \qquad (11.1)$$

where voltage = V, current = I, and resistance = R.

As shown in Figure 11.1, the concept of ohmic heating is quite simple. The passage of electric current through an electrically conductive food material obeys Ohm's law ($V=IR$); and heat generation due to the electrical resistance of the food, is given by:

$$P_{heat} = I^2 R \qquad (11.2)$$

The design of ohmic heaters is governed by the electrical conductivity of the food. Since most food materials contain a considerable amount of free water with dissolved ionic species, the conductivity is high enough for a heating effect to occur.[17]

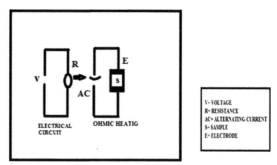

FIGURE 11.1   The concept of ohmic heating.

11.3   ELECTRIC FIELD

In order to generate heat in an ohmic heating system, an electric field must be applied to the food. The electric field (voltage distribution) is a function of the electrode and system geometry, electrical conductivity, and also of the applied voltage.[17] The electric field is determined by the solution of Laplace's equation:

$$\nabla(\sigma.\nabla V) = 0 \qquad (11.3)$$

where σ is the electrical conductivity, and ∇V is the voltage gradient.

Equation (11.3) has been obtained combining Ohm's law with the continuity equation for electric current.[26], and it differs from the usual form of Laplace's equation, because σ is a function of both position and temperature:

$$\nabla^2 V = 0 \tag{11.4}$$

The dependence of position is because foods are not necessarily homogeneous materials; the limiting scenarios being food-containing particles (e.g., vegetables soup) and that of a reasonably homogeneous liquid (e.g., orange juice). The relation of σ with temperature is usually well described by a straight line.[40]

$$\sigma_T = \sigma_{ref} \cdot \left[1 + m \cdot \left(T - T_{ref} \right) \right] \tag{11.5}$$

where σ_T is the electrical conductivity at temperature T, σ_{ref} is the electrical conductivity at a reference temperature, T_{ref}, and m is the temperature coefficient.

The main parameter in ohmic heating is its electrical conductivity behavior that depends on temperature, applied voltage gradient, frequency, and concentration of electrolytes.[29]

Icier and Ilicali[31] conducted ohmic heating studies on tylose and minced beef and reported that the temperature values of the samples inside the test tube were close to each other. The maximum temperature difference evaluated at different locations during heating was 2°C. The salt components, acids, and moisture are highly effective in increasing electrical conductivity, whereas fats, lipids, and alcohols decrease it.

11.4 HEATING MECHANISM IN OHMIC HEATING

Ohmic heating is based on Ohm's law.[36] It is a process by which the passage of an electric current through a conductor, in case of food, heat is generated within it due to its electrical resistance. The basic model of ohmic heating system consists of two plates (electrode) between which the food flow and the electric field is applied to the electrode. The simple illustration of ohmic heating system is shown in Figure 11.2.[12]

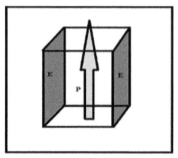

FIGURE 11.2 The principle of ohmic heating.

In ohmic heating, an electrical current is applied to the food. Owing to the food's resistance to the applied alternating current, heat is generated within it. This is due to the mechanism of movement of ions in the food and collision with their surrounding molecules, which then release energy in the form of heat.[12] Since the heating happens within the food and is not dependent on any hot surface, the heating pattern in ohmic heating is more uniform than in conventional heating technologies.[35] Ohmic heating is therefore suitable for the processing of viscous or multiphase foods.

Ohmic heating system should basically consist of generator/power supply to produce electricity, electrodes connected to power supply. Electrodes must be in physical contact with the substance to be heated. Electric field strength ($V \cdot cm^{-1}$) depends on the distance between electrodes. An electrically isolated thermocouple is used to measure the temperature. A data logger stores data on temperature, voltage, current and time. Data of voltage and current are required to measure the electrical conductivity. pH probe, refractometer can be placed as additional devices at required location.

Ohmic heating heats the substance based on the electrical resistance generated due to the conduction of electric current. There are several factors that affect the temperature generated during ohmic heating like: (1) electrical conductivity of the substance, (2) distance between the electrode, (3) number of electrodes placed, (4) voltage gradient applied, (5) thermo-physical properties of the food, and (6) amount of solutes present in the food.[21]

The most typical configuration for the ohmic heater consists of a horizontal cylinder with one electrode placed in each extremity for a batch process. Continuous ohmic heating system will have multiple electrodes that are spaced to account for the increase of electrical conductivity of food with temperature. The product will heat during its path through the heater. Batch processes are used to blanch fruits and vegetables, heat or cook meat products, compound extraction.

In order to solve Equation (11.2), boundary conditions specific for each case must be established. The solution has been obtained by de Alwis and Fryer[17] for a static ohmic heater containing a single particle, using boundary conditions: (a) a uniform voltage on the electrodes and (b) no current flux across the boundary elsewhere. For a more general case of many different particles flowing in a fluid composed of several liquid phases (e.g., vegetable soup, where different vegetable solid pieces are dip in a fluid broth with at least an aqueous and a lipid phase), the mathematical solution for Equation (11.2) is complex and is still unknown. In these cases, the prediction of the electric field has been based on semi-empirical models.[26]

11.5 ELECTRICAL CONDUCTIVITY

Electrical conductivity is the measure of how easily a substance carries electric charge, and it is expressed in Siemens per meter ($S \cdot m^{-1}$). In ohmic heating, the conductivity is a measure of the mineral or ionic content. The most common ionic ingredient in food is salt (NaCl). The more is the amount of dissolved salts in a substance, the higher will be the conductivity. Seawater has a conductivity of ~ 5 $S \cdot m^{-1}$, whereas regular drinking water has a conductivity of ~ 0.0005 to 0.05 $S \cdot m^{-1}$. To be able to process food by ohmic heating, its electrical conductivity should be in the range of $0.01–10$ $S \cdot m^{-1}$ at $25°C$.[24] Ohmic heating can be used for heating food that is an electrically conducting moist material. For example, milk contains water and ionic salts, and is capable of conducting electricity, but it also has a resistance which generates heat when an electric current is passed through it. Therefore, the rate of heating is directly proportional to the square of the electric field strength and the electrical conductivity. The electric field strength is controlled by adjusting the electrode gap or the applied voltage. However, the heating rate depends largely on the physical characteristics of the food itself, especially on the electrical conductivity.[18] Electrical conductivity is an inverse of resistance through a unit cross-sectional area A over a unit distance L, or the reciprocal of resistivity.[30]

$$\sigma = (I / V)(L / A) \qquad (11.6)$$

where A is the cross-sectional area of the sample (m²), I is the current passing through the sample (A), L is the electrode gap or sample length (m), R is the resistance of the sample (Ω), V is the voltage across the sample (V), and σ is the specific electrical conductivity ($S \cdot m^{-1}$).

In ohmic heating, research data strongly propose that the applied electric field under ohmic heating induces electroporation of cell membranes of microorganism. The cell electroporation is defined as the pore formation in cell membranes because of the electric field and as a result, the permeability of the membrane is enhanced and dispersion of material throughout the membrane is achieved by electro-osmosis.[2,36] It is assumed that the electric breakdown or electroporation mechanism is prevalent for the nonthermal effects of ohmic heating.[13,27,28,50]

11.5.1   ELECTRICAL CONDUCTIVITY OF PARTICULATE FOODS

In ohmic heating of particulate foods, the electrical conductivity of fluid and solid particles must be equal.[7] Proper electric conductance management is essential to successfully apply ohmic heating.[8,22,49,60] Biss et al.[7] suggested to raise the electrical conductivity of solids by salt infusion. This could be accomplished by slow soaking or marination process or by rapid blanching in salt solution. By altering the electrical conductivity of the solid components present in the food, it is possible to heat the solids at similar rate with the fluid.

The two main factors that influence the heating rate of particles in a fluid include relative conductivities and relative volume of the system's phase.[44] Low conductivity of solid particles compared to the fluid conductivity causes lag behind the fluid at low concentrations associated with the fluid volume. However, in certain cases where the particle concentration is high, low conductivity particles may heat faster than the surrounding fluid.[51]

Davishi et al.[15] postulated that ohmic heating is most fulfilling for products with values ranging from 0.01 to 10 S·m^{-1}, with optimum efficiency in the range of 0.1–5 S·m^{-1}. There are also critical σ values below 0.01 S·m^{-1} and above 10 S·m^{-1} where ohmic heating is not applicable. This is because very large voltage values would be required to give the amount of heat required by increasing temperature considerably by the Joule effect, in case of very low or very large σ values, respectively.[31,45]

11.5.2   EFFECT OF TEMPERATURE ON ELECTRICAL CONDUCTIVITY

Icier and Ilicali[31] conducted ohmic heating studies on tylose and minced beef and reported that with increase in temperature, the electrical conductivity of the minced beef samples was raised to a critical temperature of 45–50°C and then the rate was reduced. However for the tylose samples, the conductance of electricity values was raised as the temperature was increased to 60°C. For the minced beef samples, electrical conductivity may be reduced as a consequence of chemical reactions induced by the effect of increase in temperature and the electrical current. In particular, the denaturation of the proteins may vary in the ionic movements and reduces the conductance of electricity. As the temperature is increased, electrical conductivity values are also increased. The results are similar to those reported by Icier and Ilicali[30] and Icier et al.[32] Raise in the electrical conductivity values with increase in temperature has been explicated by the reduced drag for the ionic movement. It has been described that when water is boiling, gas bubbles are generated. This development appears because of localized high current densities of various oxidation/reduction reactions (e.g., H$_2$ or O$_2$ gas).[29] Rise in the electrical conductivity during heating of biological tissue happens because of the increase in the mobility of ions and structural changes in the tissue such as expulsion of nonconductive gas bubbles, softening, cell wall proto-pectin breakdown and dropping in aqueous phase viscosity.[15]

Electrical conductivities were decreased with temperature rise after bubbling started and the heating was stopped when bubbling started.[39,61] It has been discussed that due to formation of electrolytic hydrogen bubble, fruit juices are acidic.[15]

A linear relationship between temperature and electrical conductivity of orange juice has been found.[30] Castro et al.[8] reported that electrical conductivity of strawberry-based products was increased with temperature and there was a linear or quadratic relationship depending on the type of product under test.

The conductance of electricity of fresh strawberries at various field strengths ranged from 25 to 70 $V \cdot cm^{-1}$, and it was found that the electrical conductivity was increased almost linearly with the field strength.[10] At 25 $V \cdot cm^{-1}$, they reported conductivity to be approximately 0.05 $S \cdot m^{-1}$ at 25°C and 0.55 $S \cdot m^{-1}$ at 100°C. Sarang et al.[50] reported the conductivity at 25°C of pear as 0.084 $S \cdot m^{-1}$, red apple as 0.075 $S \cdot m^{-1}$ and of golden apple as 0.067 $S \cdot m^{-1}$. They also reported that the conductivity of strawberry was increased from 0.186 $S \cdot m^{-1}$ at 25°C to about 0.982 $S \cdot m^{-1}$ at 100°C. Typical values ranging from 0.5 to 1.6 $S \cdot m^{-1}$ at 20–80°C have been described for low viscosity liquids such as carrot, tomato and orange juices by Palaniappan and Sastry.[40] Electrical conductivity of lemon juice was reported as 0.4–1.0 $S \cdot m^{-1}$ at 30–55 $V \cdot cm^{-1}$ and 20–74°C,[14] 0.38–0.78 $S \cdot m^{-1}$ for grape juice at 20–40 $V \cdot cm^{-1}$ and 20–80°C,[32] 0.15–1.15 for orange juice at 20–60 $V \cdot cm^{-1}$ and 30–60°C, 0.51–0.91 $S \cdot m^{-1}$ for peach puree and 0.61–1.2 $S \cdot m^{-1}$ for apricot puree at 20–70 $V \cdot cm^{-1}$ and 20–60°C.[29]

11.5.3   EFFECT OF VOLTAGE GRADIENT ON ELECTRICAL CONDUCTIVITY

The minced beef samples having two different fat substances were heated using voltage gradients in the range of 10–50 $V \cdot cm^{-1}$.[31] The effects of voltage gradient on the ohmic heating were statistically significant ($P < 0.05$). The voltage gradient raised the heating time to arrive at the desirable temperature that was decreased.

Hong et al.[27] investigated the changes in the heating profiles of apple juice by ohmic heating. They found that heating rate was increased proportionally to the number of electrode pairs and was highly dependent on applied voltage.

As the applied voltage gradient was increased, the ampere passing through the system reached higher values suddenly at lower temperatures. This caused violent evaporation of water in the samples and thus heating process was stopped.[24,29,49] Darvishi et al.[14] noticed that with increase in voltage gradient, the time taken to heat the lemon juice is dropped to reach the desirable temperature. Other researchers found a linear increase in electrical conductivity with increase in temperature.[8,23,26,31,50]

11.5.4   EFFECT OF SOLUTE CONCENTRATION ON ELECTRICAL CONDUCTIVITY

The study on salt concentration of the ohmic heating rate of tylose sample reported that salt content increased the time required to reach the prescribed temperature, which was decreased especially at low-voltage gradients.[31]

The conductance of white egg was high when compared to yolk and liquid whole egg. Higher conductance of electricity of white egg may be due to the presence of high water and low fat content, and hence higher ionic mobility compared to the low water and high fat content of yolk and liquid whole egg.[15]

Ohmic heating of apple and sour cherry concentrates having 20–60% was allowed five different voltage gradients (20–60 V·cm^{-1}).[28] Icier and Ilicali.[28] observed that electrical conductivities of apple and sour cherry juices were importantly involved by temperature and concentration ($P < 0.05$). Also sour cherry juice concentrates were heated faster than the apple juice concentrates for all voltage gradients applied and at all concentrations. The electrical conductivity values of sour cherry juice had been more significant than those of the apple juice for the same temperature at all concentrations and for all voltage gradients applied. This was attributed to the higher acid content of the sour cherry juice having similar insoluble solid content compared with the apple juice.

For the same voltage gradients, the electrical conductivity values were decreased with increase in concentration at a particular temperature. The electrical conductivities of juices were increased linearly due to the decrease of insoluble solid contents, where a nonlinear relationship was reported by Palaniappan and Sastry.[38] There is a decrease in electrical conductivity with an increase in solids and sugar substance of strawberry-based products.[8] They noted that for formulations of products having solid content over 20% w/w and over 40 Brix, a different design of ohmic heater may be required due to the low values of electrical conductivity.[28]

The effects of conductivity were studied on the concentration (Brix) and the temperature (20–80°C) for lemon juice. There was an increase in electrical conductivity with increasing concentration up to approximately 30 Brix, and then it started to decrease. The electrical conductivity decrease may be due to increase in viscosity of the juices with concentration, which diminishes the mobility of the ions.[13] In addition to this, the decrease in the concentration of the apple and sour cherry juices from 60 to 20% enhanced the ohmic heating rate of the juices.[28] The electrical conductivity was dependent on the viscosity of the heated solution.[29]

The reason for the differences among the effects of solids contents by various researchers may be explained by the nature of the solutes in the samples under study. Some components may influence the electrical conductivity of the sample, counting on their electrolytic features. In fruit juices, the primary solute component is the sugar having nonelectrolytic behavior.[28] As the sugar content is increased electrical conductivities of the liquid solutions is decreased. However, the acidity of the juices enhanced their electrical conductivities. Sugar substance and the quality of the other components may induce different electrical conductivities among the juice samples under study.[28]

The drag for ionic movement was increased with increase in solid substance, which might be a cause for the decreasing pattern in electrical conductivity with increasing solid content.[37] Ohmic heating of reconstituted whey solutions (in the range of 8–24% w/v solute concentrations) was from 20°C to prescribed temperatures of 30, 40, 50, 60, 70, or 80°C and by applying voltage gradients of 20, 30, or 40 V·cm^{-1}, and conventionally at water bath.[13] He described that the effect of voltage gradient was statistically significant during from 20 to 80°C, and the solute concentration was not efficacious ($P < 0.01$). With the increase in voltage gradient, ohmic heating time was decreased sharply, may be due to the increase of generation of heating rate inside the food with the voltage gradient. On the other hand, ohmic heating was more retentive at lower solute concentrations. It could be explicated by decreasing amount of free ions/charge carriers, which decrease the electrical conductivity at lower concentrations; and faster heating occurred in solutions having higher electrical conductivity. The use of ohmic heating at high voltage gradients and at higher solute concentrations could be beneficial to receive faster heating of whey solutions.[13]

Increase in the electrical conductivity during heating of the biological tissue occurs due to increase in the ionic mobility because of structural changes in the tissue like extrusion of non -conductive gas bubbles, softening, cell wall proto-pectin breakdown and lowering in aqueous phase viscosity. Higher electrical conductivity of strawberry and peach may be attributed to the softer tissues and hence higher ionic mobility in comparison to the harder tissues of apples, pineapple and pear. Also, presence of large amount of air might result in lower electrical conductivity of apple tissues.[48]

Ohmic heating of red grape juice having concentrations of 10.5, 12.5, and 14.5 Brix by applying three different voltage gradients (10, 12 and 15 V·cm^{-1}) in the temperature range of 25–80°C showed that electrical conductivity was increased as concentration and temperature were increased. The increase of concentration of red grape juice was highly significant on increasing the electrical conductivity. Because the concentration increased solid particle in red grape juice, therefore it accelerated more electric current passage through red grape juice than low concentrated juice. The concentration of soluble solids in red grape juice can explain the change in electrical conductivity. Addition of external compounds, such as sugar and citric acid, resulted in a decrease in the conductivity values, while evaporative concentration provided an increase in conductivity.[3]

Assiry et al.[6] conducted experiment on electrical conductivity of seawater during ohmic heating, and reported that pure water is not a good conductor of electricity and it has a conductivity of 0.055 μ·S·cm^{-1}. Because the electrical current is transported by the ions in solution, the conductivity increases as dissolved ionic species in water increase. Electrical conductivity of seawater was reported to be 53 (m·S·cm^{-1}). It can be observed that as the temperature neared 95°C, there was a slight decrease in the electrical conductivity especially at high concentration and high electrical field strength. This reduction could be the due to localized boiling and production of vapor bubbles that decreased the electrical conductivity of the solution.

The concentration has pronounced effect on the electrical conductivity than the temperature particularly at high concentrations. The electrical conductivity was raised as the temperature and the solute concentration increased, at all voltage gradients. The temperature dependency of electrical conductivities of whey solutions having solute densities of 8–24% was linear.[13]

11.5.5   STATISTICAL MODEL OF ELECTRICAL CONDUCTIVITY DURING OHMIC HEATING

Ohmic heating studies on tylose and minced beef proposed a linear temperature-dependent electrical conductivity equation as below:[31]

$$\sigma = BT + C \qquad\qquad (11.7)$$

Linear relationships with significant regression coefficients were found to be in the temperature range of 30–60°C. The study on electrical conductivity for the minced beef samples showed a non-linear drift with raise in temperature, as shown below:

$$\sigma = BT^N + C \qquad\qquad (11.8)$$

Darvishi et al.[15] conducted ohmic heating of whole egg, white egg and yolk. They described that conductivity of electricity was increased almost linearly with temperature with a coefficient of determination of $R2 > 0.98$.

Due to highest value of R2, the linear model may constitute the electrical conductivity of pomegranate juice during ohmic heating.[16] The linear model has also been suggested by others to describe the ohmic heating of orange juice.[30]; seawater.[6]; apricot and peach purees.[29]; red apple, golden apple, peach, pear, pineapple and strawberry.[50]; and lemon juice.[14]

The orange juice concentrates, having 0.20–0.60 mass fraction soluble solids and ohmically heated at five different voltage gradients (20–60 V·cm−1), showed a linear relationship between temperature and the electrical conductivity for each concentration and voltage gradient applied.[30] The electricity conductivity of strawberry-based pasty products was increased with increase in temperature indicating a linear or quadratic relationship, based on the product type.[10] Icier and Ilicali.[30] recommended to use linear regression equation than the nonlinear equation for the orange juice tested at 0.20–0.60 mass fraction soluble solids at five different voltage gradients (20–60 V·cm−1), because of higher R2 with linear model (y=mx+c), where m stands for slope, x stands for variable, c denotes y-intercept, and y is the required variable.

11.6   ELECTRODES

In ohmic heating, electrodes are required to convey the current to the food material to be heated.[4, 39] Various materials have been used as electrodes in different ohmic heating studies and applications, such as aluminum, carbon (graphite), platinum, platinized-titanium, rhodium-plated stainless steel, stainless steel, and titanium.[47] In ohmic heating, the electrical energy rendered to the heating cell is most preferably used only for the generation of heat; and electrochemical reactions at electrode/solution interfaces are considered undesirable. Food materials are inherently complex mixtures of several different chemical compounds. During ohmic heating, various electrochemical reactions can potentially occur. In addition, some of the products of those electrochemical reactions may initiate a number of secondary chemical reactions.[47] Metal ion may migrate into the heating medium due to electro-chemical reaction, which may alter the color and flavor of the processed food. At low-frequency (50–60 Hz) alternating currents, electrode corrosion and partial electrolysis of the heating medium were found with most of these electrodes. It was also reported that these electrochemical effects diminish with increasing frequency.[4] In stainless steel electrodes, there was hypothesis of electrochemical reactions happening at the electrode/solution interfaces.[5, 13] The stainless steel electrodes exhibited pronounced corrosion rates, hydrogen generation, and also pH changes of the heating media at all the pH values. Titanium has high corrosion resistance and biocompatibility characteristics. Researchers observed apparent electrolysis of the heating medium and electrode corrosion during ohmic heating.[4, 5, 40, 46, 56] The use of high alternating frequencies was suggested to inhibit adverse electrochemical effects.[1, 4]

11.7 APPLICATIONS OF OHMIC HEATING IN FOOD PROCESSING

11.7.1 FRUITS AND VEGETABLES AND THEIR PRODUCTS

Several strawberry-based products can be processed using ohmic heating system in Portuguese fruit jam industry.[18] High-quality product is obtained by ohmic heating of strawberry product despite the significant differences in the electrical conductivity between the products under study. Ohmic heating might be a perfect alternative to conventional blanching treatments. The ohmic pretreatment did not cause significant changes in the moisture content of the final products.[37] A study on the influence of precooking by ohmic heating on the firmness of cauliflower showed that ohmic heating combined with low-temperature precooking in saline solutions offered a viable solution to HTST sterilization of cauliflower florets.[17] A similar study was also performed with potato cubes.[17, 20] and it was concluded that ohmic pretreatment prevented loss of firmness when compared to the conventional pretreatment (50% in some cases).

11.7.2   MEAT PRODUCTS

Ohmic heating technology has been successfully used in meat industry for meat thawing process.[43] Ohmic heating technique to thaw frozen meat samples reported uniform and quick thawing.[62] Also, meat color and pH were not changed significantly and the final products achieved a good thawing quality. Ohmic heating was first used in Finland in 1970s in meat industry, but operational difficulties led to the abortion of the project. Later in the 1990s, a project for cooking liver pate and hams was carried out in France by the Meat Institute Development Association (ADIV) and Electricite de France (EDF). Combined ohmic and conventional cooking of hamburger patties has recently been patented as a new method of cooking. [10] The method is based on passing electric current through the meat patties causing internal heat generation. Ohmic heating helps in reducing the cooking time to half the time generally required in conventional cooking. It was also concluded that ohmic heating has no negative effects on the quality of cooked hamburger.[44]

11.7.3   SEAFOOD PRODUCTS

Surimi is stabilized by myofibrillar proteins from fish muscles and it is used in number of Japanese food products. The textural properties of the products treated by ohmic heating were superior to those heated in 90°C water bath. Also, an increase in shear stress and shear strain of surimi gels was found in ohmic treatment.[65] and a superior gel quality was achieved. Higher heating rates are not beneficial in surimi manufacturing.[64] Instead, slow heating rates produce stronger gels. Ohmic processing of surimi is very effective in obtaining a wide range of linear heating rates, which play an important role in the study of surimi gelation.

11.7.4   OTHER APPLICATIONS

A large number of potential applications exist for ohmic heating in food industry such as blanching, evaporation, dehydration, fermentation, extraction, thawing, solidification sterilization, pasteurization and heating of foods to serving temperature. Ohmic heating technology is used for the processing of whole fruits, berries, fruit juices, liquid egg and soups in Japan, the United Kingdom, and Northern America.[38]

11.8   ADVANTAGES OF OHMIC HEATING SYSTEM

Ohmic heating has numerous advantages over conventional heating methods. Ohmic heating heats the entire mass of the food material volumetrically because of its inherent electrical resistance.[52] Conventional heating, the time required to raise the temperature at the coldest point, may over heat the reaming liquid. However in ohmic heating, the voltage gradient applied can be rapidly controlled and hence the product temperature can also be easily controlled. Also during ohmic heating, no residual heating occurs, after the current is shut off. General advantages of ohmic heating include.[15, 23, 33, 52]:

- Rapid and uniform heating of both liquid and solid phases.
- High energy conversion efficiency (90% of the electrical energy is converted into heat energy).
- Low capital cost due to reduced moving parts.
- Can be operated for longer duration, as fouling of the system is minimized.
- Color and nutritional value of food is retained better than conventional heating due to short processing time.
- Handles high viscous product.
- Continuous production without heat transfer surfaces.

11.9   DISADVANTAGES OF OHMIC HEATING SYSTEM

A food substance that contains fat globules can be difficult to process effectively using ohmic heating, as fat globules are nonconductive and lacking water and salt, which is required to conduct current. Fat globules may be heated slowly than highly electrical conductive region, due to lack of electrical conductivity, any microorganism present in the fat globules may not be inactivated.[62]

Another disadvantage of ohmic heating is due to the electrical conductivity of a substance. Electrical conductivity increases with increase in temperature, due to faster movement of electrons and this creates the possibility of "runaway" heating. An ohmic heating that has not been cleaned thoroughly may result in electrical arcing due to protein deposits on the electrode.

11.10 DEGRADATION OF ASCORBIC ACID DURING OHMIC HEATING

Ascorbic acid keeps away diseases like scurvy, and it acts as a biological antioxidant. As humans have no capacity to produce this component, it should then be provided with the diet.[4]

Vitamin C is the least stable of all vitamins and is easily destroyed during processing and storage. The degradation rate is enhanced by the activity of metals, particularly copper and iron, and enzymes. Availability of oxygen, prolonged heating in the presence of oxygen and exposure to light are all harmful factors to vitamin C content of foods.[19, 25] In the presence of oxygen, the share of the anaerobic degradation to the total loss of vitamin C content is small or not detectable, compared to the aerobic degradation, which has a much higher degradation rate.[59]

The usage of heat treatment is the most common method for stabilizing foods, due to its capability to destroy microorganisms and still enzymes. However, since heat can affect many organoleptic properties and decrease the contents or bioavailability of some nutrients, there is an increasing interest in searching for new technologies that have the ability to reduce the intensity of the heat treatments needed for food preservation.[58]

Degradation kinetics of AA under ohmic heating conditions with stainless steel electrodes was compared with conventional heating.[4] The results indicate that at pH 3.5 kinetics of AA degradation can be described adequately by a first order model for both methods of heating, a number of electrochemical as well as secondary chemical reactions appear to have some effects on the kinetic parameters. The influence of reactions at electrode/solution interfaces has been investigated on degradation of AA in buffer medium during 60 Hz ohmic heating with stainless steel electrodes.[5]

Vitamin C degradation in acerola pulp during ohmic and conventional heat treatment revealed that higher voltages (34 V·cm−1) presented higher degradation percentage of 10%. Lowest voltage of 21 V·cm−1 showed the lowest degradation percentage of 3%. Independent of the solids content of the pulp, lower values of degradation were obtained at low voltages. A higher degradation of ascorbic acid in acerola pulp was observed using high voltages because electrolysis and corrosion of metals were enhanced at high electric fields, leading to the production of compounds that catalyze the degradation pathways of ascorbic acid in the presence of oxygen.[24]

The kinetics of ascorbic acid degradation during ohmic heating of orange juice by the application of electric field strength of 42 V·cm−1 after 3 min of heating at 90°C, the percentage of degradation was approximately 35%.[37] Lima et al..[37] applied ohmic heating to heat orange juice for 30 min at 90°C with an electric field of 18.2 V·cm−1, and degradation percentage was approximately 21%. Clearly, the literature confirms the changes in ascorbic acid degradation. This may be attributed to vitamin C degradation mechanisms. Degradation can take place through aerobic and/or anaerobic pathways, depending on a number of factors: acidity, metal ions, pH, light, humidity, water activity and temperature, amino acids, carbohydrates, lipids and enzymes, among others.[25]

The presence of an electric field had no significant effect on the ascorbic acid degradation in orange juice. Although there was electrolysis and metal corrosion when stainless steel electrodes were used, yet these phenomena did not affect the final concentration of ascorbic acid. Vikram et al.[58] compared degradation kinetics of the ascorbic acid of orange juice by conventional heating, microwave heating and ohmic heating. They noticed that the ohmic heating method eased higher nutrient holding at all temperatures when compared to other methods of heating, accompanied by infrared and conventional heating. The highest degradation was observed in microwaved food, due to uncontrolled temperatures generated during processing, that crossed 100°C arriving at 125°C and attributed to the heat unbalanced nature of vitamin C.

The study on ascorbic acid degradation kinetics of strawberry products.[8] described that the incurred kinetic parameters were same for both the conventional and ohmic heating processes leading to the decision that the presence of an electric field does not involve the degradation of ascorbic acid. Lima et al.[37] have arrived at similar conclusions.

The study on degradation kinetics of ascorbic acid during ohmic heating with stainless steel electrodes indicated that degradation rate under ohmic heating was not significantly different from that under conventional heating, except at high power (150 W), high NaCl content (1%) and low temperature (40°C) conditions. Under these conditions, the ohmic heating degradation rate was significantly greater than for conventional heating.[5]

11.11 MECHANISM OF MICROBIAL INACTIVATION BY OHMIC HEATING

Deactivation of microorganisms is significant in many industrial applications and new low-energy or energy-efficient deactivation methods led challenging ideas.[45, 51] Pertaining to product safety and quality management, microbial inactivation is a fundamental parameter to be treated in food production processes. The occurrence of undesired or high number of certain microorganisms may lead to product deterioration like substance degradation, quality losses (appearance changes, off-odors, off-taste, deterioration of color, etc.) and/or health problems including diseases and/or illnesses. Improper cooking, for instance, is believed to be the main cause of Salmonella spp. eruption.[11, 57]

11.12 ELECTROPORATION EFFECT OF OHMIC HEATING

In addition to the heating advancement, research data strongly propose that the applied electric field under ohmic heating induces cell membranes electroporation. The cell electroporation is nothing but the pore formation in plasma membrane because of the presence of an electric field and as a result the permeability of the cell wall is increased and diffusion of material during the cell wall is attained by electro-osmosis.[2, 54] It is assumed that the electric collapse or electroporation mechanism is prevalent for the nonthermal effects of ohmic heating.[34, 53]

Regardless of being a prokaryote or eukaryote, all living cells contain cell membranes. These membranes comprise proteins and lipids. Prokaryotes usually have extra layer outside the membrane called the cell wall. At low frequencies (50–60 Hz) and high field strengths (> 100 V·cm−1) that is commonly related to ohmic heating, the naturally poriferous cell walls can permit the cell membrane to build up charges leading to the formation of disruptive pores. Since the cell membranes have particular dielectric strength that exceeds the electric field, electroporation occurs.[66] The nonconducting strength of a cell membrane is associated to the amount of lipids (acting as an insulator) present in the membrane itself. The pores could be of different sizes depending on the electric field strength, and could be sealed off after sometime. Exposure beyond limits causes cell death due to the overflow of components inside the cells through the pores. Thus, electroporation is extremely detrimental to a cell and would increase the deadly effects of thermal misuse that is already present from the ohmic heating.

In ohmic heating, the electric field seemed to have both direct and indirect consequences on the cell wall, and the materials inside the cell were oozed out to the culture medium. The exudates appeared to be comprising of protein, amino acids, nucleic acids and related material. It is concluded that below 50°C same concentrations of oozed out material were noticed in the yeast supernatant, under conventional or ohmic heating.[42]

Nevertheless at temperatures above 50°C, the exuded materials' concentration was higher compared to conventional group (P<0.01) and that the protein rate (oozed out per unit temperature increase) was significantly higher (P<0.01) with ohmic heating than with conventional heating. The influence of the electrical field within ohmic heating might have increased the rate of electroporation, thereby leading to excess exudation and cell death. It was also observed that the amount of exuded protein was increased significantly as the electric field increased from 10 to 20 V·cm−1.[42]

Lower D- and z-values were observed for the inactivation of Escherichia coli and Bacillus licheniformis when submitted to ohmic heating.[55] However, Palaniappan and Sastry.[42] found no difference between the effects of ohmic and conventional heat treatment on the death kinetics of yeast (Zygosaccharomyces bailli), under identical heating histories. Milk viable aerobes and S. thermophilus inactivation have been accessed.[63] According to many researchers microbial death is high in ohmic heated food than the conventionally heated food. They also resolved that the A. niger's critical treatment time was decreased when the electric field strength was increased.

11.13 SUMMARY

In ohmic heating processes, an alternating current is allowed through the food thus generating heat within the food. The electrical conductivity, which is one of the important parameters in ohmic heating, increases with increase in temperature. Though there are many varieties of metals used as an electrode, titanium offers high corrosion resistance and biocompatibility characteristics. The degradation rate of ascorbic acid is less in ohmic heated product when compared to other conventionally heated product.

KEYWORDS

- Advantages of ohmic heating
- alternating current
- ascorbic acid
- current
- degradation
- electric field
- electrical conductivity
- electricity
- electrochemical
- electrode
- electrode corrosion
- electrolysis
- electroporation
- firmness
- fruit and vegetable
- fruit and vegetable products
- heating mechanism
- kinetics
- juice
- meat products
- microbial inactivation
- microbial inactivation
- ohmic heating
- orange juice
- resistance
- seafood products
- solute concentration
- stainless steel
- temperature
- titanium
- voltage
- voltage gradient

REFERENCES

1. Amatore, C.; Berthou, M.; Hebert, S. Fundamental Principles of Electrochemical Ohmic Heating of Solution. *J. Electroanal. Chem.* **1998,** *457,* 191–203.

2. An, H. J.; King, J. M. Thermal Characteristics of Ohmically Heated Rice Starch and Rice Flours. *J. Food Sci.* **2007,** *72*(1), C84–C88.
3. Assawarachan, R. Estimation Model for Electrical Conductivity of Red Grape Juice. *Int. J. Agric. Biol. Eng.* **2010,** *3*, 52–58.
4. Assiry, A. M. Effect of Ohmic Heating on the Degradation Kinetics of Ascorbic Acid. PhD thesis. The Ohio State University, 1996, p 213.
5. Assiry, A.; Sastry, S. K.; Samaranayake, C. Degradation Kinetics of Ascorbic Acid During Ohmic Heating with Stainless Steel Electrodes. *J. Appl. Electrochem.* **2003,** *33*(2), 187–196.
6. Assiry, A. M.; Gaily, M. H.; Alsamee, M; Sarifudin, A. Electrical Conductivity of Seawater During Ohmic Heating. *Desalination* **2010,** *260*, 9–17.
7. Biss, C. H.; Coombes, S. A.; Skudder, P. J. The Development and Application of Ohmic Heating for the Continuous Heating of Particulate Foodstuffs. In *Process Engineering in the Food Industry;* Field, J. A., Howell, J. A., Eds.; Elsevier: London, UK, 1989; pp 11–20.
8. Castro, I.; Teixeira, J. A.; Vicente, A. A. The Influence of Field Strength, Sugar and Solid Content on Electrical Conductivity of Strawberry Products. *J. Food Process Eng.* **2003,** *26*, 17–29.
9. Castro, A.; Teixeira, J. A.; Salengke, S.; Sastry, S. K.; Vicente, A. A. Ohmic Heating of Strawberry Products: Electrical Conductivity Measurements and Ascorbic Acid Degradation Kinetics. *Innovative Food Sci. Emerging Technol.* **2004,** *5*, 27–36.
10. Castro, I. A. C. Ohmic Heating as an Alternative to Conventional Thermal Treatment. PhD thesis, University of Minho, 2007, pages 210.
11. Cho, H. Y.; Yousef, A. E.; Sastry, S. K. Growth Kinetics of *Lactobacillus acidophilus* Under Ohmic Heating. *Biotechnol. Bioeng.* **1996,** *49*(3), 334–340.
12. Coronel, P. M.; Sastry, S.; Jun, S.; Salengke, S.; Simunovic, J. Ohmic and Microwave Heating. In *Engineering Aspects of Thermal Food Processing*; Simpson R., Ed.; CRC Press, Boca Raton; 2010; pp 73–81.
13. Cristina, S. C.; Moura, D. R.; Vitali, A. D. A. A Study of Water Activity and Electrical Conductivity in Fruit Juices: Influence of Temperature and Concentration. *Braz. J. Food Technol.* **1999,** *2*, 210–216.
14. Darvishi, H.; Hosainpour, A.; Nargesi, F.; Khoshtaghza, M.H.; Torang, H. Ohmic Processing: Temperature Dependent Electrical Conductivities of Lemon Juice. *Mod. Appl. Sci.* **2011,** *5*, 210–216.
15. Darvishi, H.; Khoshtaghaza, H. M.; Zarein, M.; Azadbakht, M. Ohmic Processing of Liquid Whole Egg, White Egg and Yolk. *Agric. Eng. Int.: CIGR J.* **2012,** *20*, 224–230.
16. Darvishi, H.; Khostaghaza, M. H.; Najafi, G. Ohmic Heating of Pomegranate Juice: Electrical Conductivity and pH Change. *J. Saudi Soc. Agric. Sci.* **2013,** *2*, 101–108.
17. De Alwis, A. A. P.; Fryer, P. J. The Use of Direct Resistance Heating in the Food Industry. *J. Food Eng.* **1990,** *11*(1), 3–27.
18. De Alwis, A. A. P.; Pryer, P. J. Operability of the Ohmic Heating Process: Electrical Conductivity Effects. *J. Food Eng.* **1992,** *15*(1), 21–49.
19. Deman, J. *Principles of Food Chemistry*, 2nd ed. Van Nostrand Reinhold: New York, 1990; pp 350–351.
20. Destinee, R. A. *Ohmic Heating as an Alternative Food Processing Technology.* Master of Science Thesis, Kansas State University, 2002, pages 223.

21. Food and Drug Administration-Center for Food Safety and Applied Nutrition (FDA-CFSAN). *Kinetics of Microbial Inactivation for Alternative Food Processing Technologies-Ohmic and Inductive Heating*; FDA-CFSAN, Washington DC; 2000; pages 112.

22. Fryer, P. J.; Li, Z. Electrical Resistance Heating of Foods. *J. Food Sci. Technol.* **1993**, *4*, 364–369.

23. Ghnimi, S.; Flach, M. N.; Dresh, M. Evaluation of an Ohmic Heating Unit for Thermal Processing of Highly Viscous Liquids. *Chem. Eng. Res. Des.* **2008**, *86*, 627–632.

24. Giovana, D. M.; Steven, S.; Ligia, D. F. M.; Isabel, C. T.; Sastry, S. Ascorbic Acid Degradation and Color Changes in Acerola Pulp During Ohmic Heating: Effect of Electric Field Frequency. *J. Food Eng.* **2014**, *123*, 1–7.

25. Gregory, J. F. Ascorbic Acid Bioavailability in Foods and Supplements. *Nutr. Rev.* **1993**, *51*(10), 301–313.

26. Hayt, W. H. *Engineering Electromagnetics*, 4th ed.; McGraw-Hill Book Co.: New York, 1981; pp 45–52.

27. Hong, H. D.; Kim, S. S.; Kim, K. T.; Choi, H. D. Changes in Heating Profiles of Apple Juice by Ohmic Heating. *Hanguk Nongwhahak Hoechi* **1998**, *41*(6), 431–436.

28. Icier, F.; Ilicali, C. Electrical Conductivity of Apple and Sour Cherry Juice Concentrates During Ohmic Heating. *J. Food Process Eng.* **2004**, *27*(3), 159–180.

29. Icier, F.; Ilicali, C. Temperature Dependent Electrical Conductivities of Fruit Purees During Ohmic Heating. *Food Res. Int.* **2005**, *38*, 1135–1142.

30. Icier, F.; Ilicali, C. The Effects on Electrical Conductivity of Orange Juice Concentrates During Ohmic Heating. *Eur. Food Res. Technol.* **2005**, *220*, 406–414.

31. Icier, F.; Ilicali, C. The Use of Tylose as a Food Analog in Ohmic Heating Studies. *J. Food Eng.* **2005**, *69*, 67–77.

32. Icier, F.; Yildiz, H.; Baysal, T. Polyphenoloxidase Deactivation Kinetics During Ohmic Heating of Grape Juice. *J. Food Eng.* **2008**, *85*, 410–417.

33. Kim, H. J.; Choi, Y. M.; Yang, A. P. P.; Yang, T. C. S.; Taub, I. A.; Giles, J.; Ditusa, C.; Chall, S.; Zoltai, P. Microbiological and Chemical Investigation of Ohmic Heating of Particulate Foods Using a 5 kw Ohmic System. *J. Food Process. Preserv.* **1996**, *20*, 41–58.

34. Kulshrestha, S.; Sastry, S. K. Frequency and Voltage Effects on Enhanced Diffusion During Moderate Electric Field (MEF) Treatment. *Innovative Food Sci. Emerg. Technol.* **2003**, *4*(2), 189–194.

35. Li, F-D.; Zhang, L. Ohmic Heating in Food Processing. In *Mathematical Modeling of Food Processing;* Farid, M. M., Ed.; CRC Press: Boca Raton, FL, 2010; pp 659–672.

36. Lima, M.; Sastry, S. K. The Effect of Frequency and Waveform on the Electrical Conductivity: Temperature Profiles of Turnip Tissue. *J. Food Process Eng.* **1999**, *22*, 41–54.

37. Lima, M.; Heskitt, B. F.; Burianek, L. L.; Nokes, S. E.; Sastry, S. K. Ascorbic Acid Degradation Kinetics During Conventional and Ohmic Heating. *J. Food Preserv.* **1999**, *23*, 421–434.

38. Lima, M. Food Preservation Aspects of Ohmic Heating. In *Handbook of Food Preservation*, 2nd ed.; Rahman, M. S., Ed.; CRC Press: Boca Raton. FL, 2000; pp 741–745.

39. Palaniappan, S.; Sastry, S. Electrical Conductivity of Selected Juices: Influences of Temperature, Solid Content, Applied Voltage, and Practical Size. *J. Food Process Eng.* **1991**, *14*, 247–260.

40. Palaniappan, S.; Sastry, S. K. Electrical Conductivities of Selected Solid Foods During Ohmic Heating. *J. Food Process Eng.* **1991**, *14*, 221–236.
41. Palaniappan, S.; Sastry, S. Ohmic Heating. In *Control of Foodborne Microorganisms;* Juneja, V. K., Sofos, J. N., Eds.; Marcel Dekker: New York, 2002; pp 451–460.
42. Pereira, R.; Martins, J.; Mateus, C.; Teixeira, J. A.; Vicente, A. A. Death Kinetics of *Escherichia coli* in Goat Milk and *Bacillus licheniformis* in Cloudberry Jam Treated by Ohmic Heating. *Chem. Pap.* **2007**, *61*(2), 121–126.
43. Piette, G.; Brodeur, C. Ohmic Cooking for Meat Products: The Heat is on. *Le Monde Alimentaire* **2001**, *5*(6), 22–24.
44. Piette, G.; Buteau, M. L.; De Halleux, D.; Chiu, L.; Raymond, Y.; Ramaswamy, H. S. Ohmic Heating of Processed Meat and its Effects on Product Quality. *J. Food Sci.* **2004**, *69*(2), FEP71–FEP77.
45. Piette, G.; Dostie, M.; Ramaswamy, H. S. Is there a future for ohmic cooking in meat processing? *Can. Meat Sci. Assoc. News* **2001**, 8–10.
46. Reznick, D. Ohmic Heating of Fluid Foods. *J. Food Technol.* **1996**, *5*, 250–251.
47. Samaranayake, C. P. Electrochemcial Reactions During Ohmic Heating. PhD Thesis, Ohio State University, 2003, p 200.
48. Sarang, S. S. Safety of Low-Acid Foods Containing Particulates Processed by Ohmic Heating. PhD Dissertation, Ohio State University, 2007, p 100.
49. Sarang, S.; Sastry, S. K.; Gaines, J.; Yang, T. C. S.; Dunne, P. Product Formulation for Ohmic Heating: Blanching as a Pretreatment Method to Improve Uniformity in Heating of Solid-Liquid Food Mixtures. *J. Food Sci. E: Food Eng. Phys. Prop.* **2007**, *72*(5), E227–E234.
50. Sarang, S.; Sastry, S. K.; Knipe, L. Electrical Conductivity of Fruits and Meats During Ohmic Heating. *J. Food Eng.* **2008**, *87*, 351–356.
51. Sastry, S. K.; Palaniappan, S. Mathematical Modeling and Experimental Studies on Ohmic Heating of Liquid-Particle Mixtures in a Static Heater. *J. Food Process Eng.* **1992**, *15*, 241–261.
52. Sastry, S. K.; Palaniappan, S. Ohmic Heating of Liquid-Particle Mixtures. *Food Technol.* **1992**, *46*(12), 64–67.
53. Sensoy, I.; Sastry, S. K. Extraction Using Moderate Electric Fields. *J. Food Sci.* **2004**, *69*, 7–13.
54. Smith, K. E.; Medus, C.; Meyer, S. D.; Boxrud, D. J.; Leano, F.; Hedberg, C. W. Outbreaks of Salmonellosis in Minnesota (1998 through 2006) Associated with Frozen, Microwaveable, Breaded, Stuffed Chicken Products. *J. Food Prot.* **2008**, *71*(10), 153–159.
55. Sun, H. X.; Kawamura, S.; Himoto, J. I.; Itoh, K.; Wada, T.; Kimura, T. Effects of Ohmic Heating on Microbial Counts and Denaturation of Proteins in Milk. *Food Sci. Technol. Res.* **2008**, *14*, 117–123.
56. Uemura, K.; Noguchi, A.; Park, S. J.; Kim, D. U. Ohmic Heating of Food Materials—Effect of Frequency on the Heating Rate of Fish Protein. In *Developments in Food Engineering—Proceedings of the 6th International Congress on Engineering and Food*; Toshimasa Yano; Ryūichi Matsuno; K Nakamura (Eds.); Blackie Academic and Professional Press: London, 2004; pp 310–312.

57. USA-CDC (United States of America) Multistate outbreak of Salmonella infections associated with frozen pot – United States. *MMWR Morb Mortal Weekly Rep.*, 2007–2008, *57*(47), 1277–1280.

58. Vikram, V. B.; Ramesh, M. N.; Prapulla, S. G. Thermal Degradation Kinetics of Nutrients in Orange Juice Heated by Electromagnetic and Conventional Methods. *J. Food Eng.* **2005**, *69*, 31–40.

59. Villota, R.; Hawkes, J. G. Reaction Kinetics in Food Systems. Chapter 2, In *Handbook of Engineering;* Heldman, D., Lund, D.; Marcel Dekker Inc.: New York, 1992; pp 432.

60. Wang, C. S.; Kuo, S. Z.; Kuo-Huang, L. L.; Wu, J. S. B. Effect of Tissue Infrastructure on Electric Conductance of Vegetable Stems. *J. Food Sci.: Food Eng. Phys. Prop.* **2001**, *66*(2), 284–288.

61. Wang, W. C.; Sastry, S. K. Changes in Electrical Conductivity of Selected Vegetables During Multiple Thermal Treatments. *J. Food Process Eng.* **1997**, *20*, 499–516.

62. Wang, W. C.; Chen, J. I.; Hua, H. H. Study of Liquid-Contact Thawing by Ohmic Heating. *Proceedings of IFT Annual Meeting and Food Expo*, Anaheim, California, 2002, pages 230.

63. Yildiz, H.; Baysal, T. Effects of Alternative Current Heating Treatment on *Apergillusniger*, Pectin Methylesterase and Pectin Content in Tomato. *J. Food Eng.* **2006**, *75*, 327–332.

64. Yongsawatdigul, J.; Park, J. W. Linear Heating Rate Affects Gelation of Alaska Pollock and Pacific Whiting Surimi. *J. Food Sci.* **1996**, *61*(1), 149–153.

65. Yongsawatdigul, J.; Park, J. W.; Kolbe, E. Electric Conductivity of Pacific Whiting Surimi Paste During Ohmic Heating. *J. Food Sci.* **1995**, *60*(5), 922–935.

66. Yoon, S. W.; Lee, C. Y. J.; Kim, K. M.; Lee, C. H. Leakage of Cellular Material from *Saccharomycies cerevisiae* by Ohmic Heating. *J. Microbiol. Biotechnol.* **2002**, *12*, 183–188.

INDEX

9 781774 630525